BARTŁOMIEJ MALETA

BEEKEEPING
IN HARMONY WITH NATURE

THE EVOLUTIONARY SOLUTION
TO THE VARROA PROBLEM

TRANSLATED BY STEPHEN MARTIN

Northern Bee Books

Beekeeping in harmony with nature
The evolutionary solution to the varroa problem
Copyright © Bartłomiej Maleta

Translated by Stephen Martin

Published 2025 by Northern Bee Books,
Scout Bottom Farm,
Mytholmroyd,
West Yorkshire
HX7 5JS (UK)
Tel: 01422 882751
Fax: 01422 886157
www.northernbeebooks.co.uk

ISBN 978-1-9192004-0-8

Design and artwork DM Design and Print

AT THE CROSSROADS, OR THE PATH TO BEEKEEPING WITHOUT TOXIC SUBSTANCES / 310

Tools, characteristics, properties of bees, resistant traits / 314

The transition period to beekeeping without treatments / 347

SUMMARY / 375

BIBLIOGRAPHY / 378

INTRODUCTION TO THE ENGLISH EDITION

When writing this book I wasn't planning to publish it in English. I believe there are quite a few very interesting books on the subject in this language. By my work I wanted to show Polish beekeepers the vast knowledge on broadly understood natural and sustainable ways of apiculture were the needs of keepers meet the needs of the bees. Even a few years ago this knowledge was missing in Poland, as it wasn't discussed at bee conferences and in other media by the beekeeping authorities. I believe that we can manage our hobby or professional apiaries in ways that, still can be close to intensive and very profitable, but do not harm the bees and their adaptation processes. That would lead to the possibility of future generations of beekeepers to enjoy keeping healthy, resilient and notably varroa-resistant bees without using chemicals or other methods of controlling them or the pathogens they transmit. Honey bees can perfectly do it on their own – if we just let them, but maybe help them a little bit at the beginning.

The English version of this book would not be created if it wasn't for Prof. Stephen Martin from the United Kingdom who invested a lot of time and effort in the translation, edition and proofreading of the text – as in convincing me that publishing this book in English is an endeavour worth taking. I'm grateful for that.

I hope this book brings a different perspective on the western honey bee adaptations, health problems and varroa-resistance. Most of the knowledge and information included in my book are very universal, but some refer to our Central-Eastern European and Polish perspective. This continental-European outlook on the western honey bee problems might also be interesting and valid for some of the readers. In Poland alone there was in 2024 around 2.4 million western honey bee colonies! It is about three quarters of the estimated number of colonies managed in the United States of America, which is about thirty times bigger than Poland. It is therefore the perspective of a region, which is very urbanized with industrial agriculture and highly overpopulated with honey bees where horizontal transmission of pests and pathogens is a sad norm. An environment in which virulence of pathogens is easily growing and honey bee resistance mechanisms are difficult to breed and maintain in the population. The only way I see of getting off this vicious circle is to start a joint effort of resistance breeding and increase the efforts to improve our landscapes for pollinators.

The English edition of the book is in general a direct translation of the Polish one. In a few places however I have made minor changes. These are mostly some updates of the story and data. I also reconstructed some parts of the book where I thought they were too detailed or my personal perspective might not be needed. This included for example cutting some of the individual sentences (or adding them when I thought necessary), or some wider reconstruction of the "My apiary" chapter. This was originally written a few years ago, and since then, the situation has changed, so I wanted to present fair coverage of the events, as well as thoughts and reflections that come from them. I've also added a more detailed description of Dr Ralph Büchler's method based on so called biotechnic procedures. In the Polish version it was only mentioned briefly, but I think it's worth deeper consideration.

I hope that in this book readers will find at least a few answers to their questions.

BARTŁOMIEJ MALETA (MARCH 2025)

NOTE FROM THE TRANSLATOR

I became aware of this book after a photo request from the author. When I received a copy of this beautiful book, I translated Chapter 6 using online translators and was so impressed with the quality and depth of the information I discussed with Bartek about producing an English version. He was naturally reluctant due to the vast amount of work that had already gone into to producing the Polish version. Hence, over the 2024 Christmas period I used google translate and Deep L to translate the entire book. This took several days, but then changing the translated text into something beekeepers would understand was a much longer process since many of the beekeeping specific words did not translate well and some were very funny. But further proof-reading between Bartek and myself hopefully resolved most of these issues. We also thank Grant Elliott, Rhona Toff, Steve Riley, Joe Ibbertson and Shan/Clive Hudson for their roles in final proof-reading.

I believe that this book is timely and contains much valuable information for beekeepers around the world wanting to stop chemical control of their varroa population. Only then will this allow the bees to adapt to the mite as nature intended. It has been a labour of love to help translate this book and I have learnt so much along the way. I now hope that many other beekeepers will be able to benefit from Bartek's insights and hard work involved in researching and writing of this book.

STEPHEN MARTIN (MARCH 2025)

THROUGH THE PROFESSOR'S EYES

American physicist and philosopher, Thomas Samuel Kuhn, in his theory of scientific revolutions, defines a paradigm as "a universally recognized scientific achievement providing model solutions that gain acceptance by the majority of the community." He claims that educating students involve instilling in them paradigmatic solutions to the most important problems. In his opinion, however, a necessary condition for significant progress are scientific revolutions, when paradigmatic, commonly accepted solution is replaced by another, completely different one, even though the latter are difficult to accept at the beginning.

The beekeeping community has been struggling for over 40 years against the varroa mite, which is based on the paradigm of controlling the mites with chemical treatments may have saved commercial beekeeping. Today, however, we are not making any progress in solving the varroa problem, and the problems are getting worse. Maybe it's time for a new paradigm? Or at least use alternative methods?

In this context, this book by Bartłomiej Maleta, is worth reading. The study, supported by the reading of several hundred references and the author's many years of experience in his own apiary, introduces the reader to the issues related to the so-called organic, natural, or Darwinian beekeeping.

The foundations for this type of bee management were laid by Prof. Thomas Seeley. It involves not using chemicals and keeping colonies in conditions as naturally as possible.

It is worth adding that in 2020, there were 2.7 million bee colonies managed using the natural method, which means that on a global scale their number has increased fivefold since 2007, currently accounting for 2.9% of the worlds' bee colonies. It should also be remembered that all bee populations resistant to varroa arise because of (Darwinian) evolution mechanisms, i.e. only where treatment is not used or has been abandoned. Attempts to obtain resistant bees without stopping treatment, using genetic markers, have so far been unsuccessful. Does the future belong to natural beekeeping? We don't know that. However, many believe not.

Bartłomiej Maleta's book is therefore an example of putting the cat among pigeons and will probably spark heated debate. But when trying to change a paradigm, such reactions are standard. Many of the ideas presented in this book will probably be considered controversial, especially since the author does not hide his fascination with natural beekeeping. It often happens that adversaries do not have sufficient knowledge about the problem to discuss it fully. This book will prove to be very helpful, regardless of the position we take in this discussion. It contains a lot of information about the life of honey bees in natural conditions, their biology, evolution, health conditions, ability to survive, and especially the mechanisms of varroa resistance, along with the description of mite-resistant populations. We will also learn about many other bee diseases, the bees' nutritional requirements and environmental threats, especially those resulting from the use of pesticides. The author does not focus only on the main topic but goes much further and deeper. Particularly valuable are the footnotes, of which there are almost 400. I would only like to mention that in his enthusiasm for natural bee breeding, Bartłomiej Maleta did not avoid the risk of forcing open doors, and he also did not avoid certain inaccuracies, especially when referring to evolution, breeding genetics, including artificial insemination and inbreeding, or when he interprets the role of regional populations. However, he has presented many phenomena on this topic in an accessible and correct way.

What will be the future of natural beekeeping? Of course we don't know. Although Bartłomiej Maleta is an enthusiast of the natural method, he himself points to many obstacles to its introduction in Europe, including Poland.

He is aware that it is difficult to stop mite treatments overnight. He also considers the chances of conducting natural beekeeping without loss of honey production or the co-existence of treated and untreated colonies. Particularly valuable in this context is the chapter 'The transition period to beekeeping without treatments', in which these issues are discussed.

The book will also be useful for those who are not involved in natural beekeeping. Many can use the information contained therein by implementing only some elements of natural beekeeping. Today, even a partial increase in resistance to varroa and reduction in the use of chemical treatments is already a great success. To effectively combat varroa, let's look towards natural methods. We replace natural breaks in brood rearing by using trap-frames and artificial swarming methods.

Explaining the phenomena of the 'mite bomb' and the horizontal and vertical transfer of mites, the author provides valuable knowledge to those who struggle with mite re-infestations in areas with high colony densities. When writing about selection, he also raises an important problem in contemporary zootechnics: because of the antagonism between artificial and natural selection, farm animals can turn out to be much weaker. Therefore, more and more attention has been paid to improving functional characteristics, i.e. those related to their long-term health. Modern apiary management has distanced bees from living in accordance with their natural needs, impairing their survival abilities, which is why properly developed functional characteristics are very important for them. For this reason, the information contained in the book is also valuable material for creators of bee breeding programs. In turn, any comments regarding the specific environments in which the apiary should be located become important in the context of activities aimed at rebuilding the architecture of the agricultural landscape and grasslands around the apiary.

The entire chapter on the construction of a hive may be interesting for amateur beekeepers running small apiaries. Another problem of Polish beekeeping is the adulteration of wax. Contaminated wax is difficult to remove from circulation. To meet this threat, as if on request, the author indicates methods of obtaining clean wax in the foundation-free beekeeping chapter. Considerations on the advantages and disadvantages of keeping bees in small or standard cells are also valuable, especially in hives where there are combs made of cells of both sizes next to each other. In turn, the issue of excessive or preventive use of chemicals is important not only for the condition of bees, but also for public health (just watch commercials on TV).

An interesting addition are descriptions of famous beekeepers, whose achievements can be an inspiration for many younger beekeeping enthusiasts. The ethical considerations contained in the book also deserve special mention since they are currently rarely discussed not only in beekeeping but also zootechnical textbooks.

Finally, Bartłomiej Maleta's publication stands out in terms of editorial and figure design. Carefully selected photofigures, drawings, and pictorial diagrams, along with a clear system of references and footnotes, provide an excellent testimony to the author and editors. Therefore, despite a few controversial fragments, I believe that it is a book worth having in your home library.

LUBLIN, 10-12 2023

JERZY DEMETRAKI-PALEOLOG

INTRODUCTION

> *If insects had not been developed on the face of the earth, our plants would not have been decked with beautiful flowers, but would have produced only such poor flowers as we see on our fir, oak, nut, and ash trees, on grasses, spinach, docks, and nettles, which are all fertilised through the agency of the wind.*

CHARLES DARWIN[1]

The book you are holding in your hand is an attempt to look at the health of the western honey bee species (*Apis mellifera* L.) from an evolutionary, or as it is sometimes called a 'Darwinian' perspective. I still believe that people who know much more about it than me should write about adaptive or evolutionary processes in beekeeping. However, few publications deal with these issues. I have the impression that we are still revolving around the same problems: what type of pesticide to put in our hives this year, what varroa-treatments to choose to control their populations. Meanwhile, many publications, written from different perspectives, presenting interesting and valuable alternative methods of beekeeping can be found. Their authors point to the ability of bees to survive in beekeeping situations without the beekeeper needing to use toxic and biocidal substances.

---

1   Ch. Darwin, The Origin of Species, published P. F. Collier & Son, 1909, New York, p. 211.

One such publication is the wonderful work by American Prof. Thomas Dyer Seeley from Cornell University (USA), who discusses the history of wild bees in the 21st century in '*The Lives of Bees: The Untold Story of the Honey Bee in the Wild*'[2]. Also of interest is the book by the Welsh beekeeper Dr. David Heaf: '*Treatment Free Beekeeping*'[3].

I have focused on the problems that I consider the most important or interesting. That's how this book was created, a book that I would like to read if I were at the beginning of my beekeeping adventure. I started working on the book a few years ago. I needed time primarily to deepen my knowledge and gain experience. The chapters I wrote then had to be almost completely changed now primarily because scientific and popular science publications appeared that shed new light on many issues. Who knows, maybe if I had waited a few more seasons, many fragments that were based on our current knowledge would have to be verified and rewritten?

The text is not free from repetitions, but at least some of them are intended. I decided to do this if I thought it was worth returning to a certain topic, for example to signal or emphasize the connection with the issue I am mentioning in the current section. This is not a book about my experiences (I still think I don't have enough of them). It would be difficult for me to talk about any of my own successes in this field, especially the ones connected to apiary economy. For now, I don't have them, or they are small, but I decided to mention my own apiary and selection history to show my personal perspective. However, I know some Polish beekeepers who use beekeeping or breeding methods like mine, with better results. However, I tried to present the extensive knowledge about the ability of bees to survive, accumulated thanks to the work of other beekeepers, scientists and bee lovers whose experience are much richer than mine.

However, this is not a textbook on apiary management or even on natural beekeeping. My goal is to popularise knowledge about natural methods of keeping and breeding bees and to show a different perspective. However, understanding the content contained here requires at least a basic knowledge of bee biology and apiary management. I believe that reading a few books about bees and beekeeping or knowing the basics of practical apiary management will make it easier to understand the various details in the book.

---

2   T.D. Seeley, The Live of bees. The Untold Story of Honey bee in the Wild, Princeton University Press, 2019.

3   D. Heaf, Treatment Free Beekeeping, Northern Bee Books, 2021.

Whether my arguments will prove convincing to the reader is a completely different matter. I am aware that some ideas may seem controversial and may not be accepted by many for various reasons. However, there are also beekeepers who operate in certain shades of grey and need guidance to choose their own path. One that will best suit their worldview and beliefs. I hope that this book will help them, even if they reject all the practical conclusions from the book. I also tried to separate my opinions and views from the facts and findings of various studies, although I probably did not always succeed. Considering the basics of beekeeping based on evolutionary principles, I am trying to compromise and to reconcile the "natural needs"[4] of honey bees with the needs of the beekeepers and other people, because we all benefit from the work of honey bees. Apart from beekeepers, few people are aware of how bees are kept and how honey is produced.

Presently not all beekeepers see something wrong with the current practice of intensive beekeeping. In my opinion, modern apiary management has distanced bees from living in accordance with their "natural needs" and forces new adaptations. This is what this book will also be about.

Sometimes I had to resort to simplifications for a few reasons. Firstly, many of the issues discussed are difficult and complex. To convey their complexity, it would probably require a work of the size of the Encyclopaedia Brittanica[5]. Secondly, I believe that certain simplifications make it easier

---

4   I use quotation marks because species really have no needs, whether natural or unnatural. Disputes about what a specie really is continues to this day. Since we are unable to adopt a clear definition of a species, we are even less able to determine what the natural "needs" of a particular set of organisms are. Each individual organism has its own needs, again to simplify, survival and reproduction. In the case of *Apis mellifera* the matter is further complicated by the fact that we could be talking not only about the needs of the swarm, but also of the individual bees in the swarm. However, there is no room for such detailed considerations here. I think most readers will intuitively understand what I mean. Therefore, I used a simplification to indicate the fact that in the process of evolution, the organisms we artificially classify have developed specific adaptations to the biological niches they occupy. If the living conditions of these organisms are consistent with their adaptations, we can simply say that they live in an environment that is consistent with their adaptations, and in this sense, we can assume that their "natural needs", i.e. those shaped during evolution, are being met. I leave aside the fact that the process of evolution continues, which means that no creation is final. Therefore, the "natural needs" of a species we observe today often differ from the needs of their ancestors and may be completely different than those of their descendants (just like the environment in which they lived was different from the current one). Moreover, the evolutionary success of organisms may be supported not only by the so-called natural adaptations. After all, thanks to intensive breeding methods, the honey bee species has achieved enormous evolutionary success and has taken over virtually the entire planet. At the same time, a large proportion of honey bee colonies in many geofigureal areas have little chance of surviving and reproducing without beekeepers.

5   If we enter the words "*Apis mellifera*" into the search engine for scientific research, www.scholar. google.com, we will get several hundred thousand results. Although some of them probably only mention honey bees, a significant percentage refers directly to research on the adaptations of honey bees. This information is written faster than we can read it.

to convey and understand difficult issues. Thirdly, detailed considerations would take us away from the main themes and the book would become less interesting. I only hope that the simplifications I have used in the text are not too far-reaching. I wanted to provide as reliable and objectively the knowledge in this book, although some of the content was provided with subjective commentary[6] resulting from my way of understanding beekeeping and nature. I tried to ensure that all information was supported by sources, preferably scientific ones. I also partly rely on the experience of beekeepers who, when presenting facts, do not always indicate their scientific sources and therefore it is sometimes very difficult to find them. So, I hope that I have not made any distortions, if you, the reader, have a different opinion, please let me know. If I have made such a mistake, I assure you that it was not intentional.

Some people say that when we talk about bee health, we should focus primarily on nectar and pollen plants. I agree with this. However, in the book I will rather focus on the principles of bee-friendly apiary management and bee selection, i.e. the so-called natural beekeeping. I use the term "so-called" because Dr. David Heaf, not without reason, claims that the term "natural beekeeping" is an oxymoron. The moment we decide, as Heaf says, to put our bees in the box, our actions are already characterized by a certain unnaturalness (he calls it: Universal Scale of Increasing Artificiality in beekeeping[7]). So, the question remains: how far into the depths will we go? It seems that if apiary management is carried out in a sustainable manner[8], it will allow us to meet the needs of both bees and beekeepers, as well as consumers of bee products (e.g. honey, bee bread or mead), and all of us who benefit from the work that bees do as pollinators.

---

6   Perhaps some people will consider my subjective selection of data and issues as cherry picking, which means the selection of evidence or statistical data to prove one's own ideas, while omitting other arguments unfavourable to the presented arguments. I do not deny subjectivity in the selection of topics. However, I have the impression that the generally available literature on apiary management ignores data proving the ability of bees to independently "solve" their own problems," so I could use such an accusation myself.

7   http://www.dheaf.plus.com/warrebeekeeping/essence_sustainable_beekeeping.pdf

8   What does the term "sustainable economy" mean? According to the definition of "sustainable development", it is the idea of socio-economic development, assuming a type of activity that, while meeting the needs of society, will not limit the developmental opportunities of future generations, for example by destroying the natural environment. The shape of today's economy (including intensive agriculture) is far from this definition. We incur a huge debt in the environment, using up (often irretrievably) the planet's resources or destroying its delicate balance (for example by generating climate change). We assign this debt to future generations.

For me, natural beekeeping is a type of beekeeper's activity that involves maintaining and supporting the natural adaptability of bees. In other words, we keep and breed bees in such a way that they can survive without our help. This type of beekeeping also allows swarms, that escape from the apiary, to cope on their own in the often harsh external environment we have created for them. It is beekeeping that supports the natural behavioural characteristics and abilities of bees needed to fight predators, parasites and pathogens. Even in those cases in which specific apiary management techniques are advanced on the Universal Scale of Increasing Artificiality.

Today, there is a common belief that bees' biggest problems are environmental pollution and diseases, mainly varroosis, caused by the parasitic mite ***Varroa destructor***. In my opinion, the biggest problem with western honey bees is that they are hindered from going through the process of natural selection, and breeding typically excludes the natural adaptations they need to survive.

The name of method of keeping bees that I practice translated directly from Polish would be 'beekeeping without treatment' or 'beekeeping without chemicals'. The English term: treatment-free beekeeping sounds nicer, but it is more capacious. The word "treatment" can be understood as "cure", but it also means a way of proceeding or management, so it can also encompass other beekeeping practices. For this reason, natural beekeepers are constantly looking for new terms that would better and more accurately describe the essence of the issue. Therefore, the terms "bee-centric beekeeping" (apicentric beekeeping), "bee-friendly beekeeping" or "nature-based beekeeping" have all appeared. They all have a common denominator: they put the needs of bees first. That means that the bees should not be adapted to our beekeeping methods, but the management should be adapted to the needs of the superorganism. If the bees are healthy, all human needs will also be met. One of the British representatives of natural beekeeping, Phil Chandler, known as the 'barefoot beekeeper', likes to repeat: "Natural beekeepers don't have all the answers, but at least they ask the right questions".

What do I understand by the term "natural beekeeping" in practice? Here are my personal ideas that I try to follow in my daily work:

1. Bees can take care of themselves.

2. Bees often "know" better than the beekeeper what is good and appropriate for them.

3. Natural selection is the best way to adapt members of the honey bee species to survive on their own.

4. What's good for the bees is always good for the beekeeper, especially in the long term.

5. What is good for the beekeeper, especially in the short term, is not necessarily good for the bees.

6. The most economically efficient bee colonies are not necessarily the best adapted (and usually are not) to survive on their own.

7. Most bees' characteristics may have been necessary for survival at some stage in the species' evolution. Some they still need, and others they may need in the future.

8. Populations of the *Apis mellifera* can cope with every currently occurring disease (including varroosis), although not every colony will be able to survive the disease.

9. Systemic control of bee diseases using toxic and biocidal agents in the long-term weakens the bees' adaptation to cope with their health problems on their own and so disturbs their co-adaptation with the environment.

10. Carrying out economical, efficient and sustainable beekeeping is possible but requires us to introduce changes in everyday practice and above all, increase respect for the evolutionarily adaptations that have shaped the bees.

For most beekeepers, their profession is also a passion. A passion that unites the entire community. I am convinced that they keep bees in accordance with their best will and knowledge. I do not question the positive motivations of most of the beekeeping community, but one question still bothers me: do some beekeepers lack perspective and understanding of certain processes that take place with their bee population?

One of the greatest naturalists of all time, recognised as the father of the modern science of ecology, Alexander von Humboldt, said: "Man can only act upon nature, and appropriate her forces to his use, by comprehending her laws, and knowing her forces to his use, by comprehending her laws and knowing those forces in relative value and measure"[9]. I hope that this book will contribute to a better understanding by beekeepers of at least some of the laws of nature. I discover them anew every season and constantly become acquainted with new discoveries.

9   Wulf A, The Invention of Nature. The Adventures of Alexander von Humboldt, the lost hero of Science, John Murray Publishers, 2015; Alexander von Humboldt (1769-1859), founder of modern geofigurey, traveller and ecologist.

BEE HEALTH

Chapter 1

BEE HEALTH

," *In a few different venues, I've given a presentation called Making varroa into an Ally. The first is of Sir Albert Howard, the British agricultural scientist who gave voice, direction, and an intellectual framework to the modern organic farming movement. After a lifetime of work, one of his principle conclusions was that pests and diseases should always be seen and welcomed as friends and allies, not as adversaries to be destroyed. Their real purpose, he concluded, was to indicate where our methods, crops and livestock are weak and unbalanced, and to show the pathway to restored health and vitality.*

KIRK WEBSTER[10]

An alternative view

When I started running an apiary, I came across a book about bee diseases. Today I don't remember its author or title. However, I remember perfectly how sad and depressing the impression it made on me was. I wondered how on earth with so many threats from pathogens and parasites are honey bee colonies still able to live. I had the impression that they were all sick and doomed to a quick end. Meanwhile, bees have been living on our planet for millions of years. Although in many regions of the world, especially where agriculture is practiced on an industrial scale, their condition is deteriorating, fortunately, nothing yet indicates the extinction of honey bees. Even if they were to disappear from some continents (which also seems very unlikely), the species would probably survive in other regions, where it would create healthy, stable and self-sufficient populations.

---

10 https://kirkwebster.com/nature-has-all-the-answers-so-whats-your-question-and-a-page-from-a-treatment-free-beekeeping-diary/

It is hard to resist the impression that modern beekeeping treats the problem of bee diseases one-sidedly. In this book, I will try to present it from a different perspective than the one that is applicable to commercial beekeeping. I propose a perspective of supporting the health of the bee population. This approach seems coherent and logical, especially when we look at it from an evolutionary perspective (some also call it a Darwinian perspective). Instead of pondering over ways to directly combat the effects of pathogens, I will try to focus on describing the natural mechanisms of the bees' immunity, methods of supporting them, and tips on selection. I also have the impression that in the 21st century we have stopped paying attention to the complexity of many issues, and we are not always interested in deeper cause and effect relationships. Meanwhile, the causes of many diseases are complex, and it is often difficult to see their connection with the symptoms we observe.

The most dangerous diseases

We know very well that the situation of honey bees requires attention. The condition of the natural environment is deteriorating, pollution with toxic substances is increasing, ecosystems are impoverished, climate change is causing water shortages, the food base is shrinking, increasing urbanization is annexing wild nest sites. In addition, new diseases are appearing that affect bees, although some beekeepers seem to deny this; they claim that bees are healthy and productive like never before[11]. So, if they are healthy, why do they require constant treatments? Health according to the definition from the Polish Scientific Publishers Encyclopaedia (PWN) is "a state of a living organism in which all functions run properly"[12]. So, I ask the question: does the bee superorganism function properly if the bees require constant care from us, and if treatment is omitted, in most cases they die?

Honey bees, like almost all living organisms, suffer from viral, bacterial, fungal, and parasitic diseases. In the context of our considerations, it is

---

11   Seeley disagrees. He says: "That was the model [of intensive beekeeping focused on obtaining large amounts of honey - B.M.] that worked perfectly well before *Varroa* and the viruses that it spreads. But now, to my mind, our bees are sick. I think they are sick almost all the time, but we are still pushing them. If a human being was as sick as most of the bee colonies are, I don't think we would be asking them to be as productive as they would be if they were healthy"; https://www.beekeepingtodaypodcast.com/bee-hunting-with-dr-tom-seeley-archive-super-episode-s4-e48/, accessed: January 17, 2025.

12   PWN Encyclopaedia, https://encyklopedia.pwn.pl/haslo/zdrowie;4000909.html.

necessary to mention the most common and life-threatening ones.

The most dangerous of them all is the disease caused by the parasitic mite *Varroa destructor*, which belong to arthropods (class: arachnids, subclass: mites)[13]. The bee mite is an invasive organism that has shaped its own biology in co-adaptation with another species of bees. It is believed to have arisen because of evolutionary diversification from the species *Varroa jacobsoni*, which is a parasite of the oriental honey bee (*Apis cerana*). Currently, the mite occurs on all continents, except Antarctica[14].

Varroosis causes the death of bee colonies through the premature death of the workers parasitised as pupae. We can also find dead female varroa mites on the bottom board of the hive, photo by author.

---

13 Parasitoformes are a taxon of arachnids classified as mites. Over 15,000 species have been described so far, although their actual number may reach 200,000. Most lead a predatory lifestyle, many are parasites, often hematophagous. Parasites belonging to the parasitoforms are often of great concern of medicine and veterinary medicine, especially as vectors of pathogenic microorganisms, https://pl.wikipedia.org/wiki/ Dr%C4%99cze.

14 In June 2022, *V. destructor* individuals were found in bee colonies in one of the port towns of Australia, a continent that was previously considered free from this parasite [https://www.outbreak.gov.au/current-outbreaks/varroa-mite, accessed: 30 January 2024.]; Today we know that it was not possible to nip the invasion of varroa in the bud.

In many regions of the world, varroa has determined how apiaries are managed. Its presence in Poland first was recorded in 1980 and since then it has been the cause of huge losses in apiaries every year.

Directly or indirectly, *Varroa destructor* causes the destruction of a huge number of bee colonies around the world. It feeds on the bee, but does not actually kill it: its existence, however, causes many negative consequences that contribute to the death of individual bees and, consequently, the collapse of the entire colony. Some beekeepers believe that the honey bee is unable to develop the mechanisms necessary to survive in the presence of the parasite, and therefore believe that miticide treatments are needed to save the species from extinction. This assumption is completely false, as evidenced by numerous examples of honey bee populations functioning normally despite the presence of *Varroa*.

Another serious bee disease is nosema, a parasitic fungal disease. *Nosema apis* and *Nosema ceranae*, were formerly classified as protozoa, but now are classified as fungi[15]. Both species parasitise the digestive system of the honey bee. While the first (*N. apis*) is a form that has coexisted with the honey bee for a long time, and as a result its virulence is very limited, the second (*N. ceranae*) is an invasive species that still leads to large losses

---

15   Currently, the organisms are assigned to the genus *Varimorpha*. In the book, I will use their old taxonomic name. Those interested are referred to the series of articles: Wszystko o nosemozie [Everything about nosema], K. Pohorecka, A. Bober, "Pszczelarstwo" 2021, 12; 2022, 1; 2022, 3, 2022, 4.

in *A. mellifera* apiaries. *Nosema apis*, by causing digestive problems, can contribute to a significant weakening of the adult bees. It also often causes diarrhoea, which during wintering can end with the bee colony dying out or at least weakening it so significantly that the economic use of the colony in the next season will not be possible.

Nosema ceranae causes, among other things, acute malnutrition in adult bees due to significantly limiting their ability to absorb food, which in turn very quickly leads to the death of the bee from hunger and exhaustion. This form of the disease (similarly to varroosis in the acute phase) can, within a few weeks, lead to the "de-population" of the hive (i.e. the sudden death of the workers in the hive) and the collapse of the entire colony. Weakened bees (e.g. by mites or an improper diet) are more susceptible to diseases, but all bees, or at least most of them, are carriers of *Nosema* parasites (one or both species at the same time). The spores (endospores) of nosema can be transferred through physical contact, honey, bee bread, wax, other organisms encountered in the hive. In principle, similarly to mites, we can assume that no hive is free from nosema.

The source of American foulbrood (AFB) is the *Paenibacillus larvae* bacteria. This disease, although dangerous, is probably the most demonized of all, for two reasons. Firstly, and as it seems most important, it is the only disease whose control in Poland is required by law. Failure to report a suspicion of foulbrood to a veterinarian may result in criminal or administrative liability for the beekeeper. Administrative restrictions related to the designation of so-called infected areas cause additional difficulties in conducting beekeeping activities. This type of regulation may introduce, for example, a ban on the transport of colonies from an infected area or the sale of apiary products. While for an amateur beekeeper this will be a relatively minor issue, for a professional beekeeper it may significantly limit their ability to conduct business. Secondly, foulbrood is very easily transmitted; if left untreated, it is believed to spell a death sentence for the bee colony (at the latest by autumn).

Meanwhile, it is worth noting that this disease would be relatively easy to control if the beekeeper removed all the brood (some also suggest a two-day fast) and relocated the bees to a clean hive.

Beekeepers from other countries report cases of self-healing of bees without the intervention of a beekeeper or cleaning of frames with brood infected with foulbrood (or deliberately given to bee colonies as part of a breeding program or research)[16].

In Poland, diagnosing the disease is most often associated with the necessity of eliminating the bee colony, in accordance with the decision of the District Veterinary Office, disinfecting the hive and burning all the combs. Fortunately, this usually does not apply to the entire apiary but just the infected hive. However, it sometimes happens that veterinarians allow the treatment of colonies by resettling the bees to new hives then only the brood and combs are disposed of (the hive bodies are either disinfected or destroyed).

The name "foulbrood" accurately reflects the course of the disease: bacteria that rapidly multiply in sealed brood cause its decomposition (to the stringy, rotten and stinking phase). Bees, trying to clean the decomposing larvae, transfer the bacteria to other cells, and in this way an epidemic can develop rapidly within the bee colony. Meanwhile, it turns out that the spores of the bacteria causing American foulbrood are by no means strange with their presence detected in several dozen percent of the apiaries examined. Some beekeepers (e.g. Michael Bush) even claim that there is no hive without spores of this bacterium. They are exceptionally difficult to destroy, because it requires high temperature and appropriate pressure.

European foulbrood (EFB) is a disease caused by several species of bacteria and affects both open and sealed brood. Similarly to American foulbrood, it ends with the death of larvae and pupae. Since the disease is caused by various, commonly present strains of bacteria, its development is almost always caused by external factors, such as cooling of the brood or lack of food.

The most common fungal diseases of bees are Chalkbrood that affects larvae and Stone brood, which can affect both brood and adults. Both

---

16   Heaf cites, for example, the experiment of a beekeeping inspector, Les Crowder, from New Mexico, who placed infected frames with clinical signs of American foulbrood in a hive; the bees cleaned the frame each time, preventing the development of the disease. The research conducted in 1937 is mentioned by J. Woyke, Hodowla pszczół odpornych na warrozę [Breeding of bees resistant to varroa], "Pszczelarstwo" 1988, 11. One of the studies suggests that the high mortality rate in the case of American foulbrood is closely related to the conditions of beekeeping in apiaries, while in natural conditions this disease should not cause increased mortality of bee, Ingemar Fries et al., Vertical transmission of American foulbrood (Paenibacillus larvae) in honey bees (Apis mellifera), "Veterinary Microbiology" 2006, 5. However, there are sceptical voices that self-healing does not occur: bees clean dead pupae, which is why we do not observe foulbrood symptoms, however in the event of a negative environmental factor (e.g. hunger, prolonged cold), the disease will return.

diseases cause the death of larvae, which are infected with spores and mycelium. These diseases can stop on their own, especially if weather conditions improve and a harvest appears. Stone brood is also contagious to humans (fungi of the *Aspergillus genus*, which cause respiratory diseases, can develop in our bodies).

Viral diseases of bees include primarily: Deformed Wing Virus (DWV); Sacbrood Virus (SBV); Chronic Bee Paralysis Virus (CBPV); Slow Paralysis Virus (SPV); Acute Bee Paralysis Virus (ABPV). All of these are quite common throughout the world and are most often spread by parasites (especially *Varroa destructor*), and bees (e.g. during robbing). Undoubtedly, viral infections are most dangerous to colonies in a weaker condition, especially those where their natural protective immunity barriers have been breached. For example, high DWV infection is most often associated with a high level of varroa invasion of the bee brood.

The disease called Colony Collapse Disorder (CCD) is less and less talked about today. It was very prominent for a few years, around 2007, when in many apiaries in the United States, the phenomenon of mass disappearance of colonies began to be noted.

Bees infected with **DWV** can have deformed wings, photo by M. Adamczak.

It happened that often in the middle of the season all the bees disappeared from their colonies, leaving behind hives full of brood and food supplies, which was an unusual phenomenon. Normally, the disappearance of colonies caused by diseases happens at the end of summer and in autumn or early winter. After some time, the typical symptoms of CCD began to be noted less and less often. Since then, the mortality rate of bee colonies in the United States has oscillated around 30% each winter, and during the entire season it is estimated at around 40%. Currently, however, this is not associated with CCD. The causes of the disease have not been fully identified to this day. The symptoms (sudden and massive loss of flying bees, leading directly to the "de-population" and death of the entire colony) are often attributed to other diseases - diagnosed and testable, such as varroosis or nosema.

The mass bee die-offs connected to this specific phenomenon have been recorded primarily (or perhaps exclusively?) in North America, a region where modern and highly regulated agriculture is practiced. It has often been suggested that CCD was caused by compounds used to protect crops from pests. This was especially true for neonicotinoids (used as insecticides), which can affect the development and functioning of insect nervous systems. Ultimately, however, the connection between CCD and neonicotinoids has not been unequivocally confirmed, although there have been many theories and hypotheses about this phenomenon.

Investigating the causes of bee health problems

Why are bees unable to live on their own today? Beekeepers, when asked about this, most often cite varroa and environmental changes (primarily its contamination) as the main causes. However, few people go deeper and draw attention to the increasing limitations of bees' adaptations. Even if the respondents mention evolutionary adaptation, it is rather to state that this process no longer applies to bees today. According to beekeepers, changes in the world have gone so far that bees must either be "kept on a drip" or face the fact that they will die.

I do not intend to trivialize the problems associated with environmental changes. However, I believe that naturally living populations can adapt to these changes quite easily and quickly. It seems that they have a very important role to play today. They can indicate the direction in which beekeeping should go. Observing the surroundings of bees that manage to survive and those that die can suggest what conditions we should create around apiaries, as well as how to organise our apiaries to best serve the bees. By studying natural nest sites of honey bees, we can rediscover the principles of building hives.

By looking at the characteristics of long-lived, wild or feral colonies, we can understand what traits we should strengthen in the breeding process. We should therefore stimulate the creation of wild populations to help us balance selection and understand what is happening with the bees[17]. We can support wild colonies by creating nest sites (taking care of old trees, especially those with holes, cavities, hanging beehives or boxes, placing hives in the forest), and accepting swarming from some of our colonies.

I treat specific diseases as a symptom rather than a cause of the poor health of bees. The immune systems of organisms living in some kind of balance with the environment and adapted to it usually cope with most pathogens[18]. In the case of bees, however, we are dealing with an accumulation or even synergy of unfavourable factors. On top of all the problems that bees suffer from, caused by environmental changes, industrialisation of agriculture and intensive breeding methods, there are the further problems of invasive organisms such as *V. destructor* and *N. ceranae*. Bees must "learn" to cope with these new threats by developing the appropriate behaviour and immune responses. Then there would be a chance for the development of a specific balance in the process of co-adaptation.

Pathogens live among us. Healthy organisms have very different immune mechanisms, which are an effective defence against pathogenic microorganisms. A combination of external factors is most often responsible for the development of clinical symptoms of a disease. Therefore, we often simplify the problem, assuming that the disease manifests itself when the pathogenic organisms grow unchecked. Few people wonder why we observe pathological changes in one organism and not in the other, although contact

17   Cf.: interview with prof. J. Tautz, Aby zrozumieć świat pszczół miodnych, musimy badać je w naturze [To understand the world of honey bees, we need to study them in nature], "Pszczelarstwo" 2022, 11.

18   Even if some specific colonies succumb, populations persist and are rather stable in natural conditions. This problem must also be viewed from the perspective of the adoption of evolutionary survival strategies by organisms. It is believed that insects have adopted the so-called R-strategy, which means they produce numerous offspring, but only a small percentage of these individuals survive to a reproductive age (as opposed to the K-strategy, characteristic of mammals and birds, in which offspring are relatively few but are cared for and a significant percentage survive to maturity). In the case of honey bees (and other social insects), the adoption of the R-strategy is not unequivocal, however, because in a season they usually produce 1-3 swarms. However, the mortality rate of these swarms is relatively high and can reach over 80% (data from Seeley's studies).

with the pathogenic factor was comparable. It would be worth looking deeper for the causes. An organism that does not function properly is susceptible to various pathogenic factors. Whereas those that are more resistant to the pathogen at a given moment has a competitive advantage under their local conditions (environmental, climatic, habitat).

There is a stable population of free-living bees in Sherwood Forest, UK. According to Powell, this oak tree, which is around 700 years old, has been inhabited by bee colonies for almost 30 years, photo by J. Powell.

Also important are the prevalence of a given pathogen, the immunity and behavioural adaptions of the organism, the season, humidity, sunlight, colony activity, a varied diet, the presence of appropriate genes (and their expression).

Living organisms also can heal themselves and self-organize their life processes. Killing a specific pathogen or parasite will not change anything in the long term if the organisms do not develop the ability to control it. The situation is certainly not improved by bringing in genetically alien bees from distant regions. Their transport (this applies to queen bees, entire colonies, as well as nucs/packages) is forcing a change in their environment. They are transferred from an environment in which they can feel good to a new environment where they must adapt (different climate, nectar, landscape, co-occurring organisms, etc.). Together with the bees, we also transfer potentially more virulent pathogens, but also other microbes. We transport them from bee ecosystems, in which the bees may have already undergone co-adaptation processes, to places where such processes may not have taken place. This mechanism works both ways. On the one hand, foreign pathogens transferred to a new environment can spread in organisms that are not adapted to this (e.g. smallpox brought to America). On the other hand, the incoming organisms may not be adapted to local pathogens (e.g. the impact of tropical bacteria on newcomers from other geofigureal areas). In this situation, a holistic approach is needed, where appropriately conducted selection will go hand in hand with bee-friendly management methods. In addition, to the requirements set by apiary management, the needs of bees must also be considered. Meanwhile, one may get the impression that the welfare of bees is often considered only to the extent that it allows for high productivity of apiaries. Bees in apiaries have no right to show weakness. Replacing the queen bee (read: killing the old one) in a colony that does not develop in accordance with the beekeeper's expectations or is less productive than the others is treated as an example of so-called good beekeeping practice, and often even acting for the good of bees.

It is like saying that replacing a hen killed for broth with a newly bought chicken is an action for the good of the hens. In the beekeeping community dealing with natural breeding, it is even said that keeping honey bee colonies in apiaries helps with the protection of the bees in the same way as keeping hens on farms helps protect the birds. Meanwhile, replacing a queen bee is *de facto* killing the entire organism. **After all, it means changing the genes of all the bees in the hive, creating the superorganism.**

In my own way, I can understand the justification for these practices in commercial apiaries, which are the source of income for the beekeeper and their colony, which does not mean that I consider them right or worth imitating. However, I am not fully able to understand this type of activity in amateur apiaries. Hobby beekeepers are often outraged by the words that they are guided by profit, not by the welfare of the bees, and at the same time they are able to bring queen bees from far away just because the breeders promise that their bees have better efficiency than their local ones. As we know, nature can often surprise us. It often turns out that something that we consider a weakness becomes a symptom of health. For example, as described many times in scientific publications, it can be the longevity of organisms that consume low-calorie food or even periodically suffer from hunger. This regularity has been found not only among humans or mammals in general, but also among honey bees[19]! It also turns out that an improper diet of bees (e.g. feeding them sugar or invert syrup) can have negative effects. It causes changes in many body functions, as well as the

---

19   Bajda et al., Rewersja procesu starzenia u pszczół miodnych (*Apis mellifera*)? [Reversion of the aging process in honey bees (*Apis mellifera*)?], "Medycyna Weterynaryjna" 2013, 1; W. Wu et al., Responses of two insect cell lines to starvation: autophagy prevents them from undergoing apoptosis and necrosis, respectively, "Journal of Insect Physiology" 2011, 6.

expression of genes different than in the case of bees eating natural food[20].

It can also result in changes in the structure of the bee's intestine (which undoubtedly also affects the reproductive capacity of the microbes inhabiting the intestines)[21].

This is how poor our ability to assess long-term population health is. Scientific research indicates that human choices regarding lifestyle or diet can leave a mark even on subsequent generations. All because of epigenetic factors[22], which although they do not change the genetic code can affect the characteristics passed on to subsequent generations[23]. Changes in gene expression are initiated by certain specific environmental impulses (direct environment, stress, food, lifestyle, addictions, etc.). They can have an impact both on turning off (or silencing) specific genes, but also on strengthening their expression. Thanks to epigenetics, we can understand why identical twins can suffer from different diseases, or why animal clones differ from their parents (the famous example of a cloned cat that has a completely different coat colour than her genetically identical "parent"). In the beekeeping community, the most well-known example of epigenetics is the raising of queen bees by worker bees. The environment (a special bee cell, called a queen cell), and above all, diet, mean that any of the female larvae can become a queen bee.

We do not know all the mechanisms that govern nature, but those that we currently know teach us humility. How can we know then whether the changes to the bees' living environment that beekeepers make will not have unexpected consequences? Even if they were subtle[24]?

---

20 M.M. Wheeler, G.E. Robinson, Diet-dependent gene expression in honey bees: honey vs. sucrose or high fructose corn syrup. Scientific Reports 2014, 4; S.A. Ament in., Mechanisms of stable lipid loss in a social insect. The Journal of Experimental Biology 2011,11.

21 G. Mirjanic, Impact of different feed on intestine health of Honey Bees, XXXXIII International Apicultural Congress, Apimondia, Kijów, Ukraina, 2013.

22 Epigenetics is a science that studies the relationship between the genetic code and the living environment (it describes the mechanisms of modification of the expression of the DNA code).

23 See: P. Mazur, T. Jagielski: Dziedziczenie grzechów [Inheritance of sins], "Wiedza i Życie" 2012, 8; O. Orzyłowska-Śliwińska: Pilnuj genów [Watch your genes], "Wiedza i Życie" 2016, 1.

24 In the naturalist community, the words of Wendell Berry (American writer and environmental activist) are often repeated, which are a good complement to this thought: "We cannot know what we are doing until we know what nature would be doing if we were doing nothing".

CHANGE IN THE GENETICS OF BEE POPULATIONS DUE TO BREEDING WITH STRONG „POSITIVE SELECTION"

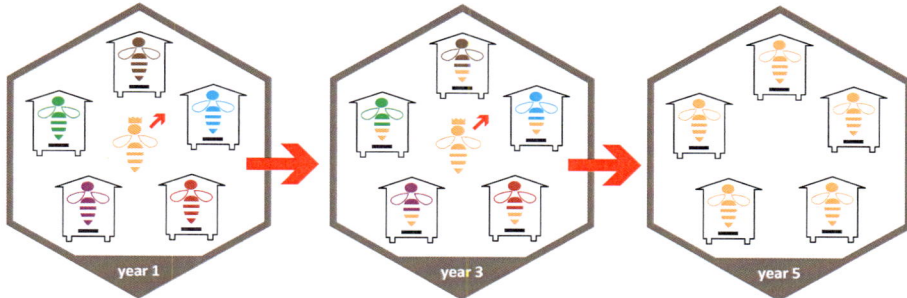

The constant replacement of the queens with the daughters from a single breeder-queen may, over time, contribute to a significant reduction in the genetic diversity of the bees in the apiary. Beekeepers are convinced that this is a good practice from an economic point of view, but it may result in a reduced ability of the population to resist various threats from parasites and pathogens, figure by Mariusz Uchman

The bee colony is, on the one hand, an exceptionally strong and resistant organism, and on the other hand, delicate and sensitive. On what do we base our belief that the stress it is subjected to will not affect its health? This problem is still poorly understood. Perhaps we are still not fully able to indicate the direction of breeding? The weaknesses of bees prove to us, however, that the path we have followed so far was not the right one.

Darwinian beekeeping

Evolutionary perspective

The gradual decline in the condition of bees has motivated many practicing beekeepers and scientists to consider a systemic solution to the problem.

Western European researchers, Peter Neumann and Tjeerd Blacquière, suggested in 2016 that the causes behind reducing bee health may be both (inappropriate for the species) methods of conducting bee management, as well as limiting the role of natural selection in apiary practice[25]. This issue was dealt with in more detail by Prof. Thomas Dyer Seeley from Cornell University, who suggests that the health problems of bees should be viewed through the prism of evolution. The western honey bee species evolved over millions of years in a specific ecological situation. Living organisms are subjected to various stimuli of the natural environment and are shaped

---

[25]   P. Neuman, T. Blacquière, The Darwin cure for apiculture? Natural selection and managed honey bee health, "Evolutionary Applications" 2016, 11.

in specific conditions and inter- and intra-species dependencies. When the ecological situation changes, some of the un-adapted organisms die, and the population rebuilds itself based on those that have acquired the tools/abilities to cope with changing conditions.

In the process of evolution, many subspecies[26] of the honey bee have developed, differing in structure, behaviour, and colour. Each of them has adapted to the local conditions in which they evolved. Modern beekeeping has changed many of these relationships. Honey bees that previously lived in distant tree holes or rock crevices suddenly found themselves on the ground and had to start coping in living in large groups (apiaries). The development of intensive agriculture and environmental changes have distanced bees even further from the conditions in which they lived during thousands of years of evolution. Thanks to the high plasticity of the population and the ability of individual colonies to shape their own habitat, they were able to function quite well, until dangerous invasive species appeared in their environment.

Bees in nature and bees in apiaries

In the early 2000s it was a popular belief in the United States that wild populations ceased to exist due to the invasion of the varroa mite[27]. In mid-2002, Seeley disproved this belief when he found that the wild population of the Arnot Forest, near Ithaca, New York, was doing just as well as it had been before the varroa invasion[28]. The density of occurrence was similar with an average of one colony per square kilometre of forest. Thus, bees were therefore able to survive in nature without any help from humans, while almost all beekeepers - including Seeley - struggled with serious problems in their own apiaries.

How do the living conditions of bees in nature and in apiaries differ? Let us list them after Professor Seeley[29].

---

26   It is commonly accepted to call them races, but from a biological point of view this term is not accurate.

27   In that part of the world, the mite appeared around 1992.

28   T. D. Seeley: Honey bees of the Arnot Forest: a population of feral colonies persisting with Varroa destructor in the northeastern United States. Apidologie 2007, 1-2: T. D. Seeley: The Lives of Bees. The Untold Story of Honey Bee in the Wild, Princeton University Press, 2021.

29   T. Seeley: Darwinian Beekeeping: An Evolutionary Approach to Apiculture, "American Bee Journal" 2017, 3. The author lists 20 differences between the living conditions of bees living in the wild and in apiaries (he already lists 21 in the book) - I have grouped some of them, I have omitted others as less important.

Thomas D. Seeley bee hunting, photo T. Seeley

1.   Bees from apiaries, unlike wild bees, are not adapted to local conditions. Buying queens, packages or nuclei from distant areas leads to "dilution" and displacement of adaptation mechanisms. It happens that these honey bees react less well to local environmental stimuli, thus forcing beekeepers to constantly intervene. Migratory beekeeping can have an equally negative impact on the local population, especially if bees are transported in large numbers and over long distances.

2.   Bee colonies in nature live at a distance from each other, usually at least several hundred meters apart, while in apiaries they are crowded. As a result, they must compete for food, they are more often prone to robbery; they also experience problems with reproduction (merging swarms, drifting queens flying into the wrong hive). According to Seeley, the most unfavourable

consequence of bee crowding is a change in the epidemiological (epizootic) situation, i.e., above all, the ease of transferring pathogens between hives (horizontal transmission).

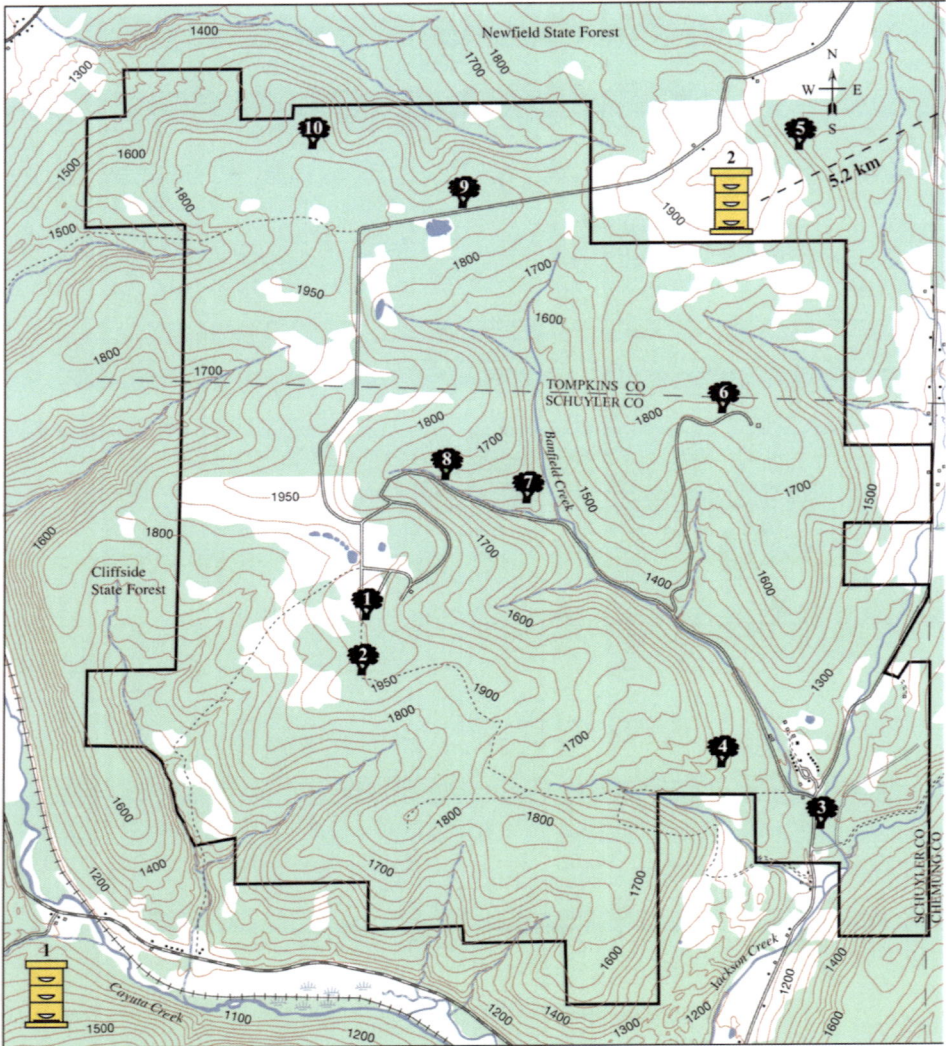

Map of Arnot Forest, showing the trees inhabited by bees found by Seeley in August and September 2011, figure by T. Seeley.

Comparison of the environments in which honey bee colonies lived in the wild (and sometimes still do) and those in which they live currently as managed colonies.	
Environment of evolutionary adaptedness	Current circumstances
Colonies genetically adapted to location	Colonies not genetically adapted to location
Colonies live widely spaced in the landscape	Colonies live crowded in apiaries
Colonies occupy small cavities (volume)	Colonies occupy large hives
Nest cavity walls have propolis coating	Hive walls with no propolis coating
Nest cavity walls are thick (good insulation)	Hive walls are thin (poor insulation)
Nest entrance is high and small	Hive entrance is low and large
Nest has 10-25% drone comb	Nest has little (<5%) drone comb
Nest organization is stable	Nest organization is often altered
Nest-site relocations are rare	Hive relocations can be frequent
Colonies are rarely disturbed	Colonies are frequently disturbed
Colonies deal with familiar diseases	Colonies deal with novel diseases
Colonies have diverse pollen sources	Colonies have homogeneous pollen sources
Colonies have natural diets	Colonies sometimes have artificial diets
Colonies are not exposed to novel toxins	Colonies are exposed to insecticides and fungicides
Colonies are not treated for diseases	Colonies are treated for diseases
Pollen not trapped, honey not taken	Pollen sometimes trapped, honey often taken
Combs not moved between colonies	Combs often moved between colonies
Beeswax is not removed, cappings are recycled by bees	Beeswax (cappings) is removed during honey harvests
Bees choose larvae for queen rearing	Beekeepers choose larvae for queen rearing
Drones compete fiercely for mating	Queen breeder may select drones for mating
Drone brood not removed for mite control	Drone brood sometimes removed for mite control

Table by M. Uchman based on 'The Lives of Bees', by T. Seeley.

In nature, the most virulent strains, killing the carriers (hosts), die along with them, and thus their populations evolve towards becoming less virulent. In apiaries, however, the virulent strains can very easily get into other hives[30].

3. Wild bees, for the most part, occupy relatively small tree hollows, and in apiaries they are kept in much larger hives (for the purpose of increasing honey production). Professor Seeley found bee colonies in tree hollows as large as a dozen or so litres; most nest sites had a volume of 20-60 litres. For comparison, the full Langstroth 10 frame box is around 42 litres. The large cubic capacity of modern hives reduces the tendency of bees to swarm, fewer swarms mean weaker natural selection that reduces the long-term selection for healthy population.

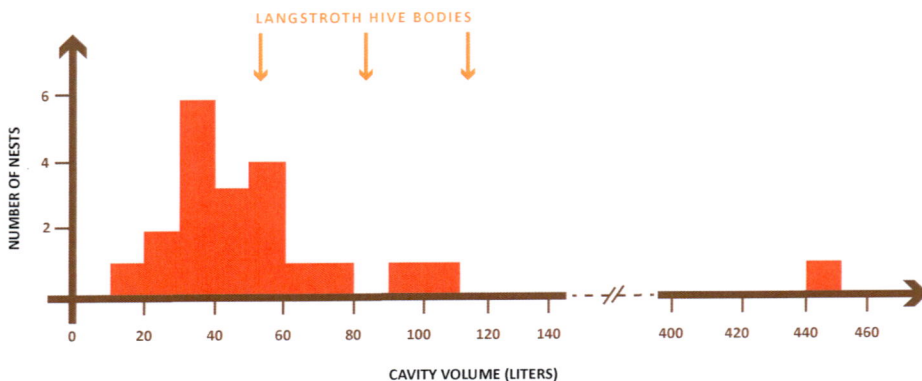

Studies of wild nests conducted by Seeley show that most nest sites have a capacity of 20-60 litres. The volume of the smallest hollow occupied by bees was less than 20 litres. One of the nests was in a very large hollow (a large, rotten tree) with a volume of over 400 litres. The bees did not occupy the entire volume, but only a relatively small part of it, at the highest point, figure by M. Uchman based on Seeley's data.

4. Bee colonies living in wild nests propolise their walls a lot, whereas bee colonies living in hives do not produce as much propolis.

---

30 Pathogens can also influence the behaviour of bees that facilitates the transmission of pathogens (e.g. drifting), which will be additionally reinforced by the small distances between hives in the apiary - A.C. Geffre, Honey bee virus causes context-dependent changes in host social behaviour, "Proceedings of the National Academy of Sciences" 2020, 4.

Bees in nature have completely different conditions than in apiaries photos by T. Seeley.

The inner surface of the nest in a rotten log stimulates the bees to produce propolis, photo by author.

5. The walls of wild nest sites (tree hollows) are usually much thicker than in hives used as standard in apiaries. It has been found that the heat losses of a wild bee colony are 4-7 times smaller than in the case of a colony living in a single-walled wooden hive[31]. Bees in hives must therefore expend much more energy[32].

---

31 The study suggests that small bee colonies are more likely to survive in tree holes than in hives, and that the reproductive success of mites is reduced; D. Mitchell: Ratios of colony mass to thermal conductance of tree and man-made nest enclosures of *Apis mellifera*: implications for survival, clustering, humidity regulation and *Varroa destructor*. International Journal of Biometeorology 2016, 9.

32 The energy requirements of bees in the Langstroth hive have been estimated by scientists, D. Cook et al., Thermal Impacts of Apicultural Practice and Products on the Honey Bee Colony, "Journal of Economic Entomology" 2021, 4. Scientists suggest that warming the hives could eliminate one of the factors of bee colony stress, but this is also possible in other cases, e.g. because of not using queen excluders.

6. Unlike most wild nests, hives usually have large entrances, located low above the ground. A large opening is more difficult for guards to defend, and its location (close to the ground) aids rodent invasion or makes it difficult for bees to perform winter cleansing flights. In addition, hives standing directly above the ground have higher humidity than a colony located a few meters above the ground.

7. Wild-living colonies - unlike managed colonies, have many drone combs. Reducing the number of drones makes it harder to pass on the genes of the healthiest colonies.

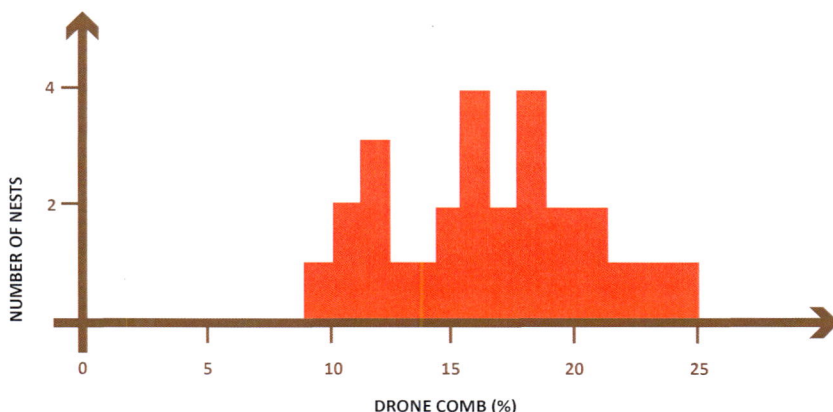

Professor Thomas Seeley, during his studies of wild nests, found that most often they contained 10-25% drone combs, figure by M. Uchman based on Seeley's data.

8. Wild colonies have complete freedom in building combs, and they also rarely experience outside interference. This is a guarantee of good work organisation. Bees in hives, on the other hand, are often subject to inspections; sometimes frames are replaced into a different place in the hive. This has a negative impact on the thermoregulation of the nest and may negatively affect egg laying by queens or pollen storage. An experiment conducted in 1963 showed that on the day of the colony inspection, bees had 20-30% less weight gain in comparison with the bees who were not disturbed[33]. It turned out that the inspections have a real impact on the organization of the bee colony's work.

---

33 S. Taber, The effect of disturbance on the social behaviour of the honey bee colony. American Bee Journal 103, 8, 286-288.

Bee colonies in unmanaged beehives live in a way like bees in nature as interference in their lives is significantly less frequent, photo by Treebeekeeping Brotherhood.

9. Bee colonies living in hives are often moved to new locations, especially in migratory type of beekeeping. Each time, they are forced to explore the new area, i.e. learn landmarks and search for food sources etc. An experiment conducted in 1975 showed that moving colonies to a new location for a week resulted in less weight gain (compared to colonies that were not moved)[34].

10. Wild bee colonies usually use a more diverse food base than those transported to monoculture crops. Those transported, despite the abundance of food, can often suffer from malnutrition (lack of a varied diet). Unlike wild-living bees, those in apiaries are fed artificial food. It has been noticed that workers consuming some pollen supplements can have health problems[35].

---

34 F.E. Moeller, Effect of moving honey bee colonies on their subsequent production and consumption of honey. Journal of Apicultural Research 1975, 14, 127-130.

35 The idea of pollen feeding is controversial among practicing beekeepers. Many note that the administration of pollen supplements (artificial protein food) stimulates bees to develop, especially during periods of dearth. On the other hand, J. Ellis from Florida State University states that the research on this subject is not conclusive; his analysis of scientific studies did not show that pollen supplements contributed to improving the health or the rate of colony development. He mentioned it many times in various episodes of the podcast: "Two Bees in a Podcast", https://entnemdept.ufl.edu/honey-bee/podcast/, accessed: October 31, 2022; see also: E.R. Noordyke, J. Ellis, Reviewing the Efficacy of Pollen Substitutes as a Management Tool for Improving the Health and Productivity of Western Honey Bee (*Apis mellifera*) Colonies. Frontiers in Sustainable Food Systems 2021, 11.

11. Bees in apiaries are subjected to various treatments, while wild colonies are not treated. Treatments disrupt the "arms race" between bees and their pathogens and parasites. Therefore, the mechanism of natural selection related to the development of resistance to diseases (pressure factors) is weakened. Seeley claims that in this situation it is not surprising that most bees living in North America and Europe have not yet developed increased resistance to bee pests such as varroa. Chemical treatments also have a destructive effect on the bee's microbiome. Today, bees are in constant contact with toxic substances that were previously unknown to them, especially those from the group of pesticides. They can interact with each other (also with preparations administered by beekeepers), sometimes causing a synergy effect, which can increase their toxicity.

12. Wild-living bee colonies do not lose the wax from the nest. Rebuilding combs is a large energy expenditure for the colony and in the case of a more primitive economy, based on cutting out honeycombs, the bee colony must make an additional effort to rebuild the nest.

A wild bee nest looks completely different from that found in a hive normally used in apiaries, photo by P. Kasztelewicz.

13. Unlike most colonies in apiaries, those living without human interference decide for themselves which larvae will be used to raise young queens. During one experiment, it was shown that bees, when raising emergency queens, did not choose larvae at random, they preferred those that came from specific paternal lines[36].

14. Drones selected for artificial insemination do not compete with others and therefore do not have to prove themselves in the mating process. Free-flight insemination promotes the selection of the male line of genes responsible for healthy and strong drones.

Observations and experiments

The differences between the living conditions of bees in nature and in apiaries allowed Seeley to put forward several hypotheses: the distances between nest sites are good for the health of bees; smaller cubic capacity may increase the tendency of colonies to swarm, which serves to limit the number of parasites in the colony. The professor then conducted research to verify his ideas.

In the first experiment, 12 hives with bees (the first group) were arranged in the same way as they are normally arranged in an apiary, i.e. at a direct distance from each other, with their entrances facing the same direction.

The same number of colonies from the second group were placed at distances of 30-80m from each other. After two years of the experiment, during which varroa was not controlled in any of the colonies, it turned out that only five of the 24 colonies survived: all in the group where the hives were separated by a greater distance. Interestingly, these were the only colonies that swarmed in the second year of the experiment. Those that did not produce a swarm collapsed in late summer or autumn because of high varroa levels. It turned out that swarming allowed the colonies to maintain the infestation rate at the level that they could tolerate. In the group where the hives were placed close to each other, all the colonies died, even those that produced swarms.

---

36   R.F.A. Moritz. Rare royal colonies in honey bees, *Apis mellifera.* Naturwissenschaften 2005, 92.

Seeley used queen bees in two characteristic colours for the experiment, in one part of the apiary the queens were dark, in the other part, intensely yellow. While because of free-flight insemination of queens and their female offspring (worker bees) had different colours, the drones retained the colours of their mothers. This allowed the professor to study the drifting of drones between colonies. It turned out that in colonies placed close to each other the drones mixed between hives, which was not observed in the hives in the second group. He therefore concluded that spacing the hives with greater distances allowed the elimination of bees drifting.

Meanwhile, in the group of hives set close together, in two colonies that swarmed in the second season, the varroa mite population stabilized at a relatively low level during the summer, but it increased rapidly at the end of the season, when neighbouring colonies began to die off. The colonies set close together infected each other with mites. In those that were separated by a few dozen meters, however, this tendency was not observed. The phenomenon of the so-called mite bomb (varroa bomb) consists of the fact that bee colonies weakened by the increasing level of mite infestation are robbed, and some of the infested workers fly into other hives, which causes the mites to infest nearby colonies (those that rob weakening colonies or those that allow bees carrying mites into their colonies)[37].

---

37 Seeley described the phenomenon of mite bombs in: Mite bombs or robber lures? The roles of drifting and robbing in *Varroa destructor* transmission from collapsing honey bee colonies to their neighbours [D.T. Peck, T. Seeley, published in "PLoS ONE" 2019, 6]. The experiment showed that robbing was most important for rapidly weakening colonies, where the mite levels were very high – so the transfer of parasites were from weakening colonies (robbed) to those that were robbing (at the time of colony collapse and robbing, the infestation began to decrease in those, while in other colonies it grew rapidly). Other conclusions resulted from the experiment of K. Kulhanek et al., Accelerated *Varroa destructor* population growth in honey bee (*Apis mellifera*) colonies is associated with visitation from non-natal bees, Scientific Reports 2021, 11. The highest increase in mite infestation was observed in the colonies that allowed the entry of foreign bees into the hive (weak nest defence), but not in the hives robbing others. It was also suggested that the use of protection against robbing at the exit (specially constructed nets - robbing screens) can significantly help to maintain a low mite infestation of colonies, e.g. significantly reduce the increase in the infestation of mites in colonies moved to areas of good forage in autumn. Conclusions from both experiments suggest then, that the transfer of mites during robbing may be either from robbing colony to robbed one, or the other way round – depending on the situation. Seeley does not deny that drifting bees also played a very large role in the spread of mites (before colonies with a large infestation level began to collapse).
In turn, another study suggests that the increase in stress caused by parasites and viral infections, combined with the impact of viruses on the cognitive abilities of bees, led to an increase in drifting among worker bees, and thus to an intensification of the phenomenon of spreading parasites between bee colonies, G. DeGrandi-Hoffman, Are Dispersal Mechanisms Changing the Host-Parasite Relationship and Increasing the Virulence of *Varroa destructor* (Mesostigmata: Varroidae) in Managed Honey Bee (Hymenoptera: Apidae) Colonies?, "Environmental Entomology" 2017, 8. John Kefuss, in turn, claims that the phenomenon of mite bombs is associated primarily with colonies in which resistance selection is not carried out (primarily treated ones). When using the so-called Bond

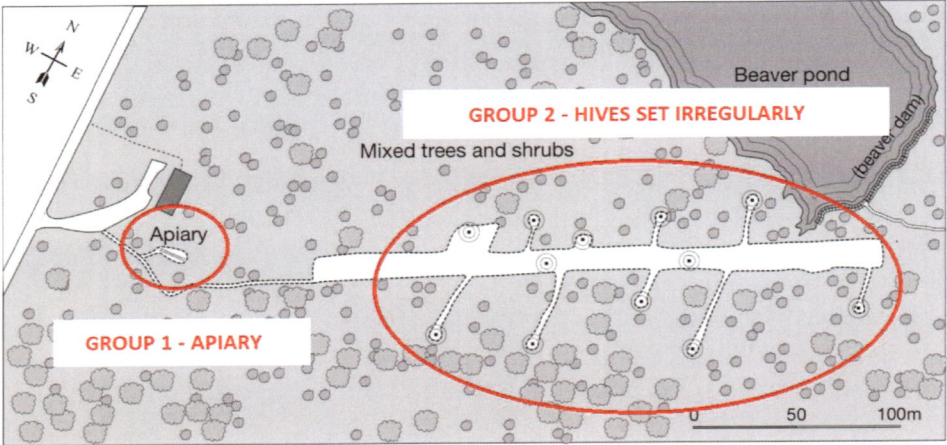

GROUP 2 - HIVES SET IRREGULARLY

Mixed trees and shrubs

Beaver pond

(beaver dam)

Apiary

GROUP 1 - APIARY

0   50   100m

The hives from the first group were set up as in a standard apiary, and the hives from the second group at 30-80 m from each other, photo by T. Seeley.

Another experiment shows that mites from a colony weakened under the influence of excessive levels of varroa infestation can infest bee colonies even within a radius of 1.5km[38]. In my opinion, however, Seeley's research suggests that the effect of a mite bomb does not necessarily have to affect all colonies in the area, if they are not too close to each other. This means that

---

test, i.e. subjecting the colony to natural selection, the non-resistant colony dies, then the threat it creates is short-lived because the non-resistant colony is dead. More than mite bombs, the breeder is afraid of not eliminating "bad genetics" (lines) from the apiaries, regardless of whether they are not resistant, sustain the bees' maladjustment to threats, or are simply unproductive. According to Kefuss, the matter is simple: it is enough not to breed from colonies that have a large mite infestation. In this way, the problem of mite bombs would be solved very quickly.

38   E. Frey. Invasion of *Varroa destructor* mites into mite-free honey bee colonies under the controlled conditions of a military training area. Journal of Apicultural Research 2015, 4.

colonies weakened and dying, with a high level of mite infestation, will have the most severe impact on colonies located close by, i.e. in the same apiary.

It is worth adding that some beekeepers who do not treat bees, e.g. Solomon Parker[39], deny the phenomenon of the mite bomb (they do not notice the relationship between one colony weakening and finally collapsing in the apiary and the situation in the neighbouring hives). Other beekeepers (e.g. David Heaf or John Kefuss) claim that the effects of the mite bomb can be minimized if the bees develop mite resistance. Resistant bees may give up robbing weakening colonies or not allowing bees infected with mites into their hives or simply quickly get rid of mites from the hive if their number increases. In this context, it is worth mentioning that Seeley did not use bees selected for resistance in these studies. These were bees from commercial farms, typical of apiaries where standard management is carried out.

In the second experiment researchers verified the hypothesis concerning the positive influence of a relatively small nesting volume (nest cavity) on the health of the colony. Again, two groups were compared, each containing 12 colonies. This time, the variable was to be the volume of the nesting cavity. The first group was placed in one Langstroth box with a capacity of 42 litres. The second group had boxes added to them as the colony grew.

One group of bees was left in single boxes, the other had boxes added successively as the colony grew, photo by T. Seeley.

The differences began to become apparent in the spring of the second year of the study. It turned out that out of 12 colonies in small hives, 10 swarmed, some several times. Meanwhile, only two colonies swarmed from large hives. The mite population was regularly checked during the experiment. In the second year of the experiment, the mite population in the small hives was lower and remained relatively unchanged until autumn, when it turned out to be around 2%. The number of mites in large hives

---

39   I mention S. Parker in the chapter Bees and People.

increased throughout the year, and colonies began to collapse at the turn of August and September. At that time, in four of the small hives, the mite population was growing rapidly, which was caused by robbing the collapsing large hives (and thus by a mite bomb). Seeley considered this to be his mistake, because both groups were placed too close to each other (they were separated by 60 m). However, this mistake allowed the observation of the bomb phenomenon. From the group located in large hives, only two colonies survived the winter after the second season (those that swarmed), while from small hive group eight of the 12 survived. In the small-hive group, those that collapsed were those that robbed the collapsing colonies in autumn, and where the infestation increased rapidly. It is worth adding that no honey was obtained from small hives, while over 200 kg was collected from large ones. However, these harvests were paid for by the deaths of the colonies.

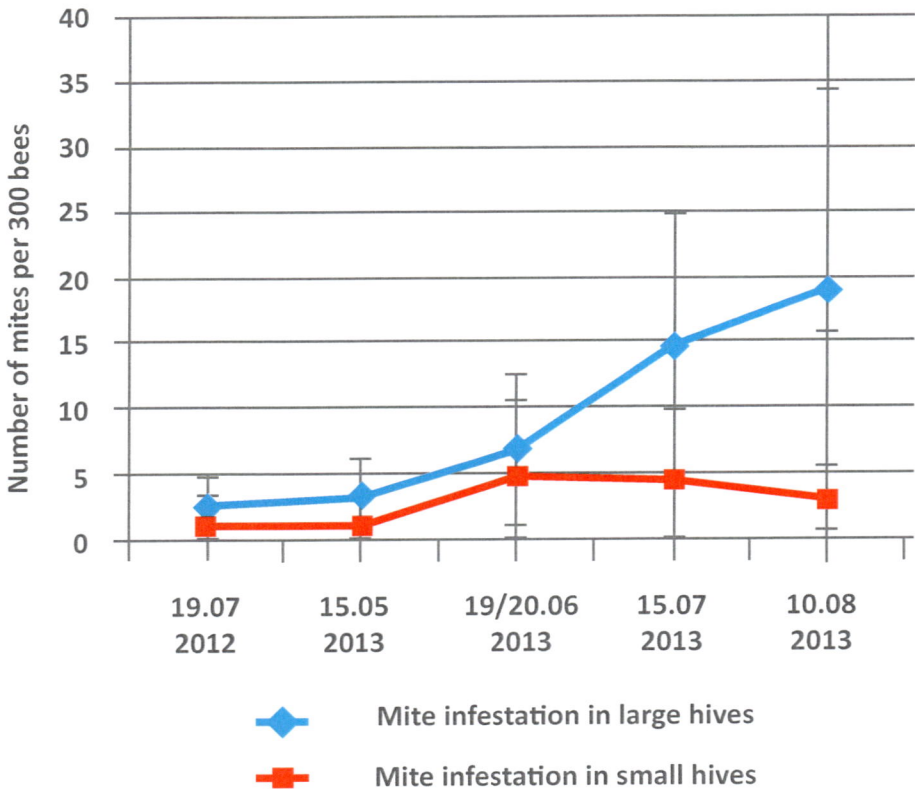

In the second season of the experiment, differences in the degree of infestation of colonies with varroa mites began to be revealed. In the first group (large colonies), the infestation increased to the level of 10%, in the second it remained at a low level until the end of the season, figure by M. Uchman based on T. Seeley

Darwinian beekeeping principles - Conclusions

Thomas Seeley claims that colonies of western honey bee species can survive in nature without human help. I think most of the scientific community would agree with this thesis. However, the possibility of running a profitable apiary without using varroa treatment is rather questionable. Seeley formulated several principles, thanks to which it would be possible to increase the chances of survival of apiaries in conditions where "chemicals" are not used. For understandable reasons, they apply more to amateur apiaries than commercial ones[40]. They can be used in the so-called transition period, i.e. intended for conducting basic selection of bees. As selection progresses and bees become more resistant, it would also be possible to intensify apiary management (e.g. systematically increasing the volume of hives or gathering more colonies in one place). The principles of Darwinian beekeeping are simple in their assumptions, although in some cases they are not easy to implement. Let us get to know the most important of them:

1. Keep locally adapted bees in your apiary. This can be done by anyone, regardless of whether you are a professional or amateur beekeeper. Bees should be adapted to the local climate, floral sources and local ecological situation. The best solution is to buy bees from local beekeepers, catch swarms or raise queens from bees from your own apiary. The importance of "localizing" bees for their health is also emphasized by Polish scientists.

   Some claim that for the condition of bees to improve noticeably for practicing beekeepers, it would be enough to limit the migration of bee colonies for just two years. This is one of the key conditions for maintaining the health of the population, and it would be easy to meet, but despite this, beekeepers are not convinced to implement such practice. In the pursuit of better harvests, both breeders[41] and professional beekeepers, and even amateurs, compete in bringing foreign bees into their apiaries,

---

40 Seeley says, "I'm not saying it's appropriate for everybody. I think it's probably only appropriate for a certain segment of the beekeeper world. It's a segment that I think we could respect more and give more credence to people that want to do it on a small scale. Want a colony of bees that's something they enjoy as a piece of nature, not as a production unit and I think that's valid.";
www.beekeepingtodaypodcast.com/bee-hunting-with-dr-tom-seeley-archive-super-episode-s4-e48, accessed January 18, 2025.

41 To be more precise, this mainly concerns so-called amateur queen producers not breeders.

which is supposed to solve their problems (most often small honey harvests or having too defensive bees). In the long term, however, this action does not bring the expected effects, and only causes an increased burden on the health of the bees.

2. Space the hives as far apart as possible. The best solution for the health of bees is to copy nature. In the past, wild bees occupied nest sites at least several hundred meters apart. The most primary and primitive form of beekeeping was based on these principles. However, efficient apiary management requires a compromise between the health of the bees and the needs of the beekeeper. Seeley believes that with every meter the risk of bees drifting decreases, and it can be prevented by placing the hives about 30m apart. With increasing distance, also the probability of robbing decreases, although in this case an even greater distance is necessary. However, we can undoubtedly organize an apiary in a compromise way. You can place the hives in smaller groups on a larger number of bee yards. Swedish breeder and retired professional beekeeper, Erik Österlund, suggests that no more than 6-8 colonies should be placed in one spot. It seems that it is worth coming to such a group of bees to perform the necessary work and inspections, at the same time, the group is small enough that the risk of cross-infection between colonies is lower and affects only a smaller group.

    We can also take care to place the hives as far apart as possible in each place, with small entrances facing different directions. This reduces the likelihood of drifting bees and makes the guards' job easier.

3. Keep your bees in small hives. Professor Seeley advises using a hive no larger than a full Langstroth box (42 liters) as a nest, and a medium (28 liters) super as a honey box. The honey harvest will then be smaller, and the bees will most likely swarm. For amateur beekeepers who want healthy bees free from so-called chemicals, and count on harvests only for their own use, this could be a compromise. In such a hive configuration, the probability of maintaining a low level of mites increases. Seeley adds that since his discoveries were popularized in the US, the so-called small hive beekeeping movement has become popular among amateurs.

The widest possible spacing of the hives is to prevent the bees from drifting and to limit robbing. In our conditions, this is not guaranted that the bees will survive, photo by author.

4.   Make sure that the inside walls of the hive are rough. Such surfaces will encourage the bees to cover them with propolis, which will provide them with a shield against pathogenic micro-organisms. It is worth mentioning that by constantly scraping, cleaning, burning (or working on smoothing the walls of the hive in other ways) we are doing the bees a disservice. In this way, we are only permanently removing the natural protective barriers of the bees.

5.   Use hives that provide adequate thermal insulation.

6.   Place the hives as high above the ground as possible. Bees prefer their nest to be located higher (at a greater distance from the ground) for several reasons. Firstly, because there is less humidity in such a nest cavity. Secondly, such a location protects them better from predators and other pests. In natural conditions, a higher nest was also more reliable protection against bears or rodents. In apiaries, beekeepers usually take care of such things. Regardless of the location of the hive, however, it is certain that bees manage perfectly well without "landing boards", "runways"

(or "mouse platforms" as some natural beekeepers ironically call such structures), since bees are fully capable of landing on the vertical wall of the hive.

German Einraumbeute hive (with the size of the Dadant frame vertical) with walls well insulated with reeds, photo by D. Heaf.

7. Allow bees to use 10-20% of the combs for drone cells.

8. Do not change the nest organization. During inspections, each frame should be returned to its place. Seeley also advises not to separate the brood using empty frames as part of anti-swarming treatments.

9. Minimize the frequency of moving hives. Research recommends that if it is necessary to move a bee colony to another location, wait for the period of no flow, then interference in the work of the bees will have the least severe effects.

10. Refrain from treating bees against varroa.

Seeley gives this last piece of advice with a warning. He believes that if the apiary is not properly organized, refraining from treating bees can lead to a situation where evolution, instead of favouring resistant bees, can support virulent mites and pathogens. This applies to a situation in which they can kill the colony, but do not die with it, as is most often the case in nature. Due to the high density of hives in the apiary, weakened colonies are robbed, and virulent pathogens and parasites reach the nearby colonies. The professor suggests closely monitoring the level of mite infestation and reacting when it reaches several percent. After exceeding this threshold, the colony should be treated or killed (Seeley calls this "pre-emptive killing").

He claims that in this case our actions are in line with the "evolutionary direction"[42]. First, we eliminate the colonies that do not have sufficient resistance to the parasite and (with high probability) would collapse. Secondly, we reduce the risk of a mite bomb in the apiary almost to zero (in nature, this phenomenon can also occur but, due to the distance of the nest sites, it is of little importance to the entire population. Seeley claims that if we do not perform these procedures, colonies that show a certain level of resistance will be overwhelmed by such a scale of mites that they will not be able to cope. After treatment, the queen should be immediately replaced with one from a colony that shows the ability to keep the mite population in check. This serves to systematically increase the resistance of one's own apiary[43].

---

42 The term "evolutionary direction" may seem inappropriate, because evolution does not have a set direction, but follows the pressure of the environment. In our apiaries, the "evolutionary direction" will be the one we know from our practice: pathogens and parasites become virulent and gain a specific advantage in an arms race with the bees, whose adaptations are systematically weakened. The evolutionary direction will depend on various circumstances, including the broadly understood internal and external environment. The idea of Darwinian beekeeping assumes the creation of such conditions as to reverse the current evolutionary direction in a direction that will serve the health and adaptation of bees. The statement is therefore another simplification, a reference to the direction that evolution would probably take (we can assume this based on cases in various parts of the globe) in conditions close to natural ones (i.e. where human interference is limited to a minimum).

43 This type of selection is sometimes referred to - after Kefuss - as a soft-Bond test. It is the same breeding philosophy that was adopted by, for example, Erik Österlund or Randy Oliver.

David Heaf does not fully agree with the justification for pre-emptive killing, at least in relation to the conditions in his environment. In the Welsh county of Gwynedd, where resistance of bees to varroa is relatively common, the problem of mite bombs does not exist. By killing a colony - Heaf argues - in which the mite infestation is increasing, we are essentially acting against nature, not with it. It happens sometimes that some colonies in which the level mite infestation increases manage later to reduce it and survive.

According to David Heaf, this happens, for example, when a colony (despite problems) swarms, or activates some kind of hygienic self-cleaning mechanism. It happens that colonies in which the number of mites increases periodically, later remove them from their colony. If the bees show resistance (or tolerance) to the pathogens carried by the mites, they have a chance to successfully survive the winter, even if the mite infestation is relatively high.

When treating colonies in which the mite infestation is increasing, one must be aware that any type of pressure exerted on the mites and pathogens may, in the long term, lead to an increase in their virulence or pathogen resistance. For this very reason, Seeley suggests pre-emptively killing the colony. Then we eliminate the entire colony along with the parasites and micro-organisms present in it. If, however, we treat the colony, some of the parasites and pathogens will certainly survive, despite the use of substances that are harmful to them. These pathogens and parasites are the most dangerous[44], which is why it is a better solution to kill the bee colony along with the entire population of the bees. For many people, however, this is not only ethically unacceptable, but also uneconomical. Therefore, both arguments must be weighed. Seeley claims that he decides to treat colonies when he can immediately replace the queen with better genetic material (i.e. not in the next season, but immediately after the treatment). Otherwise, he commits pre-emptive killing of his colony.

As a professional beekeeper for whom work in the apiary was a source of a significant part of income, Erik Österlund did not practice pre-emptive killing; in such a situation in the previous years, he always treated the colonies and replaced the queen. After the treatment, he placed a frame of brood in the hive from a colony that has remained in the apiary untreated for the longest time. According to him, this is a kind of injection of healthy bacterial microflora for the colonies that have been exposed to miticides

---

44   They show a type of resistance to the treatment (to the substance used); if this feature is encoded in the genes of the organism, they can pass it on to their offspring. In this context, it is difficult to consider a treatment that resulted in the elimination of about 90% of parasites as a success.

(biocidal) substances. Such a practice can have an impact on compensating (at least partially) the harmful effects of the treatment.

Undoubtedly, the direction Darwinian beekeeping is taking can help to develop a kind of balance between bees and their parasites and pathogens. First, we should shape our apiaries in such a way that, together with the non-resistant colonies, we can destroy the most virulent strains of pathogens and parasites. Meanwhile, most beekeepers, by organizing their apiaries according to the principles of intensive beekeeping, and above all by constantly intervening in a kind of "arms race" between the honey bee and bee pests (and the pathogens they carry), contribute to the growth of pathogen virulence. The longer this situation persists, the more difficult it may be to break the vicious circle.

Fighting against varroa - a road to nowhere

Impact of treatments on the ecological system

Polish law lists the western honey bee (*Apis mellifera*) as a farm animal. However, the discussion on its status is still ongoing. Some believe that it is a domesticated organism and, according to this group of people, bees are unable to survive without human help, while others treat bees as wild organisms. Thomas Seeley claims that bees are partially domesticated animals, as a large part of their population is dependent on humans[45], but at the same time, humans have not managed to eliminate from their genome the characteristics necessary for independent survival[46].

The situation of the honey bee is complicated because, unlike other farm animals, its life cannot be separated from the natural environment[47]. Although theoretically it would be possible, for example keeping them in

---

45 Seeley argues that this is largely due to the conditions in which we keep the bees.

46 I share this opinion. Bee populations from Poland and continental Europe seem to be largely domesticated, but representatives of the same species also live in other regions around the world where the situation is completely different. If we take the entire species into account, it seems to be closer to the wild organism.

47 No legal regulations or statutory definitions will change this. Many animals, including those that are difficult to consider fully domesticated (e.g. cervids, foxes, etc.), can be kept in boxes, farms, etc., providing them with the necessary nutrients and largely controlling their living conditions. Even fish, whose breeding conditions are regulated, can be isolated in water tanks. I am far from claiming that this is good for the animals. In some cases, the conditions in which animals are kept raise strong objections and are also an epidemiological threat. The blame for this should be placed on modern agriculture, in which efficiency counts. The same agriculture (via veterinary services) postulates the constant need to carry out treatment for these animals, because they suffer from many diseases.

huge indoor buildings, but then keeping honey bees would completely lose its meaning, because they are primarily pollinators. Obtaining honey and other apiary products is a side effect of the basic task we set for bees. Isolation would mean giving up the benefits that honey bees provide us with that we consider the most valuable[48]. We can therefore try to supervise bee nest sites effectively, but we are not able (to the same extent as we do in the case of other animals) to control the contacts of bees with various species of fauna and flora, including those that may be harmful to bees (e.g. toxic plants, predators or pathogenic microorganisms). While we can effectively (or poorly) prevent, for example, contacts between pigs from different farms or with wild boars, we are unable to prevent our bees from contacting other insects, both those of its own and other species.

We also cannot fully control their selection and reproduction. Queen bees fly out of their nest cavity on mating flights and are inseminated by several (sometimes even several dozen) drones. For some time now, attempts have been made to control selection using artificial insemination methods, but on the scale of the entire population, this is a marginal phenomenon, which is additionally associated with other problems (not only technical or logistical, but also, for example, inbreeding). The assessment and control of bee characteristics is also much more difficult than in the case of pigs or cattle.

The history of the last few decades of the fight against varroa mite, proves that this is not a path that leads to a systemic solution to the problem. Although beekeepers have learned to better control the parasite infestation of bees, it does not always bring the expected results. Despite, as it seems, increasingly effective treatments, the average mortality of bees remains at a constant, high level. Some scientists claim that immediately after varroa mite appeared in Poland, the death of an untreated bee colony most often occurred in the third or fourth year after infection. The colonies therefore lived longer and died with a higher level of infestation than is the case today. Currently, the average bee colony is dying no later than at the end of the second year from the last treatment, and often already in the first autumn (this was the case in my apiary in 2014).

---

48   This is a simplification for the purposes of considerations: no organism or species has a centrally determined "task". However, we often tend to assign them specific roles, which most often results from economic practices, and sometimes also from philosophical or religious concepts. Organisms whose tasks or roles we are unable to define, we treat as pests (often having no qualms about killing them). Meanwhile, it seems that the only "task" of all living organisms is to survive.

What is more, several dozen years of conducting apiary management based on pesticides has resulted in bee colonies dying with mite infestation levels three times lower than previously seen. Such data is given, for example, by Dr. Maciej Howis[49]. These opinions are also shared by Prof. Paweł Chorbiński, as well as scientists from abroad, e.g. Prof. Jürgen Tautz from the University of Würzburg.

Most beekeepers with more experience claim that at the beginning of the invasion, the problem of varroa for the entire season could be solved with just one treatment. Today, some practitioners also assure that they use only one dose of so-called chemicals per season, which effectively allows them to keep the bees alive. However, when we ask for details, they start to present a long list of apiary procedures. It turns out that most often they use isolation of queens to force a brood-less period, sometimes they use brood removal (cutting out drone brood to limit the invasion of drone brood) throughout the season or remove all brood altogether, disinfect all elements of the hive at least once per season (or more often), disinfect stored combs in fumes of various substances, sterilize apiary tools and - finally - every two or three seasons they burn all frames (sometimes other elements of the hive, e.g. bottom inserts).

Some additionally use acids or oils, not considering them as so-called chemicals, despite their decidedly biocidal nature. What's more, some beekeepers also claim that they do not treat bees, but only fumigate colonies to check the level of infestation. Of course, there are also those who use so-called chemicals sparingly and do not perform a large part of the above-mentioned procedures. However, they periodically incur quite large losses. It seems that over the last few decades in the fight against the mite, it has only led to the fact that varroa is killing bee colonies much faster than before.

The problem is therefore much more difficult to solve and requires much more work than it did several decades ago. When I asked Professor Thomas Seeley about the reason for this situation, he did not hesitate to point to the increase in the virulence of pathogens, primarily DWV. Since the invasion, the problem has been getting worse, because intensive apiary management encourages the evolution of pathogens towards an increase in

---

49   M. Howis, Aspekty biologiczne rodziny pszczelej – relacje pomiędzy *Apis mellifera* a *Varroa destructor* przy stosowaniu zabiegów ograniczających populaację pasożyta [Biological aspects of the bee colony - relationships between *Apis mellifera* and *Varroa destructor* when applying treatments limiting the parasite population], University of Life Sciences, Wrocław 2012.

their virulence[50]. Living organisms, including bees, when exposed to strong pressure can evolve very quickly[51]. However, it must be remembered that we do not mean major evolutionary changes (e.g. concerning the anatomy of insects). The honey bee has evolved for several dozen million years and during this time has developed many immunological mechanisms and specific behaviours that serves to maintain population health. During such a long period, bees have had to deal with a specific species of brood parasites or so-called ectoparasites more than once. It would seem, therefore, that it would be enough for them to use previously developed immune mechanisms to fight any new threat. Varroa may of course be more dangerous than many other parasites that bees have encountered during their evolution. In regions where bees have undergone a process of natural selection after the invasion of Varroa destructor, the threat is incomparably lower than in our economic region. There are places in the world, although not free from the *varroa* mite, where the colonies don't suffer from varroosis[52].

Paradoxically, varroa is dangerous for bee colonies only where beekeepers constantly fight it. Therefore, it is not a threat that could prove fatal to the entire species. Looking from a broader perspective, I am ready to defend the idea that the problem of varroa and the disease it causes is primarily an economic problem, not a biological one[53]. Let us also look at one of the examples of bees' rapid adaptation to another threat, which is passed over in silence today. This concerns another mite, the tracheal mite (*Acarapis woodi*), which according to many people in the first half of the 20th century almost led to the extinction of honey bees in the British Isles, and at the end of the last century it decimated apiaries in North America.

In both countries, the problem ceased to exist after a few years. Today,

---

50   Seeley claims that if the induction of disease increases the chance of parasite transmission, there will be an evolutionary increase in its virulence, D.T. Peck, T.D. Seeley: Mite bombs or robber lures? The roles of drifting and robbing in *Varroa destructor* transmission from collapsing honey bee colonies to their neighbours, "PLoS ONE" 2019, 6. This happens in the conditions of modern apiary management, where bees are not immune. This should not happen in the case of bees that do not succumb to the disease and/or with large distances between nest sites.

51   A.S. Mikheyev et al, Museum samples reveal rapid evolution by wild honey bees exposed to a novel parasite, Nature Communications" 2015, 8.

52   The parasite is there (colonies are infested by varroa), but we cannot speak about the occurrence of varroosis (a disease it indirectly causes). The mite lives in the hive and feeds on bees, but the health situation of the entire superorganism is not much different from that of colonies infected with, for example, wax moths.

53   By this statement I do not deny the threat that varroa poses to individual bee colonies or larger populations in those regions of the world where bees have not developed resistance through natural selection.

many amateur beekeepers probably do not even know about the existence of tracheal mites[54].

In my opinion, the continuous treatment of a disease of farm animals, which has been permanent for several years, does not make sense. In such situations, we should look for systemic solutions, e.g. selection. Treatment would make sense in the case of a disease that affects only some animals or the entire population, provided, however, that it is for a limited period (season). Meanwhile, varroa is an organism that permanently resides in all hives. In principle, we should assume that, firstly, every bee colony is infested by the mite, and secondly, that treatments will not always stop the decline of non-resistant colonies. For this reason, beekeepers are forced to permanently perform treatments against varroa. This is also proven by an experiment conducted by the Bee Disease Department of the National Veterinary Institute in Puławy[55], where experiments are conducted on the effectiveness of various types of medicines for bees. The tests consist of administering specific preparations, according to the manufacturer's recommendations, and then verifying their effectiveness using other methods, in this case Apiwarol (amitraz). The bees are treated until the effect is achieved, i.e. if dead mites appear on the bottom of the hive. When they are gone, we assume that the colonies have been completely cured. Meanwhile, the following year, even several thousand mites are detected in some colonies (the inglorious record was achieved by a colony in which almost 9,200 female mites died after the treatments!).

What are the conclusions of this experiment? Firstly, that it is impossible to eliminate the varroa problem once and for all, even with the help of very effective treatments. Secondly, that in one apiary, and therefore under the same conditions, the difference between the number of mites in individual colonies can be very large. This means that colonies can control the varroa population at very different levels[56]. Thirdly, and finally, it is proof of how wrong the ideas of so-called control fumigation of only some bee colonies in a bee yard to check the varroa infestation in the entire apiary. It turns out

---

54 Some beekeepers from the USA still remember it very well, as the disease caused by the tracheal mite decimated their apiaries shortly after the arrival of the mite (in the 1980s).

55 K. Pohorecka, Oczekiwania versus rzeczywistość, czyli skuteczność zwalczania inwazji Varroa destructor a liczebność populacji roztoczy w następnym sezonie [Expectations versus reality, i.e. effectivness of combating *Varroa destructor* invasion and the size of the mite population in the next season], Pszczelarstwo 2020, 10.

56 This diversity provides a basis for further selection of bee resistance. Selection is effective primarily where there is high variability of traits in the population.

that the level of infestation of one colony can be even ten times higher than in a neighbouring one.

I therefore believe that in a situation of constant and widespread threat caused by the presence of the *Varroa destructor* in bee colonies, it is necessary to direct our actions towards evolutionary or adaptive changes. Otherwise, we will remain in a vicious circle, condemning ourselves to a constant arms race. However, with this distinction, that on one side of the barricade we will place *Varroa destructor* (and the pathogens it carries), and on the other beekeepers and the entire chemical and pharmaceutical industry, and not the bees with their abilities and immune systems.

Varroa will constantly adapt to the pressure of the treatment, this has already been observed many times. Firstly, the development of resistance to some active substances has already occurred[57]. Behavioural changes were also observed: the life cycle of the mite changed in such a way that it spends much less time on adult bees and more time in the brood cells reproducing which avoids exposure to most chemical treatments[58].

As a result, the more we use so-called chemicals, the more difficult it is to defeat the pest and consequently we go in circles. In doing so, we have contaminated the bees' living environment and for decades we have prevented them from developing natural biological adaptations to combat varroa. The chemicals we use in the apiary penetrate the beeswax, the combs are melted down and the wax returns to the hives in the form of foundation. In this way, toxic substances (along with their metabolites) have been present in the hives for several decades. However, their concentrations in the hives are not high enough to threaten the existence of pathogens and bee parasites[59]. In turn, subjecting pathogens to constant exposure to toxic substances promotes the development and consolidation of their resistance, also in combination with other factors, may even cause an increase in virulence. It is worth noting that after only a few years of discontinuing treatment against varroa, a certain balance is established between the honey bee and the mite (and the pathogens it carries). The most important change that then occurs

---

[57] Various types of mite resistance to treatments have been found with respect to their ability to metabolize some of the substances used in apiaries, D. Sammataro, The resistance of varroa mites (Acari: Varroidae) to acaricides and the presence of esterase, International Journal of Acarology 2005, 31; P. Rosenkranz, Biology and control of *Varroa destructor*. Journal of Invertebrate Pathology 2010, 1; N. Morfin, Surveillance of synthetic acaricide efficacy against *Varroa destructor* in Ontario, Canada, The Canadian Entomologist 2022, 4.

[58] The acquisition of such skills is the result of unintentional natural selection, and not conscious learning by the mites to avoid exposure to chemical treatments.

[59] They create stress zones.

is a significant reduction in the virulence of pathogenic micro-organisms, and the development of a more effective immune response in bees.

It is obvious that all living organisms are subject to the laws of nature, with the laws of evolution at the forefront. Also, when man interferes with the ecological system, e.g. through chemical treatments or biotechnical procedures, it may be necessary to maintain this artificial balance indefinitely[60], but it is not a balance that we can intuitively call natural, where co-adaptation of organisms takes place without human intervention. The procedures applied systematically caused a shift in the centre of gravity, which is why this system can be compared to the movement of a pendulum or the stretching of a spring. Moreover, we observe the "spring return" effect in apiaries every few years - after all, a very large population of bees die periodically in some regions. A spring stretched to its limits must eventually return to its natural state.

Toxic substances can have a significant impact on the health of both individual bees and the entire population. Not only scientists but also ordinary beekeepers noticed the reduced fertility and longevity of queen bees. For most of them, however, this is not a big problem, because the beekeepers have adopted the practice of replacing their queen bee no later than in their third year of life. Most often, queens are therefore killed before their health problems appear[61].

---

60   I refer to the common understanding of these concepts. If we look at the ecological system from a natural perspective, we will see that equilibrium does not exist. Each organism realizes its own selfish goals: to survive, to satisfy more and less basic needs, to reproduce (with the emphasis on shifting the ecological system towards itself, in such a way as to realize the goals most fully). These actions do not have to be (and in most cases are not) undertaken "consciously". None of the organisms cares about the good of the species; consciously or not, it cares about its own good (sometimes evolution develops mechanisms in which it "pays" for organisms to take care of others to survive). However, if we do not delve into the individual relationships, we may have the impression that ecological systems strive for a kind of equilibrium in which different organisms coexisting in the same environments have "learned" to live next to each other in different relationships, to benefit from each other's presence, and at the same time to defend themselves against threats.

61   Here I will refer to the book by P. Wohlleben, The Hidden Life of Trees, published by Greystone books, 2016, considered controversial (I share this opinion). Both the principle of anthropomorphising of trees and forest adopted by the author are controversial, as well as the way of presenting the issue (the language of emotions described by some reviewers as downright infantile, and at the same time presenting phenomena occurring in the forest in a way detached from the objectivity or rationalism that should characterize popular science works or reporting). However, the book is interesting for another reason: it shows the evolution (or maybe even revolution?) in the approach to nature, departing from the philosophy of nature treated purely in terms of utility. Wohlleben says that since trees from such productive forests do not have the opportunity to grow old - because they go to the sawmill at the age of one hundred - the negative effects they have on health are almost unnoticed. He also notes that when he started his professional career as a forester, he knew as much about the secret life of trees as a butcher knew about the feelings of animals. Modern forest management is concerned with the production of timber, i.e. cutting down trees and then planting new ones.

Refraining from a systemic control of bee diseases could bring many benefits in the long term. The most important ones include:

> Natural selection of a population of bees that would be adapted to the presence of pathogens and parasites (reducing bee mortality due to diseases);

> Reintroduction of *Apis mellifera* to nature, i.e. building up self-sufficient wild populations of honey bees supporting the natural adaptations of farmed bees;

> Reduction of workload and costs of running apiaries related to the purchase and administration of medicines or other chemical substances;

> Improving the quality of bee products.

Principles of natural selection - ecological tolerance

Some beekeepers believe that under apiary conditions, when using unnatural nest sites (hives), we cannot speak of natural selection. They claim that the tools of Darwinian selection don't really apply[62]. Such an understanding of natural selection is not in accordance with biological knowledge.

With a certain amount of goodwill in the interpretation of words and the recognition that nature has some right or proper state (however, this is not a view consistent with science), we could accept this as true, if we consider

---

He claims therefore that when reading professional journals, one may get the impression that the good of the forest is worthy of attention to the extent that it is necessary from the point of view of its optimal exploitation. Doesn't an analogy to considerations about apiary management come to mind?

62   For example, German researcher Torben Schiffer believes that in the conditions of apiary management we cannot speak of fully natural processes - including the process of natural selection. It is not only about coping with disease factors, mainly varroa, or external factors (e.g. climate, nectar flow system). Schiffer believes that bees that live in a frame hive lose the ability to build combs that are correct from a biological point of view, often starting halfway up (attaching the combs to the walls) instead of from the top, which makes it difficult for them to survive the winter. I assume, however, that any controversy regarding the understanding of the term "natural selection" in this case results rather from translation inaccuracies than from replacing their strict understanding with a specific ideology.

bees as exclusively wild animals and our economic relationship with them is over. But the world of bees does not end with natural tree cavities or tree-beehives. In our geofigureal region (Poland), bees living in hives dominate, and I wouldn't even dare to say that 2% of them occupy fully natural nest sites (and probably it's 10 times less).

Natural selection is one of the key mechanisms of evolution, which leads to changes in populations in a completely unplanned way, and as a result, promotes the adaptation of organisms to different environments and conditions. The process directly affects individual organisms; they do not live in a vacuum, but in the environment of biotic (living organisms) and abiotic factors (non-living elements of the environment). In this sense, entire ecological systems are subject to evolution, and the process of natural selection promotes the co-evolution of different organisms within the ecosystem, shaping the relationships between them[63]. Organisms that cope with specific ecological conditions and situations can pass on their genes to new generations. Genetically diverse populations (which are also constantly subject to mutations and other evolutionary processes), along with expansion into new biological niches or changes in current environmental conditions, can undergo differentiation into new ecotypes, subspecies and species. As a result of the long-term process of evolution, organisms can therefore change completely in relation to their own ancestors (or other evolutionary lines) and only genetic studies will show their kinship. This process does not have to take place only in the places that most of us have in mind when we say "nature". It affects organisms both in the jungle and on the steppe, in the city, in the henhouse, the barn and finally in the hive or other wild bee nest cavity. We will also deal with the process of natural selection when beekeepers carry out treatments against pests with the help of toxic substances.

In this case, however, the parasite-pathogen-host system is in some way disturbed or rather expanded by the presence of another organism: the human beekeeper.

---

63   If we tried to place the same biocentric systems in different types of biotopes, the results of the evolution process might be different (and this is often the case).

Evolution does not promote the "best" organisms (the strongest, fastest, cleverest, etc.). It also does not clearly favour the best-adapted organisms, although these undoubtedly achieve the greatest reproductive success[64]. What is really important for organisms is to prove to be sufficiently adapted to survive and reproduce. So, we are not talking about the **best**, but **sufficient characteristics** to effectively pass on their genes to the new generation. Thus, it is difficult to talk about evolutionary success in the case of drones, which live until the end of the season and die of hunger and cold exiled from the hive before wintering. Those that die immediately after fertilizing the young queen are successful. Organisms also bear the costs of maintaining their adaptations. They must invest certain resources in maintaining the trait. It is therefore worth maintaining it only if there is an environmental pressure. On the other hand, if the pressure decreases, and the trait is no longer needed (in specific ecological conditions), it may disappear and often will[65].

Traits and characteristics are therefore not constant, given once and for all, and they are not subject to evolution to a state of perfection, but only to a level that is **sufficient** for the organisms that are equipped with them to survive and achieve reproductive success.

For the purposes of these considerations, let us assume that we will speak of surrendering honey bees to the process of natural selection when a specific balance is formed with limited human participation. I therefore assume that natural selection (elimination of individuals as part of the evolutionary process), related to resistance to parasites

---

64   The process of evolution also promotes/favours the best (highly specialized) organisms in their fields. The entire mass of organisms are however simply individuals of "average level of perfection" (in the context of a given trait) and this factor very often determines their evolutionary success. They have a different balance of characteristics, and their adaptations are/can be a response to a wider range of pressures or environmental changes (they are better adapted than outstanding organisms). Highly specialized organisms are often adapted to specific biological niches, so a change in environmental conditions can lead to depopulation or even extinction of a given species. The evolutionary formation of organisms depends on many factors. The process will be diverse in various types of organisms – for example, predators or organisms at the top of their food chains will evolve differently, than parasites or herbivores, plants, fungi or microbes, etc. The rule is quite simple, however: if you are sufficiently adapted to your niche, you will have better probability to survive. The puzzle begins at other levels, e.g. what does the phrase "sufficiently adapted" mean in conditions that are constantly changing. This process becomes more complicated in the case of social insects (all social organisms), because there is a pressure factor related to the distinctiveness of individuals (workers) versus the entire colony. Honey bee workers are not at all obedient cogs in the superorganism system (at least not always), they have their own needs and are often guided by their own interests (more than the interest of the colony, e.g. rebels), interview with Dr K. Kuszewska; "Radio Warroza" [Radio Varroa]; https://www.warroza.pl/2022/04/zbuntowane-pszczoly.html), accessed: November 1, 2022.

65   On islands where there are no predators, some birds evolve towards losing the ability to fly, as flying is extremely energetically expensive.

and pathogens, occurs where humans do not intervene in a specific arms race between organisms. It is true that the dividing line is artificial and not everyone will accept its designation in this place, and the very formulation of this definition may be considered incorrectly selected. However, such treatment of the evolutionary process in the apiary seems to reconcile the management of the economy with the preservation of the environmental adaptation of bees, which they need to survive independently. This means that in a situation where a beekeeper suddenly disappears (illness, death, lack of time, abandonment of beekeeping, etc.), such a method of running an apiary would allow a large part of the bee population to survive without humans. Of course, each of us has the right to define these phenomena on our own (and sometimes our view will not be where the science of biology strictly defines it). Some believe that feeding bee colonies that have not collected food for themselves excludes natural selection (i.e. then the death of individuals would have nothing to do with evolutionary processes). Others will state that performing artificial splits (nucleus colonies) also puts natural selection in question. The definition I have proposed is undoubtedly artificial and subjective, and it also tries to relate biological concepts to the reality of apiaries, artificially eliminating some phenomena of the process, so it may not be satisfactory for those who want to operate within strict definitions or terms. I hope, however, that it will be understandable[66]. However, I believe that standard apiary management procedures (inspections, feeding, making splits, taking honey, etc.) interfere to a much lesser extent with the ability of bees to survive independently outside the apiary[67]. The way of creating a new colony (swarm, artificial swarm, package, split), although it may have an impact at the start, it does not seem to be crucial from the perspective of developing resistance to the mite (and maintaining it), proper immunological response to pathogens, or the ability of bees to survive in subsequent seasons.

---

66   We are dealing with two extreme views with the spectrum in between. The first view assumes that natural selection occurs only in natural conditions (it is excluded by any interference, including even the use of an artificial nest cavity for the nest, such as a roof space). The second, which seems to me to be truer from an evolutionary point of view, assumes that the process of natural selection does not consider whether the pressure was created by geological factors or evolution, or whether it is the result of human activity. I try to set my own definitional boundary in a place that will allow for the management of apiaries while maintaining the direction of evolution. This will allow for the development and maintenance of their ability to survive independently (and therefore without human care) in the current environmental conditions.

67   Before *Varroa destructor* appeared, the ability of most bee colonies to survive without human care was not questioned. Let's assume that we return to such a situation where we run apiaries as before the mite invasion, ignoring the presence of varroa.

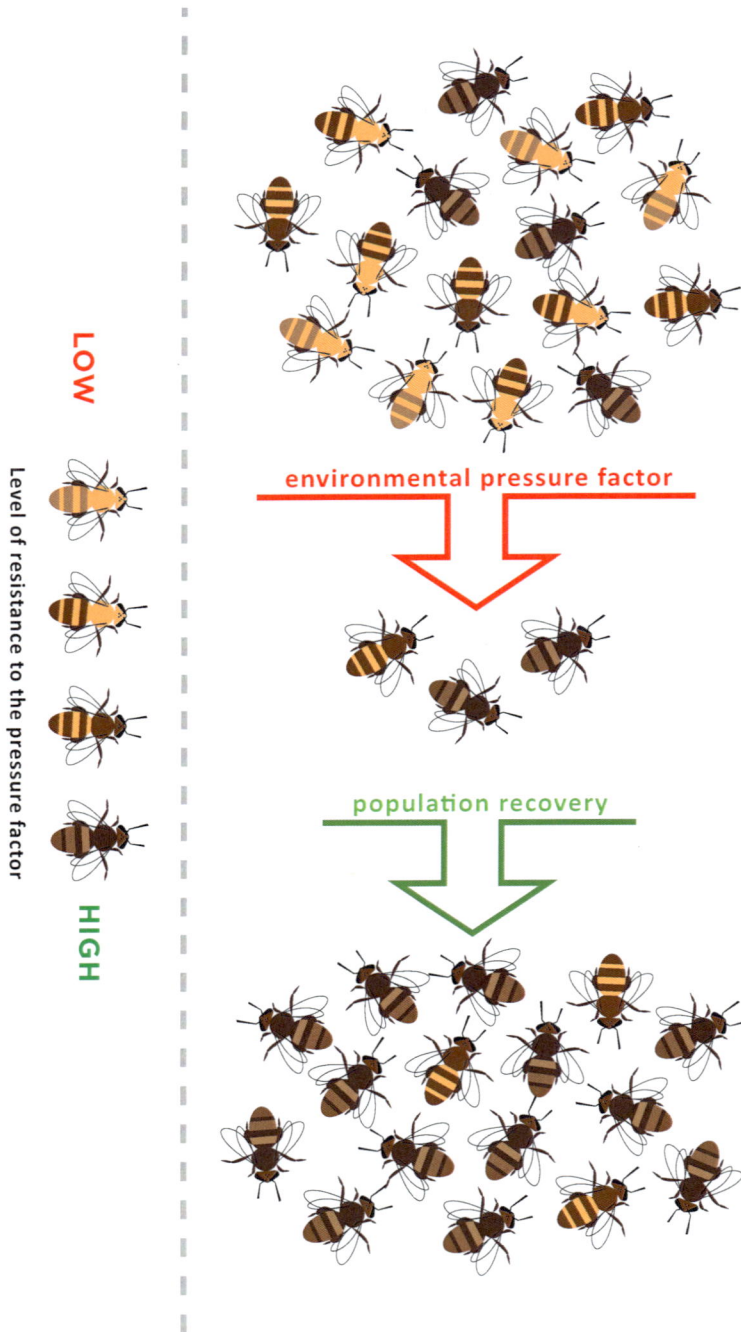

LOW

Level of resistance to the pressure factor

HIGH

environmental pressure factor

population recovery

The so-called evolutionary bottleneck. In every population diverse in terms of genetics and phenotypic traits, we will find individuals with different resistance to environmental pressures (low or high temperature, presence of pathogens or parasites, poor food availability, etc.). As a result of the pressure, part of the population dies out, but only individuals with high (sufficient) resistance survive, based on which the populations are rebuilt, figure by M. Uchman.

Although in the long-term evolution can contribute to the creation of completely new characteristics, abilities or even a complete reconstruction of the anatomy of organisms, in the case of the development of resistance to some environmental pressure factor (e.g. temperature, humidity, availability of a specific type of nest site or food, but also a pathogen or parasite) we are talking about a much simpler mechanism. It assumes that the population has already developed specific tools for coping with the new parasite or pathogen, but not all individuals have these tools. Natural selection eliminatesorganisms that are not resistant and leads to reproductive success of those that can develop an appropriate response to the new threat. It should be noted that if the pressure factor exceeds the tolerance capacity of every individual, the population will become extinct.

Let's look at this evolutionary mechanism. Living organisms have their own ecological tolerance range, beyond which they die. In biology we distinguish few zones/niches. The Latin term *optimus* (optimum) means, in reference to the ecological niche of a species, excellent living conditions and the possibility of easy reproduction and growth. A worsening (*peius*) niche allows organisms to survive but with reduced growth, while lethal (*pessimum*) niche is an environment where organisms exist on the edge of survival.

If representatives of a species start reaching places where they do not have optimal development conditions (driven, for example, by competition for food or nest sites or escaping from predators), individuals that tolerated the selection pressure factor less well will sooner or later be eliminated from the gene pool. They may survive for some time in a lethal niche, but they will eventually die without passing on their genes. As a result, those for which local conditions are not lethal, will begin to reproduce. After some time, populations may differentiate between neighbouring environments, and genetic crosses or mutations may occur that will make it easier for the organisms to survive. Then we will be able to talk about the isolation of a new ecotype or subspecies. Due to genetic differences within a population, the same environment for one individual of the same species may be a lethal zone, and for another in a worsening niche or even an optimum niche. Within a given population, subspecies adapted to different niches (climatic conditions, microflora, nest sites, predators or parasites, etc).

In the longer term, within populations occupying different niches may undergo such large changes that we will be able to speak of the emergence of a new species[68].

Thanks to evolved adaptations (e.g. the ability to shape their own nest site or gather food for times of hunger or winter), colonies of the same species (*Apis mellifera*) can survive in extreme conditions. They live in hot and dry semi-desert climates, in marshy areas of a temperate climate, and in the far north, where winters are exceptionally frosty and can last up to six months.

Within the species *Apis mellifera*, we can distinguish over 20 basic subspecies of honey bees[69]. Some of them differ greatly from each other, both in terms of morphology (colour, size) and adaptive or behavioural abilities.

For example, some African subspecies of bees cannot form a winter cluster, so they are not able to survive the winter, a period that requires the bees to warm each other and demonstrate the ability to limit heat loss.

The bees' march northwards was therefore probably determined by the pressure of cold, and stress of the environment (extreme conditions of ecological tolerance) in a way that shaped the adaptation of the northern subspecies to the increasingly shorter season, greater temperature amplitudes, the ability to warm themselves with each other, to store enough food for the winter, and finally to function without having to defecate for

---

68 Diversification (even leading to the emergence of a new species) may also occur in the same area, e.g. because of competition for food (in the case of food shortages, some individuals will start to eat food that was previously less tolerated by them, e.g. less tasty or nutritious) or sexual selection (some females will prefer males with a specific appearance or behaviour).

69 Some scientists distinguish up to 35 subspecies of the western honey bee. (The Origins of Honey Bee, "Two Bees in a Podcast", https://open.spotify.com/ episode/05iB7Jbfu2lC1bau1TpKzj, accessed: November 10, 2022. A very interesting publication has recently been published, shedding new light on the origin of the honey bee species, K. Dogantzis et al., Thrice out of Asia and the adaptive radiation of the western honey bee, Science Advances 2021, 12. Genetic studies indicate that the honey bee originated in Asia (where all other species of the genus *Apis* occur), so the hypothesis of its African origin, the most popular so far, has been questioned. The new study also confirmed the existence of seven evolutionary lines of honey bees, which were formed in different environments, then divided into subspecies. These, within the evolutionary lines, are more closely related to each other than located between evolutionary lines. Interestingly, the evolutionary line M (*A. m. mellifera*, *A. m. iberiensis*) has fewer genetic affinities with evolutionary line C (*A. m. ligustica, A. m. carnica*) than the latter, with the line O, occurring in western Asia (*A. m. anatolica, A. m. caucasica*). This means that Carniolan bees are genetically closer to Anatolian or Caucasian bees than to Central European bees (often called: European dark or black bee). Recent studies allow us to better realize the scale of the potential threat of parasites or pathogens originating from East Asia, where there are at least several known species of bee parasites of the genus *Apis*, formed over millions of years of evolution, which have adapted to coexist with eusocial bees. I have no doubt that the future will reveal the existence of previously unknown bee parasites. Will they, like *Varroa destructor*, be able to change hosts?

several months. This was also a period of co-adaptation of these bees with the local fauna and flora. Local bees therefore better tolerate the presence of pathogens characteristic of a given area. African subspecies cope better with the small hive beetle (*Aethina tumida*) than European ones, while those from continental Europe adapted better to the presence of the tracheal mite, while the island population in Great Britain and bees from the United States encountered it only later. For this reason, importing foreign subspecies or breeds of bees to our apiaries weakens local genetic adaptation. We import bees that are, or in certain conditions can be, more productive than local bees, but are often unable to fully survive on their own. They require greater care from beekeepers. Together with bees, we import pathogens and bee parasites to biological niches, burdening non-resistant populations. This is what happened with the *Varroa destructor* mite.

While specific situations can be very complex, the basic principle of evolution seems simple in principle. If an organism is under strong selection pressure, an evolutionary response can be expected. Not every species will withstand the pressure; it is estimated that many more species have disappeared from the face of the earth irretrievably than exist today.

Today, many species are at risk of extinction mainly due to the pressure exerted by humans on the environment. We are changing ecosystems, reducing the number of wild areas, putting pressure even on those that still exist, we are releasing toxic pollutants, changing the climate etc.. For decades, the number of species has been constantly declining.

Nature will always strive for balance, which is never completely stable, but rather fluid and variable. The relationships between organisms are also very complex. They are based on countless factors, such as the availability of food and water, competition between individuals of different species and within species, the presence of predators, parasites or pathogens, the state of the environment and the availability of nest sites, etc.

Before Charles Darwin created his theory of evolution, one of the most eminent naturalists of all time, Alexander von Humboldt, revolutionized the way of perceiving the natural world, creating the concept of nature as an inseparable web of life in which all organisms are interconnected[70]. However, he pointed out that this web is burdened with the risk of destruction through attempts to remove the threads that make it up. This happens

---

70 Today, Humboldt's concept seems to be a great simplification. But it is obvious, that he was already aware of the fact that one species could influence another and that he was aware of the co-adaptive processes of different species.

when the organisms connected to each other lose their sources of food or the environment is no longer able to support their existence. The richer the ecosystem, the less important bilateral relationships are (or can be) (e.g. parasite - host, predator - prey, etc.), because the organism can be influenced by new factors (e.g. in a complex ecosystem, a predator can have more species of prey to choose from). The more diverse the web of relationships, the more stable and resistant to change it is, because organisms can replace each other within the complex web. In theory, however, reducing or increasing some pressure can lead to a complete reconstruction of a given ecosystem.

Depending on the definitions and views adopted, pressure can be divided into **natural** and **artificial**. However, both have an impact on the formation of the population, because evolution does not distinguish between them. Natural selection will eliminate from the gene pool those individuals that are not adapted to the changes in a sufficient way to be able to survive and produce offspring.

Virulence, horizontal and vertical transmission

Virulence is defined as the ability of a pathogen to penetrate another organism, multiply in it, and damage its tissues. Its measure is the ability to kill the host. Mild strains either encounter an effective immunological barrier or their ability to cause disease (damage to tissue) is relatively low[71]. Non-virulent pathogens can multiply rapidly, increasing their ability to infect, but do not cause significant health issues within the population. The process of acquiring virulence is not fully understood, but we can indicate the circumstances in which it occurs. Pathogens need a host organism to achieve reproductive success. If, because of the "arms race", the host has developed protective barriers, then to achieve reproductive success, the pathogen must diversify to bypass these barriers (we are not talking about a conscious and deliberate process). Micro-organisms are simply subject to constant changes; they mutate or because of contacts with each other exchange genetic material. Since their life cycle is very short, changes can occur very quickly. Evolutionary success is achieved by those that have developed the ability to bypass the host's defences. However, their virulence

---

71   They infect the host, but the infection is asymptomatic, or the symptoms are mild and not noticeable.

will not always increase evolutionarily so that it would become more dangerous for the host – it depends on many factors and circumstances[72].

One could try to formulate general principles of the process. They would sound like this: if the rapid weakening of the host (or even its death) favours the spread of the pathogen, it will start to show increasing virulence (because this will serve its evolutionary success). On the other hand, if killing the host (or causing severe symptoms of the disease in it) will simultaneously mean the death of the pathogen or make it difficult to spread between organisms[73], the population will start to evolve towards being less virulent[74].

Micro-organisms are also subject to the pressure of chemical treatments (sterilizing agents), which also shape pathogen populations. Killing a single micro-organism is easy but killing micro-organisms *en masse* under natural conditions is extremely difficult. First, they can hide in various nooks and crannies, effectively protecting themselves from the treatment; second, they can mutate very quickly and change their own metabolism, adapting to life in the new environment. Repeated treatments or constant exposure to a toxic substance in low concentration can serve to develop resistance in organisms[75].

---

72   I. Fries, S. Camazine, Implications of horizontal and vertical pathogen transmission for honey bee epidemiology, Apidologie 2001, 5-6.

73   A sick individual, for example, will isolate himself from others.

74   Pathogens that have a high capacity to be transmitted between organisms evolve most often towards mildness (see: the pandemic caused by the SARS-CoV2 virus). Others, exceptionally deadly, most often have a low ability to transmit between organisms, sometimes they are extremely infectious, but due to, for example, the short infectious period, their ability to spread is not great (e.g. the Ebola virus).

75   Such processes occur not only in the world of micro-organisms, but also in more complex organisms. The problem also concerns the varroa mite, A. Millán-Leiva et al. Mutations associated with pyrethroid resistance in Varroa mites, a parasite of honey bees, are widespread across the USA, Pest Management Science 2021, 3; the study confirmed the widespread presence of alleles associated with resistance to pyrethroids used in beekeeping (e.g. flumethrin, tau-fluvalinate) in the mite genome. J. Woyke claims that it is possible to breed bees resistant to insecticides (by exposing bees to small doses and selecting those that survive and remain healthy). Such processes take place more easily and quickly in the world of micro-organisms, if only due to the rate of generation replacement and the possibility of easy exchange of genetic material by bacteria.

DIFFERENT ENVIRONMENTAL PRESSURE FACTORS AND THEIR INFLUENCE ON SHAPING THE POPULATIONS

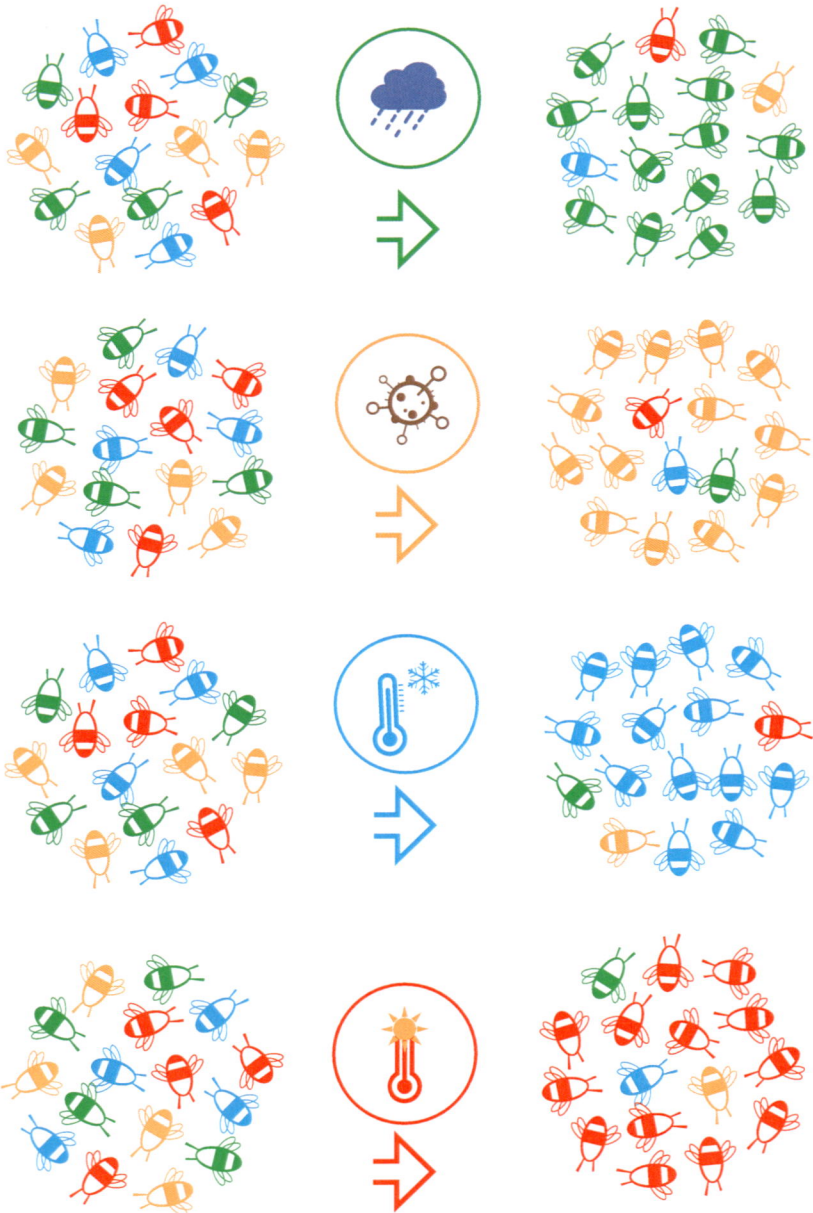

The formation of a population may take different courses, depending on many environmental pressures (e.g. humidity, presence of pathogens, cold or heat), as well as other factors (e.g. sexual selection). This process may lead to the separation of evolutionary lines, subspecies and ecotypes. Over time, it may lead to such a far-reaching differentiation of individuals that eventually we will be able to talk about originating of a new species (process called speciation), figure by M. Uchman.

We could kill microorganisms, perhaps even relatively easily, by exceeding their ecological tolerance range, e.g. by using high temperature or high concentration of a biocidal agent, such as acid. In such a situation, however, the tolerance range of not only micro-organisms but also bees would be exceeded[76]. The problem begins when the virulent organisms can kill the host, and at the same time are able to survive and transfer to a new host. Natural beekeepers therefore postulate that in apiaries, the so-called **horizontal transmission** of parasites and pathogens should be excluded or minimized, and that conditions should be created that would at most allow for the so-called **vertical transmission**.

Horizontal transmission involves the transfer of pathogens and parasites between neighbouring hosts (bee colonies). If the system favours horizontal transmission, even those pathogens that are very virulent and deadly to the host are quickly transferred in the population. Such an ecological situation can promote the maintenance or even increase pathogen virulence.

Vertical transmission, on the other hand, means that parasites and pathogens are transferred only via a vertical line, i.e. from parent directly to offspring (bee colony to swarm, queen to worker). If the ecological systems do not allow horizontal transmission (or makes it very difficult), a parasite or pathogen that leads to the death of the host dies along with it. Such a system favours the evolutionary weakening of pathogen virulence, because the most virulent strains undergo self-elimination. If the host develops resistance to the virulent strain or the pathogen becomes milder, both organisms can survive. This situation evolutionarily promotes the most resistant organisms and benign pathogens.

In this system, both strains of micro-organisms that have not developed virulence (or have limited it) can survive with a host that is not resistant to their virulent versions, as well as resistant hosts with pathogen forms that are virulent for other members of the species.

Let's apply this theory to the situation of honey bees. In today's reality, pathogens spread very quickly. What's more, the weakening of colonies accelerates the horizontal transmission of mites and the pathogens they carry (primarily through robbing and drifting). In evolutionary terms,

---

76   The problem is to select the dose of the toxic substance in such a way as to cause maximum harm to pathogenic organisms and parasites without harming the bees. In my opinion, it would be difficult to prepare a dose that would not have a negative effect on the bees and at the same time eliminate 100% of mites and pathogens. If we reduce the dose to a level relatively harmless to the bees (and repeat the treatment several times a year for several decades), undoubtedly some of the bee pests will also survive, thus causing them to develop resistance to the active substances.

pathogens therefore "benefit" from high virulence, weakening, and even killing bee colonies - the death of the host does not, in principle, prevent them from rapidly reproducing and spreading to other bee colonies.

In my opinion, the situation could change in two cases. First, if horizontal transmission were stopped, or at least significantly hindered. If bee colonies lived far apart from each other (as in nature), a large percentage of parasites and pathogens that led to the death of bees would die with them. However, this condition is impossible to meet, because it would mean the elimination of beekeeping altogether, or at best a return to the traditions of tree-beekeeping. Second and this seems possible in practice under certain assumptions, if bees began to become increasingly resistant to pathogens, the pathogens would begin to evolve towards being less virulent. This sounds suspicious and quite illogical: it means that pathogens will cease to threaten bees when they become resistant to them, everyone will agree with that. But how can this be done in a situation where pathogens are becoming increasingly virulent, and bees are by no means still not immune? The situation will seem more logical if we consider that we are not talking about a specific point in time, but about ongoing processes. A few decades ago, as scientists claim, pathogens were not as virulent as they are today. We ourselves contributed to their increased virulence via the way we run our apiaries.

In theory, the way out of this situation is very simple. It is enough to change certain rules of apiary management, and above all the priorities of bee breeding[77].

Many beekeepers, also supporters of natural beekeeping and treatment-free beekeeping, believe that by using natural selection in apiaries (allowing the death of colonies), we are promoting horizontal transmission, thus supporting the evolution process leading to an increase in the virulence of pathogens. In my opinion, this is only part of the truth, as it concerns only a certain period of selection, when the basic immunity of bees has not yet developed. At that time, many colonies may die, although if they lived in isolation, the threat would be smaller. After some time, however, the situation changes - precisely because of the processes described. The immunological resistance of the bees increases, and pathogens begin to evolve towards lower virulence[78].

---

77   See chapter 6: At the crossroads, or which way to beekeeping without toxic substances.

78   These processes are ongoing in many regions of the world, but most often spontaneously in places that have exceptionally bee-friendly surrounding conditions: where there is relatively low managed

Complete elimination of horizontal transmission of pathogens is unlikely (if not impossible), even in nature, where bee colonies are hundreds of meters apart. Sooner or later, bees will find a dead colony and rob it of any remaining honey. This is confirmed by Thomas Seeley's experiment, who, setting up an empty hive with honey in the forest, noticed that it takes bees at most a dozen or so days to find it (the bees found the same hive set up in an apiary in a few hours). Bees are in contact with each other, even when they live far apart, although then their interactions are much weaker. Keeping bees in apiaries, regardless of whether we treat the bees or not, increases the intensity of horizontal transmission. The cause of this situation is not only the bees (drifting or robbing), but also human activities (infected hands, apiary equipment, clothes, exchange of combs between hives etc.).

Some pathogens (e.g. spores causing fungal diseases, chronic bee paralysis virus [CBPV]) can also move through the air, although usually over short distances. However, we can try to limit this type of transmission, e.g. by refraining from combining or reinforcing colonies.

Disinfection of the bees' nesting site

The subject of treating bees and combating varroa relates to the issue of disinfection or sterilization of the bees' living environment (hives, frames) and beekeeping tools. Two fundamental questions come to mind: first, does it make sense at all; second, do these activities cause more harm than good?

Organisms live in complex ecosystems. There are about 2-2.5 kg of bacteria inside a human, that is trillions of living organisms. It is assumed that there are as many bacterial cells as human cells or even several times more. So, there is at least as much foreign DNA in us as our own. Some researchers treat this bacterial ecosystem as an organ of our body. Bacteria take part primarily in digestive processes and in building the immune system of organisms on at least several levels.

The bacterial ecosystems (microbiome) consist of friendly, commensal (symbiosis with another organism) and pathogenic organisms (the presence of the latter is completely natural). A specific balance (interactions) between the groups will affect the overall health of the host. A healthy ecosystem

honey bee density, and relatively large populations of wild/feral bees or on islands. We also deal with them in other regions (less friendly to bees), but there it requires active support from beekeepers (with appropriate breeding or management methods).

of micro-organisms (which, in addition to bacteria, also includes fungi) somehow drives out pathogenic organisms (e.g. because of competition for food and nest sites, not allowing their populations to grow excessively). According to some scientists, the sterilization of the human living environment and pollution by unnatural substances have led to an increase in people's susceptibility to various diseases, e.g. dermatological, gastroenterological, allergies, auto-immune problems.

Some researchers claim that bacterial microflora disorders caused by our unnatural living conditions, e.g. poor diet, may contribute to the occurrence of ailments that we would not associate with our microflora. These include heart disease, diabetes, obesity, but also depression and other mental disorders.

The role of microorganisms in the world of bees has not been fully understood[79], but we know that it is significant[80]. Scientists' research shows that in bee nest sites and on/in insects, over 6,000 species of microorganisms can be distinguished (some sources even mention 8,000). Micro-organisms (including bacteria and yeasts) take part in the process of pollen fermentation and digestion[81]. Those found in bee bread promote the production of enzymes, vitamins, organic acids and fats (which makes it easier to absorb and more valuable and preserves it longer). It has also been found that the presence of some micro-organisms can prevent the development of chalkbrood, and even completely stop the development of *Paenibacillus larvae* (causing American foulbrood)[82] and other dangerous pathogens[83]! It has also been found that in a situation where treatments disrupt the structure of the microbiome, pathogenic bacteria develop

---

79   This is a relatively new area of research, even in the field of human health.

80   M. Gilliam, Identification and roles of non-pathogenic microflora associated with honey bees, FEMS Microbiology Letters 1997, 10; Z. Lipiński, J. Wojtacka, Przełom w pozaniu biologicznej roli symbiotycznych bakterii przewodu pokarmowego pszczół miodnych [A breakthrough in understanding the biological role of symbiotic bacteria in the digestive tract of honey bees], Pszczelarstwo 2016, 2; P. Engel et al., The Bee Microbiome: Impact on Bee Health and Model for Evolution and Ecology of Host-Microbe Interactions, ASM Journals 2016, 4.

81   J. L. Fredrick, Saccharide breakdown and fermentation by the honey bee gut microbiome, Environmental Microbiology 2014, 6.

82   E. Forsgren, Novel Lactic acid bacteria inhibiting *Paenibacillus larvae* in honey bee larvae, Apidologie 2010, 1.

83   For example, the bacterium *Melissococcus plutonius*, one of those that cause European foulbrood, A. Vasques, Symbionts as Major Modulators of Insect Health: Lactic Acid Bacteria and Honey bees, "PLoS ONE" 2012, 3.

faster[84], which shortens the lifespan of the bees[85]. It also turns out that the microbiome can play a role in detoxifying environmental factors that are harmful to bees. It also helps bees develop naturally by stimulating the expression of genes related to aging and changing roles (tasks) of bees in the colony[86].

We can assume that the microbiome is at least as important for the health of bees as it is for humans. We can also say with full certainty that in modern hives, especially those that are regularly cleaned and disinfected, the micro-world is poorer than in the natural nest sites of bees, such as tree hollows, or even in old wooden hives, insulated with straw or sawdust. The external and internal environment affects the entire microbiome and shapes it for many years. Sterilization does not act selectively, during this procedure we kill not only pathogens, but also those bacteria and fungi that are useful or neutral to bees. On disinfected surfaces, pathogenic flora can develop much faster than on so-called biofilms (also called a biological membrane), composed of multicellular structures of bacteria, surrounded by a layer of organic and inorganic substances, which are in constant interaction with each other. Micro-organism communities can also be found on chitinous exoskeletons and epithelial tissues of bees.

They are protective barriers that do not allow pathogens to penetrate the bees' bodies. When treating or sterilizing the bees' environment (especially with acids), we not only destroy bacterial biofilms, but also greatly strain the physical bees' protective barriers (chitinous exoskeleton or epithelial tissues). There are known cases when bees that managed to build natural immunity and balance collapsed due to prosaic health problems after using organic acids in the hive[87].

---

84   The results of antibiotic treatment were studied primarily, but perhaps similar effects may occur after treatment with strong biocidal agents, e.g. acids, J.E. Powell, Field-Realistic Tyrosine Exposure Impacts Honey Bee Microbiota and Pathogen Susceptibility, Which Is Ameliorated by Native Gut Probiotics, ASM Journals 2021, 6.

85   The impact was minimized by administering probiotic cocktails; however, the composition of most probiotic products available on the American market does not correspond to the composition of the bee's gut microbiome (bacteria from silage or products that are fed to other animals); it should be noted, however, that their regular administration may be harmful to bees, because probiotics can lead to the reconstruction of the ecosystems of micro-organisms in the intestines of bees; interview with Dr. Aneta Strachecka, https://www.warroza.pl/2022/06/Aneta-Strachecka-odpornosc.html; accessed: November 1, 2022.

86   For example, from nurse bees to foragers, Tyrosine exposure impacts on Honey bees, "Two Bees in a Podcast", https://entnemdept.ufl.edu/honey-bee/podcast/, accessed: November 1, 2022.

87   E. Österlund, Buckfast Breeding Principles, https://www.beesource.com/threads/ buckfast-breeding-principles.365573/#post-1843975, accessed: November 1, 2022.

It is also important to remember that for the immune system of each organism to function properly, it must be constantly trained. If we start to replace it in fulfilling its tasks (e.g. through disinfection), we will consequently weaken the defence mechanisms. This is suggested by research conducted at the University of Pennsylvania (USA) under the direction of Margarita López-Uribe, in which wild bees were compared with a population of bees not immune to the pest, bred in apiaries. Both groups were infected with bacteria, then samples of the bees were taken and the expression of genes responsible for the immune response were examined. It turned out that on average the immune response was twice as high in wild bee colonies than managed bees. A comparative experiment was also conducted between the groups of wild and apiary colonies in terms of DWV pressure. It was found that in wild colonies pressure from DWV is significantly higher than in colonies from apiaries (due to high levels of mite infestation, the vector of DWV). Despite this, mortality in both groups were similar. It turned out that the survival of bee colonies is not dependent on the pathogen pressure, but on the immune response of the bees[88]. Pressure from DWV was higher in wild colonies, i.e. those whose immune system is constantly exposed to the pressure of pathogens[89].

Another very interesting study on the ways in which honey bees develop social immunity was conducted by a team led by Gyan Harwood[90]. It shows that bees create specific vaccines. Through royal jelly, workers can transfer many compounds against micro-organisms to both the queen and the larvae during the feeding. It was found, for example, that in this way they transfer fragments of cells of the bacteria[91] causing American foulbrood to the queen, and then the queen can then transfer them directly to the eggs, which stimulates the larvae's immune system to fight it[92]. This mechanism can be

88   The researcher associates the phenomenon of higher resistance to pathogens with higher genetic diversity within wild colonies, M. López-Uribe et al., Higher immunocompetence is associated with higher genetic diversity in feral honey bee colonies (*Apis mellifera*), Conservation Genetics 2017, 2; similar conclusions were also reached by S.D. Desai, Genetic diversity within honey bee colonies affects pathogen load and relative virus levels in honey bees, *Apis mellifera,* Behavioral Ecology and Sociobiology" 2015, 7.

89   Ch. Hinshaw, The Role of Pathogen Dynamics and Immune Gene Expression in the Survival of Feral honey Bees, Frontiers in Ecology and Evolution 2021, 1.

90   G. Harwood, Social immunity in honey bees: royal jelly as a vehicle in transferring bacterial pathogen fragments between nestmates, Journal of Experimental Biology 2021, 4.

91   Most often, these are components that build cell walls or other molecular structures of microbes (so-called antigens).

92   This phenomenon is called TGIP (transgenerational immune priming).

used to vaccinate subsequent generations of bees against other pathogens. This has also been confirmed in the case of DWV[93]. It is a unique system of sharing specific immunological memory by subsequent generations of the bee superorganism!

The honey bee has only started to function in sterilized hives (e.g. Styrofoam, but also wooden ones with smooth walls) in recent decades. Before that, it had evolved ecological interdependencies in tree holes full of humus (or other composted organic material), leaves, twigs and remains brought in by various animals.

Often, the nest sites are re-occupied by colonies after the original colony has disappeared, e.g. those that had collapsed for various reasons (including diseases). However, only those colonies that showed sufficient resistance survived. Bees interact within a complex ecological system with a wide variety of organisms, including other insects and arachnids. Today, we have completely changed the ecological relationships within bee nest sites. It seems that in the short term, this serves individual colonies after all, we are the ones who solve their problems, which allows them to develop faster and produce (for us) more honey. In the long term, however, this has a negative impact on the ability of the entire population to survive on their own. We are destroying the ecosystems of micro-organisms living on and in bees and the surfaces of their nest sites, which means interfering with the conditions that are conducive to the evolution of the most resistant insects.

Paenibacillus larvae spores were found in several dozen percent of the apiaries examined. Some even claim that they can be found in almost every bee habitat. Nosema spores (*N. apis, N. ceranae*) are present in most of the bees examined. Although a large percentage of apiary bees deprived of treatment do die, the species still copes with these pathogens. In my opinion, we simply must accept all micro-organisms in the bees' environment and treat them all together, as allies rather than enemies.

Meanwhile, modern beekeeping advocates the fight against microbes. It constantly recommends carrying out so-called hygiene procedures, which, according to our assumptions, are aimed at maintaining the bee colony at optimal health. However, they are nothing more than sterilization of the bees' living environment. Zbigniew Lipiński, professor of beekeeping, and Dr. Joanna Wojtacka note, among other things, that any proposals to

---

93   S. Lang et al., Context-Dependent Viral Transgenerational Immune Priming in Honey Bees (Hymenoptera: Apidae), Journal of Insect Science 2022, 1. It was found that in the case of deformed wing virus, the occurrence of the TGIP phenomenon is conditioned by many factors, including genetic, epigenetic and the route of infection of the queen bee with the virus.

use disinfectants in the prevention and control of bee diseases should be approached with great caution. They also claim that the use of such agents within the bee nest/hive leads directly to the annihilation of symbiotic bacteria. This can result in secondary malnutrition of the bees, as well as an increase in their sensitivity to infections by pathogenic micro-organisms, and therefore, among other things, a decrease in productivity. Researchers also admit that although at the start of using disinfectants (especially in spring) there may be more flights to the forage, we can also observe an increase in hygiene behaviours, etc., because of excessive, downright stressful stimulation. Later we will usually notice a sharp decline in the condition of the bees, a weakening of resistance to infections and the development of diseases, including American foulbrood[94].

It also turns out that the reconstruction of the microbial ecosystem in sterilized biological niches can take a very long time or never at all. This can be compared to the clearing of a primeval forest of all the trees. Even if we allow the area to be reforested, the return to the previous state may take hundreds of years, and some exterminated species of fauna and flora may be lost forever. This is because nest sites that are unique to very old forms of forest, necessary for maintaining some ecological relationships (e.g. holes in old trees, which are used by many species of invertebrates, now threatened with extinction), will disappear. However, this does not mean that the balance cannot be rebuilt, even if it is poorer by many species, they can still remain in healthy relationships.

To sum up this part of the discussion, I will refer to an example from my own apiary. I have only had to deal with chalkbrood twice. In both cases, the disease affected bees brought from outside, for which I changed the conditions in my hives. In colonies where I do not use so-called chemicals at all, there are no symptoms of this disease at all. The first case, which took place in the spring of 2014, is important for our considerations. At that time, I bought overwintered colonies with hives from a beekeeper who (due to his age) began to reduce the size of his own apiary. What I saw in the new Styrofoam hives terrified me (today my reaction would be completely different). On the bottom, I found a thick layer of compost, and the hive was full of channels dug by wax moth larvae. As soon as the weather turned nice, I removed all (black) frames that lack any brood or stores (honey or

---

94   Z. Lipiński, J. Wojtacka, Przełom w poznaniu biologicznej roli symbiotycznych bakterii przewodu pokarmowego pszczół miodnych [A breakthrough in understanding the biological role of symbiotic bacteria in the digestive tract of honey bees] "Pszczelarstwo" 2016, 2.

pollen), and I resettled the colony to a clean, freshly made hive. During the next inspection, I found very strong symptoms of chalkbrood, which did not want to go away. I have analysed this story many times since. Of course, I cannot be certain that the disease would not have developed in the hive in which the bees came to me. In my opinion, however, the pathogens were kept in check by the bacterial and fungal ecosystems present in the old nest site, which I replaced with a new and "clean" one.

Chemicals in the bees' environment

What are chemicals in beekeeping?

Everything is chemicals. The entire material world around us is made up of chemical compounds, both animate and inanimate. However, the term "chemicals" is commonly used to refer to those chemical compounds that are or may be harmful and do not occur naturally in nature[95]. These are e.g. pesticides[96], preservatives or artificial fertilisers. In my opinion, however, these substances can hardly be considered unequivocally detrimental. They are a kind of tool with which we can, for example, increase crop yields[97]. Hypothetically, we could therefore increase the harvest from a smaller area, and consequently give the area not covered by cultivation to other organisms, in a wild or original state. This would probably be the case in an ideal world or at least a better one. But it is not. The obstacle is the widespread and, in my opinion, completely irrational use of chemicals. Pesticides (e.g. insecticides, herbicides, fungicides, and in beekeeping also acaricides or bactericides, such as acids or oils) are used *en masse* in every type of agriculture, which disrupts the ecological relations between organisms in

---

95   For an explanation of the problem of artificial divisions and definitions of chemical compounds used in beekeeping or agriculture in its broad sense, see: J. Jaroński, Pszczelarska chemia i pestycydy [Beekeeping chemicals and pesticides] (https://www.warroza.pl/2021/03/pszczelarska-chemia-i-pestycydy.html), accessed: November 2, 2022.

96   Pesticides (Latin *pestis* 'plague', *cedeo* 'I kill') are substances used to kill organisms that we consider harmful or undesirable; they can be of artificial or natural origin. Natural ones (e.g. plant nicotinoids) are often more harmful to many organisms than those produced by humans. Pesticides are divided into many groups. The most used are insecticides (kill insects); acaricides (kill arachnids, including mites); fungicides (kill fungi); bactericides (kill bacteria, these include e.g. antibiotics).

97   We are not as unique as we think: pesticides are used by ants and termites.

basically all the surrounding ecosystems[98]. The artificial fertilizers that we use, often in irrational quantities, end up in groundwater, then in seas and oceans, causing excessive growth of certain micro-organisms, creating the so-called dead zones.

Compost on the bottom of a hive purchased in 2014 (partially cleaned), photo by author.

Pesticides, preservatives or fertilizers can also be of natural origin (found in plants or produced by micro-organisms or animals). Pesticides are therefore also substances used to kill pests, considered natural or organic (e.g. thymol, formic acid or oxalic acid).

Meanwhile, substances of natural origin can be as harmful or poisonous

---

98   Herbicides are used, among other things, to accelerate the ripening of some crops, to eliminate plants in the cracks of paving stones, sidewalks or instead of using a lawn mower.

to various organisms as synthetic ones, sometimes even more[99]. Paracelsus, a physician and naturalist, considered the father of modern medicine, used to say that everything is a poison, and nothing is a poison; the effect depends on the dose[100]. The harmfulness of a substance may also be related to the method of administration or the phenomenon of synergy (increased toxicity under the influence of contact with another substance)[101]. Compounds that are harmless to some organisms may be toxic to others, much depends, for example, on the metabolic capabilities of individual species, or even specific individuals. A substance may be extremely harmful to humans, and completely harmless to other organisms - or *vice versa*[102].

For the purposes of our considerations, I will use the term "chemicals" to refer to all pesticides used to combat bee parasites and pathogens[103]. It is therefore irrelevant whether they are of natural origin or not, or whether they occur naturally in wildlife, or even in the hive (e.g. in honey).

These substances have a diverse impact on honey bees and their environment, but all of them seem harmful from the perspective of developing an ecological balance between bees and other organisms (including pathogenic or parasitic ones). Therefore, I define chemicals as all substances that have reached the hive through beekeepers, not bees. I also include pesticides (e.g. fungicides or herbicides used in agriculture) harmful to bees, which foragers could bring from the fields into the hive in nectar, pollen or water.

---

99 Some people believe that the most toxic substance in the environment of bees or beekeeping activities is bee venom.

100 Any substance in excess can cause poisoning, for example drinking too much water can cause water poisoning, also known as overhydration hypotonic, consisting in a disturbance of the body's electrolyte balance.

101 R.M. Johnson, Acaricide, Fungicide and Drug Interactions in Honey Bees (*Apis mellifera*), PLoS ONE 2013,1; R.M. Johnson, Pesticides and honey bee toxicity USA, Apidologie 2010, 5-6; D.J. Hawthorne, Killing Them with Kindness? In-Hive Medications May Inhibit Xenobiotic Efflux Transporters and Endanger Honey Bees, PLOS ONE 2011, 11.

102 The toxicity of a substance is one issue, but equally important is whether organisms will be exposed to it. Even if a substance is extremely toxic to bees, it may not be dangerous to them because they will not encounter it (e.g. it is used on non-flowering plants or in a way that prevents it from getting into the nectar or pollen of other plants or water used by bees). Some suggest that when describing the effects of substances on living organisms, we should not talk about their toxicity, but about the risk they carry, e.g. fungicidal substances are of reasonably little harm to insects, and for this reason they are sometimes used on flowering plants. Although their use rarely leads to acute poisoning of bees, it can lead to chronic poisoning and, therefore, to the weakening and death of the colony.

103 Both the division and the definition are not very precise, but I will use them in accordance with the common, somewhat intuitive understanding of this concept.

I have no doubt that a one-time discontinuation of treatments against the varroa mite would cause over 90% of colonies to collapse within the first few seasons. This would be a problem not only economic (loss of potential income, costs of purchasing new bee colonies, possible environmental/agricultural losses caused by a decrease in the number of pollinators), but also moral and emotional. Professional beekeepers cannot afford to lose their bees. They would lose their source of income along with them, so it is hardly surprising that they reach for chemicals. What surprises me more is that for four decades the entire beekeeping community in Poland, including scientists, breeders, professional and amateur beekeepers, has not sought to introduce selection mechanisms that would eventually allow them to abandon the use of these toxic substances The situation is sometimes similar in other countries, but in some it is much better in that regard.

What are the reasons for this situation? It is difficult to answer this question unequivocally. However, if the main reason is the lack of confidence in the ability of bees as a species to cope with the mite, then I regret to say this indicates fundamental deficiencies in the education of beekeepers[104]. Many beekeepers still believe that treating varroa saves the western honey bee (species) from extinction.

The problem also has another aspect, a more economic viewpoint. Even if beekeepers, as they claim, undertake treatments for purely ethical or emotional reasons, some of their actions show that they are largely guided by convenience and concern for the size of the harvest. I am talking here about the selection of bees towards increasingly greater gentleness or efficiency, while fully accepting the lack of systemic selection of their mite-resistance. On the one hand, it is difficult to question the love of beekeepers for bees and their profession, but on the other hand, how can we explain their breeding and economic choices? Are they not guided primarily by the utility values of the bees, and not their population health? How else can we explain the fact that although many of them are aware that local bees better cope with surrounding conditions, they still import genetically foreign queens, just because they have proven to be more productive? Everything indicates that the resistance/health of bees most often loses out to economic (short-term) calculations and convenience. It is easier to reach for chemicals than to start deepening our knowledge or undertake breeding work. It is also surprising that some beekeepers (fortunately a minority) still deny the harmfulness of

---

[104] I am writing this with the full awareness that the problem is complex, difficult to solve in practice and requires a systems approach.

pesticides to bees, claiming that they have no impact on the quality of bee products[105].

I believe that the problem could be solved at a relatively low cost and effort. By a "relatively low cost" I mean that the cost will be low if we compare it with the effort and costs incurred in previous decades and those, we are facing to keep the bees alive. If we add up the annual costs and estimate the effort associated with treating bees, and then multiply the results over the last 40 years, we will realize the scale of the outlay. The costs are not only financial, but also environmental and health.

Let's not forget that in recent decades in many regions, like in Poland, the beekeeping community has not come even an iota closer to a systemic solution, but rather it has contributed to the accumulation of particularly virulent strains of pathogens. I therefore believe that with the current course of action, the costs and efforts are significantly higher than those that would have to be incurred to solve the problem once and for all.

Hard chemicals

In everyday language, we divide chemicals into hard (artificial substances, often referred to as 'synthetic pesticides'[106]) and soft, natural, also called ecological or organic compounds. In some respects, this classification seems rational, but in others it is artificial (as discussed below).

In Poland, the most popular hard chemicals are artificially created acaricides, or acaricidal pesticides. The most used active substance is amitraz (administered in the form of Apiwarol tablets that are burned in the hive). Beekeepers also use other substances, such as fluvalinate or coumaphos. Studies show that bee products contain many pesticides, including those

---

105   They argue that if pesticides were harmful, the relevant institutions would not allow their use. Meanwhile, pesticides can penetrate apiary products. The most at risk in this respect are products containing fats, mainly wax and propolis. Honey in most cases meets the requirements of food standards but is most often tested for the presence of specific active substances. As consumers, we do not have the opportunity to find out what (possible) harmful substances it contains outside the tested spectrum, e.g. the decomposition of pesticides used in fields or apiaries. Their concentrations are probably not significant, but we cannot predict what the long-term effects will be.

106   The name does not seem entirely appropriate, and to some extent may even be misleading. A synthetic substance is a compound produced by chemical synthesis. However, we may be dealing with synthetic substances, identical to natural ones. Perhaps it would be more appropriate to distinguish between: artificial substances (those that do not occur in the environment or nature) and natural substances (those that we can encounter in nature, e.g. produced by plants or animals, or produced in the environment because of natural biochemical processes).

brought from the fields[107]. They enter the hives via the pollen, nectar or water, because herbicides, insecticides and fungicides are commonly used not only in fields, but also in home gardens.

Interestingly, in this respect, statistically better results are obtained from samples taken from hives set up in cities than in agricultural areas. The results of samples taken in fruit-growing areas are particularly alarming, so it seems that it would be better to abandon beekeeping in the vicinity of orchards or at least to take the hives away for the time of tree blossoming. Toxic substances used in agriculture, similarly to acaricides administered by beekeepers, penetrate the combs. Therefore, even if we refrain from using chemicals, the production of foundation from our own wax in a closed circuit will still carry the risk of contaminating the bees' living environment with synthetic biocidal compounds.

It is not my goal to analyse each active substance (and its decomposition products) in terms of harmfulness to humans or bees, I will rather limit myself to general conclusions. In the case of many of these substances, exceptionally harmful effects have been found, sometimes even carcinogenicity is said to exist[108]. Isn't awareness of such a risk a sufficient reason enough not to use them? The manufacturers of these agents recommend wearing protective goggles or gas masks while working[109]. Perhaps a single or limited use of these drugs would be justified if the problem could be solved forever. But the treatments must be repeated constantly and without them the fate of non-resistant bees seems sealed.

The use of pesticides that target mites (i.e. acaricides), although they do not generally lead to acute poisoning, may also be harmful to bees in the long term. These pesticide residues can increase in the hive environment,

---

107 Compare: P. Johnston et al., Trudny los pszczół. Analiza pozostałości pestycydów w pierzdze pszczelej i pyłku odłowionym od pszczoły miodnej (*Apis mellifera*) w 12 krajach europejskich, [The difficult fate of bees. Analysis of pesticide residues in bee bread and pollen collected from honey bees (*Apis mellifera*) in 12 European countries], Greenpeace technical report, Greenpeace Foundation Poland 2014, 3; Mullin et al., High Levels of Miticides and Agrochemicals in North American Apiaries: Implications for Honey Bee Health, PLOS ONE 2010, 3; M.P. Chauzat, J-P. Faucon, Pesticide residues in beeswax samples collected from honey bee colonies (*Apis mellifera* L.) in France, Pest Management Science 2007, 11.

108 Amitraz is listed by the US Environmental Protection Agency as potentially carcinogenic to humans. However, the studies were conducted on mice (its carcinogenic effect on humans has not been unequivocally confirmed). Amitraz poisoning may, however, have negative effects on many body systems, see: S. Dhooria, R. Agarwal, Amitraz, an under recognized poison: A systematic review, "Indian Journal of Medical Research" 2016, 9.

109 Some people do not want to follow this recommendation just because the sight of a beekeeper in a gas mask by the hive could arouse negative associations (so-called black PR). After all, substances from the hive are considered medicine for many!

affecting the entire hive ecosystem (leading to chronic poisoning). In addition, we are not able to protect bees against the metabolites of these substances. Most acaricides penetrate the wax (they are lipophilic, i.e. soluble in fats), where they accumulate and can affect the fertility of queen bees and drones. Bees constantly function in an environment that is toxic to them (even to a small extent) which can negatively affect their health[110]. Beekeepers are also known to be poisoned with amitraz. Many of them complain of feeling unwell after treatments using these substances. Evidence for many of these ideas can also be found in the article by Prof. Lipiński and Dr. Wojtacka[111]. The authors write:

"In principle, amitraz is of low toxicity to bees. The dose that causes the death of 50% of bees after administration with food is 12 micrograms per bee, and the contact dose that penetrates the chitin cuticle is 12 milligrams per bee. Small amounts of amitraz that are found in honey are rapidly disintegrated, mainly to 2,4-dimethylformamide (DMF) and N-(2,4-dimethylphenyl)-N'-methylformamide (DMPF) molecules, which constitute almost 50% of the substances derived from amitraz disintegration and these compounds are toxic to bees. The DMPF molecule is stable in honey for at least 45 days. The presence of amitraz in honey is not always detectable. Amitraz dissolves better in wax, where it quickly disintegrates into DMPF, a substance that turns out to be more stable in wax. DMPF is the most common residue found in honeycombs. In 2010, Mullin[112] found DMPF in 60% of wax samples and 31% of pollen from North American apiaries. Beekeepers sometimes

---

110   *In vitro* studies by a team of American scientists show that amitraz, coumaphos and fluvalinate can lead to the death of bee larvae; however, most often the concentrations of pesticides in the combs were not as high as those administered during the studies in the laboratory environment (*in vitro*) and which led to the death of the larvae. However, the negative impact of lower concentrations (such as those found in the combs) on the later development of bees has not been ruled out, P. Dai et al., Chronic toxicity of amitraz, coumaphos and fluvalinate to *Apis mellifera* L. larvae reared *in vitro*, Scientific Reports 2018, 4. It was confirmed that these substances (sublethal levels) can lead to physiological, neurological, metabolic and behavioural changes in bees, N. Desneux et al., The Sublethal Effects of Pesticides on Beneficial Arthropods, The Annual Review of Entomology 2007, 52; also J.Y. Wu et al., Sub-Lethal Effects of Pesticide Residues in Brood Comb on Worker Honey Bee (*Apis mellifera*) Development and Longevity, PLoS ONE 2011, 2. It has also been found that pesticides can penetrate royal jelly, which means that worker bees can transfer small, sublethal doses to larvae or queen bees, F. Böhme et al., From field to food will pesticide-contaminated pollen diet lead to a contamination of royal jelly? Apidologie 2017, 8.

111   Z. Lipiński, J. Wojtacka, O amitrazie i nie tylko [About amitraz and more], "Pszczelarstwo" 2015, 11. In their article they also describe the effects of the substance on humans (beekeepers). They report cases of poisoning of beekeepers working with amitraz without masks and gloves, resulting in vomiting, drowsiness, loss of consciousness, slowed heart rate, and respiratory disorders.

112   C.A. Mullin, High Levels of Miticides and Agrochemicals in North American Apiaries: Implications for Honey Bee Health, PLoS ONE 2010, 3.

use such high doses of amitraz that DMPF penetrates the bee bread and royal jelly. Since amitraz is, as mentioned, relatively non-toxic to bees, but DMPF residues can cause chronic poisoning, especially in conditions of simultaneous contamination with other pesticides, it is important to introduce as little amitraz as possible into the hive environment".

However, not everyone agrees that the contamination of beeswax (and indirectly the entire environment of bees) with toxic substances is a real or significant problem for bee health (compared to others). This group of people includes even beekeepers who do not carry out chemical treatments in their apiaries daily. For example, Solomon Parker claims that, firstly, such contamination is relatively insignificant (often below the accepted standards), secondly, that it constitutes a small percentage of all the dangers threatening bees, and thirdly, that it is enclosed in the structure of wax (bees have relatively little contact with it). The removal of toxic substances in wax is a specific ability, developed during evolution, ensuring the cleanliness of the environment and bee food.

It was also mentioned that for some synthetic toxic substances found in plant nectarines where subsequently almost absent in the honey. Bees can therefore filter nectar from toxic substances, which they get rid of from their own bodies together with the wax, thanks to which the food (honey) is uncontaminated (they also give it to the larvae, free from contamination). The wax comb is therefore a specific organ of the bee colony and takes over some of the functions that the liver and kidneys perform in the mammalian body. In a world where we use huge amounts of synthetic pesticides, this type of adaptation serves bees well. So maybe Solomon Parker was right? Maybe the safest place to collect toxic substances is the bee comb[113]?

Soft chemicals

In the common understanding, term "soft chemicals" refers to substances that occur in nature, sometimes also in the hive (or in low concentrations also in honey). This is probably one of the most important arguments for their use (both for beekeepers and consumers). Beekeepers, who realized

---

113   Similar considerations, reasoning and doubts regarding lipophilic substances administered to bees in the hive (and enclosed in the wax structure) were conducted by Prof. James Ellis from Florida State University, who, however, did not reach any clear conclusions. He stated that there is a lack of research that could resolve the dilemma of whether toxic substances enclosed in the wax structure are harmful to bees, and whether they can affect the immunization of the varroa mite. See: Climate effects on Honey bee food source, "Two Bees in a Podcast" episode 123, https://open.spotify.com/episode/ 4fOD2KRhZcHFnuycZpn52Q, accessed: November 10, 2022.

how dangerous artificial acaricides can be, have been convinced to use them.

It was also recognized that soft chemicals do not pollute the bees' environment and do not contribute to the development of resistance in the varroa mite. The problem is that in the doses given to bees in the hives, the concentration of "soft" substances is much higher than those occurring in nature.

The most used organic acids to treat bees are oxalic, formic and lactic acids, which are strong pesticides, and are effective in combating mites. Acids are not soluble in fats, so they do not accumulate in wax. Their supporters also try to prove that due to the above advantages, they are not harmful to bees. However, this is not true. The use of organic acids to combat mites is not only **not** a natural procedure, but for several reasons is harmful.

Both formic acid and oxalic acid are used in laboratories for disinfection, since their biocidal effect is exceptionally strong. Therefore, acids will act much less selectively on hive micro-life (ecosystems of micro-organisms coexisting with bees) than synthetic acaricides. Their use can lead to the death of bee larvae[114]. It was also observed that in colonies in which oxalic acid was used, the amount of brood decreased compared to control colonies, and moreover, some colonies lost their queens after the treatment[115]. High concentrations of acid can also lead to the death of adult workers[116].

Opponents of the use of acids believe that after the application bees are very irritated; the daily drop of adult bees also increases because some of them (e.g. older ones or those infected with pathogens) die[117].

Some of the natural beekeepers I met during bee conferences[118] claimed that after acid treatments they found bee antennae on the bottom board of the hive, which are organs through which insects receive various stimuli. This phenomenon was explained by the fact that the antennae were amputated by the bees, which could not tolerate the irritation. It cannot be ruled out

---

114   A. Gregorc, Cell death in honey bee (*Apis mellifera*) larvae treated with oxalic or formic acid, Apidologie 2004, 9-10.

115   M. Higes, Negative long-term effects on bee colonies treated with oxalic acid against *Varroa jacobsoni* Oud., Apidolgie 1999, 4; E. Rademacher, Effects of Oxalic Acid on *Apis mellifera* (Hymenoptera: Apidae), Insects 2017, 8.

116   R. Martin Hernandez, Short term negative effect of oxalic acid in *Apis mellifera iberiensis*, Spanish Journal of Agricultural Research 2007, 4.

117   Some people even see this as an advantage, because only young and healthy bees remain in the hive.

118   For example, Heidi Herrman (Natural Beekeeping Trust) or Torben Schiffer.

that after acid treatments bees could self-harm[119]. Attention is also drawn to the adverse effect of organic acids on the condition of queen, which is also mentioned by practitioners.

Acid treatments can have a negative impact on the internal systems of bees and their epithelial cells[120], facilitating the penetration of dangerous pathogens into the body, thus posing a threat to life. The susceptibility of bees to bacterial diseases and diarrhoea also increases. Some even claim that acids can disrupt the bee communication system and, consequently, the work of the colony, because one of the ways bees communicate using pheromones, so through chemical messages[121].

Acids not only disrupt chemical messages but can also impair the ability to receive them (mainly by damaging the antennae). As a result, the bees' ability to detect threats is reduced, and their hygiene instincts are weakened. Some practitioners claim that after the use of acids, in late autumn, there are cases of colonies collapsing.

Supporters of acid treatments claim that the undesirable effects of the acids are most likely due to beekeeper error, i.e. using too high an acid concentration. In their opinion, low concentrations are completely harmless to bees.

To treat bees, substances are also used that are not intended for this purpose, such as cleaning agents or disinfectants. Their effectiveness often results from their composition, since they contain acids or other substances that are toxic to mites (e.g. various types of detergents). They may also contain

---

119 Regardless of whether bees can self-harm or not, scientific studies have confirmed the negative impact of formic acid treatment on the sensory receptors in bees' antennae; E. Tihelka, Effects of synthetic and organic acaricides on honey bee health, Slovenian Veterinary Research 2018, 55(3), the list includes many other confirmed negative effects of pesticide treatment, including "soft chemicals".

120 M. Howis, op. cit., p. 21; I. Papeznikova, Effect of oxalic acid on the mite *Varroa destructor* and its host the honey bee *Apis mellifera*, Journal of Apicultural Research 2017, 56(4). Other scientists note: "The therapeutic procedure is performed only once and should not be repeated before three months have passed. Even a double application during this period may be dangerous for the bees. A certain amount of acid is consumed by the bees, which leads to a decrease in the pH of their digestive tract content and haemolymph, and consequently shortens their life. It also causes a rapid increase in the demand for water by bee colonies. The acid is also dangerous for brood, which is why during its rearing not all forms of application can be used. Long-term effects of acid on colonies have been confirmed, resulting in a reduction in the amount of brood, even after several weeks, even strictly according to the instructions, is not indifferent to bees: M. Bykowy, P. Chorbiński, Kwas szczawiowy w zwalczaniu *Varroa destructor* [Oxalic acid in combating *Varroa destructor*], Pszczelarstwo 2021, 6.

121 The bee dance is just one of many forms of hive communication.

irritating substances that stimulate bees to clean (groom) themselves[122]. The bees simply want to remove such a substance from themselves[123].

One of the substances once used in apiaries to treat bees was 'Rapicide™' (disinfectant). According to Zbigniew Lipiński and Joanna Wojtacka, its use without proper protection may be dangerous to humans, as it may lead to damage to the skin, eyesight, irritation (or even damage) of the respiratory tract, drowsiness and dizziness. The substance is also harmful to other organisms, e.g. aquatic organisms[124]. Researchers point out that some disinfectants may be toxic to bees (especially larvae).

In recent years, various types of aromatic oils have been increasingly used to combat mites. These are natural substances produced by plants primarily to protect against pathogens (natural pesticides). They circulate in the channels of the plant, accumulating in places where external barriers are damaged, i.e. where pathogens can most easily penetrate the plant interior. Some oils are also effective against parasites (e.g. oils from healthy trees effectively protect against caterpillars of various insects), so they are elements of the plant immune system, and for micro-organisms - a deadly toxin. It turns out that some of the oils are also toxic to mites. Currently, thyme oil, i.e. thymol, is quite popular in the treatment of varroa, but beekeepers reach for menthol, clove, fir, lemon, among others[125].

Due to their decidedly biocidal nature, none of these oils remain without influence on what happens in the hive. Some oils are also completely unsuitable for use in the hive due to their toxic effect on bees (e.g. nicotinoids, which are a natural insecticide).

---

122 A similar effect is produced by acids, M. Howis et al., Uszkodzenia mechaniczne i pozycja *Varroa destructor* na denicy ula po zastosowaniu różnych środków warroabojczych [Mechanical damage and the position of *Varroa destructor* on the bottom of the hive after the use of various varroacidal agents], Medycyna Weterynaryjna 2012, 10.

123 For this purpose, beekeepers reach for the strangest substances, some use e.g. coca-cola (contains phosphoric acid). Since they cause irritation, these substances are effective in their own way, because at the same time the bees remove the pests from themselves.

124 Z. Lipiński, J. Wojtacka, Przełom w pozaniu biologicznej roli symbiotycznych bakterii przewodu pokarmowego pszczół miodnych [A breakthrough in understanding the biological role of symbiotic bacteria in the digestive tract of honey bees], Pszczelarstwo 2016, 2.

125 An attempt to document the efficacy of varroa treatment with oils was made by Czech researchers, M. Hybl et al., Evaluating the efficacy of 30 different essential oils against *Varroa destructor* and Honey bee Workers (*Apis mellifera*), Insects 2021, 11. Some of the oils showed over 70% varroacidal effectiveness. Better than thymol were oils from: peppermint, manuka bush (*Leptospermum scoparium*), oregano, Litsea tree, carrot, and cinnamon.

It should be noted that the advantage of oils is that they probably have a significantly weaker effect on the physical protective barriers of bees than acids (they do not seem to damage the epithelium of bees), and they do not pollute the hive environment (if they penetrate wax or honey, they should not be a direct problem for either humans or bees). Their strong smell, however, may interfere with bee communication. Some oils are used for this very purpose, e.g. when combining colonies. In higher concentrations, they can (e.g. thymol) cause great stress in the colony, leading to egg laying disorders, or, as some claim, even the effect of drone laying of the queens.

For most beekeepers, using toxic or biocidal substances is simply a necessary evil. But is this evil really necessary? I have no doubt that we should look for alternatives to chemicals, whether hard or soft, they should be the last resort. Meanwhile, many beekeepers use them preventively or prophylactically, e.g. in early spring, even though they have performed all the treatments in the fall and repeated some treatments in the winter (when the risk of robbing has dropped to zero). Michael Bush has one answer to the question about the principles of using chemicals to combat varroa: "you should get to know them well enough to know why you shouldn't use them"[126].

Varroosis

What kind of disease is this?

Varroosis (varroasis) is still the number one problem of beekeeping in densely populated, industrialized countries, where intensive agriculture is practiced. It was here that the invasion of the varroa bee changed the entire ecological situation of the honey bee. Until then, beekeeping was the last bastion of agriculture, where pesticides were not systematically used. Since the invasion of the varroa mite, most bees have ceased to be independent. Natural wild populations have been decimated, and in regions with a high density of managed bees, but a relatively small number of available nest sites (e.g. in Poland or even the whole of continental Europe), they have probably not fully recovered yet. There are certainly natural wild colonies

---

126 M. Bush, Ten Commandments of Beekeeping, https://www.bushfarms.com/beestencommandments. htm, accessed: January 21, 2025.

living around us, and more than we think[127], but most often they are escapees from apiaries, for whom it is extremely difficult to exert a genetic influence on the rest of the population. The situation is completely different in Africa or South America, where large wild populations of honey bees live in vast areas beyond human control.

It is quite a popular opinion that due to the invasion of varroa, bees have become susceptible to many diseases. In my opinion, the reason for this situation is that for several decades beekeepers have been keeping them alive using increasingly sophisticated methods. It also seems that varroa has allowed the selection of bees to move towards gentleness and increased efficiency.

Semi-wild apiaries where bees lived freely, almost as in nature, are gone. They used to contribute to maintaining the genetic diversity of the population, and serving to develop adaptations valuable for wild bees and favouring their survival. In the opinion of breeders, such bees were destroying their work, producing their own semi-wild drones. Since the invasion of varroa, bees in apiaries began to require care and treatment, which led to a change in the way beekeeping is practiced in basically all industrialised countries. The mite contributed to what is sometimes called the professionalisation of beekeeping. Many people consider this process to be a good direction. In their opinion, it means that beekeepers are becoming more competent, they must educate themselves to keep their bees alive. In my opinion, the direction of this "professionalisation" is not good from the perspective of bee health. This process concerns primarily honey production, the use of anti-varroa treatments and so-called hygiene procedures, which primarily lead to the weakening of various immune mechanisms of the bees.

Let us consider why bees die faster today and with a significantly lower level of mite infestation than before. Professor Jürgen Tautz already in 2008 stated that the number of mites, deadly for the colony is only one tenth of the number from decade earlier. So how did it happen that bees without care live shorter lives on average? Have mites become more virulent? If so, what does this increased virulent nature means? Do mites reproduce faster? Do they feed on a larger number of worker bees? There are many questions.

---

127 T. Seeley mentions the study of wild bees in various regions of the world in his book 'The Lives of Bees: The Untold Story of the Honey Bee in the Wild'. There are also other sources available about wild bees, such as www.freelivingbees.com. A very interesting study was conducted in Serbia, in the Belgrade region. It shows that the area is inhabited by many wild bee colonies, which, according to scientists, are probably form a feral or wild and self-sufficient population there, and are not just escapees from apiaries, J. Bila Dubaić et al., Unprecedented density and persistence of feral honey bees in urban environments of a large SE European city (Belgrade, Serbia), Insects 2021, 12.

The problem is of course complex. The health of bees depends on many factors, but the bees die for a specific reason. It is not the mesh bottom, the height of the stand, the sugar or the material the hive is made of.

Varroa originates from the Far East. As I have already mentioned, *V. jacobsoni* is a natural parasite feeding on the eastern honey bee (*A. cerana*).

During its evolution, the mite and *A. cerana* have developed a relatively balanced relationship. The natural range of western honey bees (*Apis mellifera*) is Africa, Europe and the western and south-western part of Asia. The eastern honey bee (*A. cerana*) only occurs in Asia, so although *A. mellifera* and *A. cerana* are sister species their natural ranges did not overlap[128]. The eastern honey bee has developed many adaptations to coexist with varroa; it can detect parasites in the brood cells and remove them, preventing their population from growing too large. In extreme cases, when these mechanisms fail, it leaves the nest (absconding), leaving most of the mites behind, along with the infected brood.

In the process of adaptation of the Eastern honey bee to the presence of varroa, a mechanism has developed to direct the force of the parasite's impact onto the drones. Most often, mites reproducing in the worker brood remain infertile. According to other sources, the worker brood is so delicate that when infected by the mite, it dies, thus interrupting the parasite's reproductive cycle[129]. Regardless of which explanation is correct (maybe both), mites reproducing in the worker brood do not achieve reproductive success. It is difficult for me to assess the scale of mortality of Eastern honey bee colonies due to mites, because I do not know of such data, but it seems that it is not significant (due to the long evolutionary relationship between both species).

The mites feed on the tissues of bees. Until recently, it was believed that they fed only on haemolymph (insect blood). However, research

---

128 It is estimated that the evolutionary lineages of *A. mellifera* and *A. cerana* diverged about 7-8 million years ago (K. Dogantzis), but the scale of possible later contacts between the species is difficult to estimate.

129 Studies show that several dozen percent of worker bee larvae infected with the pest die in a process like cell apoptosis, hence the phenomenon is called social apoptosis, P. Page et al., Social apoptosis in honey bee superorganisms, Scientific Reports 2016, 6; for more information see chapter 6: At the crossroads (...).

conducted by Dr. Samuel Ramsey[130] showed that the basis of their diet is the fat body, an organ of great importance for the health of the honey bee – it is their reservoir of fat, protein and glycogen (which is important, for example, for survival during wintering)[131]. Feeding by mites on the fat body of workers (so-called winter bees) most often probably means death due to the serious depletion of reserves necessary for survival. This organ also takes part in many metabolic processes (its role is sometimes compared to the role of the liver in vertebrates). Honey bees with impaired fat bodies may therefore have problems metabolizing certain toxic substances, thus even minor poisonings (with sublethal doses) may lead to the death of individual workers.

---

130 S.D. Ramsey et al., *Varroa destructor* feeds primarily on honey bee fat body tissue and not haemolymph, Proceedings of the National Academy of Sciences 2019, 1. The author takes a proactive approach to bee problems. He is currently conducting research on bee parasitic mites in Asia, in particular the Euvarroa spp. and *Tropilaelaps mercedesae*, which both can cause potential threats to western honey bees, like varroosis. *Tropilaelaps* are parasites of bee brood. Their main hosts are the giant bee (*A. dorsata*) and the Himalayan bee (*A. laboriosa*), however, the ability of the parasites to infect *A. mellifera* colonies has been confirmed. The parasites reproduce exceptionally quickly, and their life cycle takes place mainly in the sealed brood, which makes them difficult to control. *Euvarroa* spp. are parasites of the dwarf honey bee (*Apis florea*) in nature. Laboratory studies have confirmed the possibility of their existence and feeding on *A. mellifera*, and even the ability to raise offspring in the worker brood, however, no change of host in nature has been observed so far, S.D. Ramsey, Foreign Pests as Potential Threats to North American Apiculture *Tropilaelaps mercedesae, Euvarroa* spp. *Vespa mandarinia*, and *Vespa velutina*, Veterinary Clinics of North America: Food Animal Practice 2021, 3, see also: "Beekeeping Today Podcast"; https://www.beekeepingtodaypodcast.com/dr-samuel-ramsey-fall-2022-updates-55-221/, accessed: November 9, 2022. Ramsey believes that thanks to scientific work, we can be better prepared in the event of an invasion by another dangerous parasite.

131 The fat body is a tissue characteristic of insects functionally equivalent to the liver and adipose tissue of vertebrates. It is also the site of hormonally controlled changes in the intermediate metabolism of insects and acts as a reserve tissue, accumulating reserve carbohydrates, fats and proteins, while releasing trehalose, diglycerides and proteins into the haemolymph. It sometimes deposits products of nitrogen metabolism (uric acid) as well. The fat body is in the body cavity (hemocoel), is divided into a visceral part (around the insect's digestive tract) and a peripheral part, reaching around the muscle fibres to the body integuments; http://stareaneksy.pwn.pl/biologia/1468973_1.html. accessed: November 8, 2022; see also: A. Strachecka, Ciało tłuszczowe, tkanka odpowiedzialna za metabolizm [Fat body. Tissue responsible for metabolism], Pszczelarstwo 2022, 11; J. Chobotow et al., Morfologia I funkcje ciała tłuszczowego owadów z uwzględnieniem pszczoły miodnej *Apis mellifera* L. [Morphology and functions of the insect fat body with regard to the honey bee *Apis mellifera* L.], Veterinary Medicine 2013, 12.

Professor S.J. Martin believes that working on bee behaviour is the only solution to the varroa problem. Here, during research conducted on the island of Oahu, Hawaii (USA), photo by S. Martin.

Although the limitation of this organ's function is a serious threat to the life of the worker, it must be remembered that a superorganism consists of many thousands of individual insects. If we were to assume that the death of the colony was to be caused solely by feeding (excluding other negative effects of the presence of the mite), the number of mites would have to be huge. A specialist in bee pathogens and parasites, Prof. Stephen Martin from the University of Salford in Manchester, believes that there would have to be about 80,000 to kill a colony by feeding alone!

Varroa can only reproduce in the brood[132]. Female mites enter the cell containing the bee larva shortly before it is sealed. There, the adult female lays the first unfertilized egg, from which the male hatches. Then, every few dozen hours (approximately every 30 hours), more fertilized eggs arrive, from which females hatch. The mites mature until the young bee hatches from the cell. Mature females leave the cell and look for a new host, while those whose maturation cycle has not completed by that time die. The longer the development of the bee pupa lasts, the younger female mites have a better chance to mature. For this reason, mites reproducing in drone brood achieve greater reproductive success than in the case of reproducing in worker brood. In a drone cell during one reproductive cycle, the maturity necessary for survival outside the cell can be reached by up to two more young females. According to the reproductive cycle of the mite, the young male fertilizes his own sisters and dies shortly thereafter.

Cross-fertilization can only occur when two adult female mites enter a given bee cell and both lay eggs in it[133]. As a result, the mite is characterized by relatively little genetic diversity, especially in those places where it is controlled on a mass scale.

---

132 Thanks to Dr. Ramsey's research, it was also noted that varroa mites, which feed on the fat bodies of both larvae and adult bees, achieve significantly greater reproductive success. From both developmental forms of the bees, the varroa mite takes specific proteins that are incorporated into the structure of the yolk of their eggs, which means that the varroa mite can reproduce faster and with lower energy requirements. This phenomenon has been called kleptocytosis: the name refers to the process of placing "stolen" proteins directly into the egg cells of the parasite. The process thus allows to bypass the biochemically expensive and time-consuming process of protein synthesis necessary for reproduction by taking them directly from the host organism. Ramsey suggests that kleptocytosis may be one of the key adaptations that allowed for the enormous reproductive success of the varroa. He also believes that the discovery of this process will allow the development of new ways of combating the parasite in bee colonies (research is ongoing); S.D. Ramsey et al., Kleptocytosis: A Novel Parasitic Strategy for Accelerated Reproduction via Host Protein Stealing in *Varroa destructor*. [At the stage of translating the book, the work has not yet been peer-reviewed and published in a scientific journal, but the text is available at. https://www.biorxiv. org/content/10.1101/2022.09.30.509900v1.article-info, accessed: November 10, 2022.] In this context, Dr. Ramsey asks whether the period of mites' existence on adult bees (imago) should be referred to as the phoretic phase. Strictly speaking, this phase concerns the transfer of some organisms by others (using their locomotor abilities to occupy new nest sites). However, research has indicated the key importance of *Varroa destructor* females feeding on adult bees to achieve greater reproductive success.

133 Adopting such a reproductive strategy seems strange - it turns out that incest (as the basis of its own reproduction) allowed the mite to achieve much greater evolutionary success than if it had used the more common cross-fertilization process.

A. cerana colonies are smaller than those of *A. mellifera*[134], and this species is more difficult to breed and manage. When subjected to great stress, *A. cerana* tends to abandon the nest (similarly to some African subspecies of the western honey bee). Increased interference in the life of colonies usually ends with their abandoning the hive. Therefore, keeping colonies of *A. cerana* is less efficient and profitable than *A. mellifera*. For this reason, western honey bee colonies began to be transported to regions where the Eastern honey bee occurs naturally. As a result of contacts between these two species, varroa was able to shift to and exploit a new host. According to various studies, this happened at least several times. For a long time, it was believed that the mites feeding on the western honey bee and the eastern honey bee were the same species: *Varroa jacobsoni*. Ultimately, it was found that the individuals feeding on *A. mellifera* belonged to a different species: *Varroa destructor*[135].

Scientists suggest that varroa switched hosts from *A. cerana* to *A. mellifera* several times. One early switch was by the Japanese haplotype, whereas the Korean haplotype switched hosts later. It is commonly assumed that the Japanese haplotype is less virulent than the Korean one. Many beekeepers believe that without human help, bees can only cope when infested with the Japanese haplotype, so in the past, the ability of bees to survive without

---

134 Both species of bees belong to the so-called eusocial insects, i.e. those that have developed complex social relationships (characterized by, among other things, division into morphologically different castes and cooperation in caring for offspring). They belong to the genus *Apis*, in which, depending on the adopted taxonomy, 7-11 species of bees are most often distinguished.

135 D.L. Anderson et al., *Varroa jacobsoni* (Acari: Varroidae) is more than one species, Experimental and Applied Acarology 2000, 3; the researcher suggests that there are several haplotypes of the Eastern honey bee, previously referred to by the collective species name: *Varroa jacobsoni*. It was found that six of these haplotypes are larger and less spherical; due to their distinctive features, they were recognized as a separate species, which was named *Varroa destructor*. However, the study also shows that only two of the haplotypes (Japanese and Korean) changed hosts and began to parasitise *A. mellifera*, which allows us to assume that other haplotypes, both *V. jacobsoni* and *V. destructor*, are not adapted to feeding on *A. mellifera*. However, in the literature on the subject, there is a common view that *Varroa destructor* was formed after a change of host because of genetic differentiation associated with population isolation (so first the change of host *Varroa jacobsoni* would have occurred. Only then, due to different living conditions, genetic differentiation would have occurred, and a new species would have been created). It is not for me to decide which version of species diversification of the mite is true. However, the feeding of *Varroa jacobsoni* has undoubtedly been confirmed in at least one population of *A. mellifera* honey bees (on the island of New Guinea). As a result of the research, it was concluded that another change of host of the parasite occurred there, which could suggest the truth of the second of J.M.K. Roberts' hypotheses: Roberts, et al., Multiple host shifts by the emerging honey bee parasite, *Varroa jacobsoni*, Molecular Ecology 2015, 4; J.M.K. Roberts, Tolerance of Honey Bees to varroa Mite in the Absence of Deformed Wing Virus, Viruses 2020, 5. When trying to verify which of the hypotheses is true, one of the scientists specializing in the problems of honey bee parasites replied to my question: "The truth is that our scientific knowledge changes or evolves every time new evidence is revealed." Research is ongoing.

treatment in South or North America was explained by the presence of an avirulent type of parasite. However, I have not found an explanation for this difference in virulence in scientific sources. Randy Oliver[136] claims that the Japanese strain can only reproduce in drone brood, while the Korean can do so in both drone brood and worker brood. It therefore achieved greater reproductive success and began to dominate basically all over the world, including in regions where the Japanese strain initially functioned.

Meanwhile, Professor Stephen Martin believes that the differences in the mite haplotypes are minor and insignificant. No real differences have been found in reproduction abilities of both haplotypes, and both can also carry viruses that are lethal to bees. So, what explains the differences in the survival abilities of some honey bee populations infested with different haplotypes of varroa? According to the professor, it has nothing to do with the varroa haplotype.

It turns out that the secret of the bees' ability to survive in the presence of varroa comes down on the one hand to their immune response, and on the other hand to the (absence of) deadly strains of some viruses such as the deformed wing virus (DWV)[137]. Here we come to the heart of the matter: why does the presence of varroa not mean the destruction of the colony? In Africa or South America, you can find honey bee populations infected by thousands and sometimes even dozens of thousands of mites, which do not have the health problems characteristic of varroa infestation[138]. Before the era of varroa, DWV was also not a real problem for bee colonies, it was almost completely harmless to them. Only the combination (mite plus virus) causes DWV to start spreading rapidly between non-immune workers. Then an epidemic breaks out in the hive, with the number of bees rapidly decreasing, until it becomes impossible to maintain the basic life functions of the colony. Sometimes the hive empties completely of bees; sometimes in the summer the last handful of bees is finished off by robberies, and in the

---

136   Randy Oliver's blog: www.scientificbeekeeping.com.

137   This thesis is confirmed, among others, by studies conducted in New Guinea, where both *A. cerana* and *A. mellifera* occur. It is believed that both populations are infected with the *Varroa jacobsoni* mite (*Varroa destructor* is not present there). According to scientists, around 2008, in New Guinea, there was another change in the mite's host from *A. cerana* to *A. mellifera*. It should be noted that the populations were also attacked by another mite: *Troplilaelaps mercedesae*. After initial collapses in the honey bee population, the bees acquired immunity that allowed beekeepers to continue their normal activities. Scientists suggest that the immunity of bees from New Guinea results from the complete absence of the DWV virus in this isolated population, see: John M.K. Roberts, Tolerance., op. cit.

138   They are infested by varroa, but despite this do not suffer from varroosis.

winter - by frost. The process of developing an epidemic in the hive could be stopped, however. How?

Firstly, bees can slow down the development of the mite population to such an extent that the pathogens do not have time to infect the brood and workers to an extent that would avoid the collapse of the colony[139] (we call these bee colonies' resistant to varroa). In other words, the course of the disease in the hive is under control, because the basic viral vector has been blocked. Then the colony has a chance to survive the winter and start developing in the new season (even if it is not in the best condition in spring).

Secondly, bees can develop an effective immune response to transferred pathogens[140] (we call this tolerance of bee colonies). Then, even if the infestation with varroa is relatively high and the level of virus in the colony is high, the immune response mechanisms of the bees will prevent the bees from dying and the colony collapsing. For example, in the British apiaries of David Heaf or Ron Hoskins, we can find bee colonies with untreated varroa, showing a very high level of specific DWV strains (type B), to which they are not susceptible[141].

---

139   The safe level of infestation is considered to be 2-5% (depending on the time in the season, among other things), but most breeders selecting resistant bees believe this safe limit to be around 3%. Recently, however, probably due to the progressive increase in the virulence of pathogens some researchers have begun to consider infestation at a level of even 1-1.5% as potentially dangerous to the bee colony (it's viewed as the threshold at which treatment is recommended).

140   From a practical point of view, it is simply much easier for us beekeepers to focus on observable characteristics or properties of bee colonies during selection. It is easier to measure the level of colony infestation or observe the effect of various bee behaviours (e.g. uncapping brood cells, hygiene tests, etc.) than to study the immune response of bees to pathogens, since this requires specialist knowledge and equipment. Of course, the situation is complex, and evolutionary processes run in parallel. Bees that develop resistance to pathogens are most often the same bees in which we observe characteristics related to limiting mite reproduction.

141   J.L. Kevill et al., DWV-A Lethal to Honey Bees (*Apis mellifera*): A Colony Level Survey of DWV Variants (A, B, and C) in England, Wales, and 32 States across the US, Viruses 2019, 5. The DWV type B strain is considered by some scientists to be milder, but most researchers consider it to be more virulent, D.P. McMahon et al., Elevated virulence of an emerging viral genotype as a driver of honey bee loss, Proceedings of the Royal Society B 2016, 6; M.E. Natsopoulou et al., The virulent, emerging genotype B of Deformed wing virus is closely linked to overwinter honey bee worker loss, Scientific Reports 2017. 7. It is also considered to be more virulent by many Polish scientists (e.g. Z. Lipiński, K. Pohorecka). It was found that one of the main differences between the two types is that type B can reproduce not only in the cells of the bee, but also in the cells of the mite. This would mean that the bee is a biological vector of the type B virus but is only a mechanical vector of the type A virus (the viruses do not reproduce in the mite, which only transfers pathogens from infected bees to others). Despite this, the B strain (considered milder in this study) displaces the more dangerous A strain. However, some suggest that the resistance of bees in the UK may not be related to whether the strain is objectively milder or more virulent for the wider population honey bees. Local bees may simply have developed resistance to the strain of the pathogen that is dominant in the region (and in this sense it is milder locally).

Thirdly, natural selection in a larger population should mean a systematic reduction in the virulence of pathogens. However, this will be difficult to achieve if the apiary is small and located in the vicinity of such apiaries where evolutionary processes are more conducive to an increase in the virulence of both parasites and pathogens. It seems, however, that cooperation that would support the necessary and positive evolutionary changes would be possible already at the level of the beekeeping club (however, it would require limiting the frequency of bee transport and undertaking selection).

Fourthly, and finally, uncontrolled population development, at least of some pathogenic micro-organisms, should be prevented by healthy bacterial and fungal biofilms. It is difficult to say whether this will happen in the case of viruses, but such a relationship can be said about spores of the *Nosema* genus or some bacteria. And this time the conclusion suggests itself: all these processes would be most conducive to subjecting a large percentage of bee colonies to the laws of natural selection.

Varroosis reaches its critical point most often in late summer or autumn. This is the moment when the parasite population is at its highest, the queen bee egg laying is reducing, and therefore the colony naturally shrinks before winter. This is when the relative infestation of the mite increases rapidly, because the number of bees decreases, and the number of mites continues to grow. If the colony is not immune, it usually starts to fail in September or October. Colonies in which there are fewer mites, or in which some immune mechanisms are present, can survive longer.

It also happens that the effect of varroa can be enhanced by the presence of *Nosema*. This parasite also shortens the life of worker bees, thus are simply unable to survive the entire winter, especially if the colony has been seriously weakened by viruses transmitted by the mite. In the case of varroosis, early spring is also a critical period for the bees. Some colonies come out of winter very weakened. They sometimes occupy only 2-3 frames. If the weather stabilizes and spring flows are good the colony may survive and develop during the summer. However, if the weather breaks in early spring, such colonies very often collapse. Workers weakened by the winter are unable to raise a new generation of bees. In the case of a massive varroa invasion (large-scale infestation) the death of a colony can occur at virtually any time of the year. However, in the peak season (May-June), such cases are extremely rare, because the colony development is rapid with young bees appearing faster than older ones or those infected with the virus, are dying. It is worth adding that if high mite levels coincided with the peak period of brood rearing (which happens rarely, because it would mean that the colony came out of winter heavily infested with mites), it cannot be ruled out that the colony could cope with such a few mites, at least for a certain period. I recall a case of a colony at the peak of the season, with a very high level of mite infestation, that was split (an artificial swarm was made with an old queen), which was supposed to provide a break in brood production and thus stop the development of the mites and encouraging the bees' hygienic behaviours. Although the colony became very weak at that time, it came out of the crisis without any other treatments or procedures than a brood break. It also managed to rebuild the strength needed for overwintering. The bees from this colony (split into nucs the following year) survived for another two years.

Can bees cope with varroa?

The deadly nature of varroa raises the question: are bees able to stop the development of the parasite on their own? **I will answer in the affirmative, without a moment's hesitation.** However, I mean the population, not individual colonies. Probably most scientists and beekeepers, even those who do not skimp on chemicals in their apiaries, will admit that it is possible theoretically, for the honey bee (as a species) to coexist with the mite in a balanced relationship.

They will probably also admit that bees can cope with most diseases on their own. However, it seems that they will deny that this is possible in the current apiary management ways and economy concepts. I also agree with the latter, at least to some extent. The difference between the way of thinking presented and the one I represent is that instead of convincing people to use treatments, I believe that we should help bees by reorganising apiaries and ways of how we use them for honey production. However, this requires a change in the way of thinking. We need to consider what needs to be changed in the apiary and how to reorganize the selection of bees to create conditions for them, to be like those found in varroa resistant populations.

Every season, on several continents, the western honey bee proves that it is a self-sufficient species and can find a balance in its relations with varroa mite, other of its various pests and pathogens. In our country, as well as almost all continental Europe, this balance is not easy to achieve. We will not find many regions where large populations of bees could pass through the process of natural selection. Europe has a relatively high population density, and human settlements are spread relatively evenly throughout its area (apart from regions that are also not friendly to bees, e.g. high mountains etc.). Unlike both Americas, Africa or Asia, there are no huge, practically uninhabited areas in Europe where wild bees or swarms that escaped from apiaries could start the process of natural selection, avoiding inter-breeding with the vulnerable apiary populations. In the United States, the situation is different. There, the population of bees in apiaries is about 25-30% larger than in Poland, while the human population and area of the country are many times larger. The average density of bees is therefore much lower, which is good for the general health of the population. In many areas, however, the situation is like that found in continental Europe, which results from the uneven population density of this geofigureally diverse country. In addition, to densely populated regions (where the density of the bee population is relatively higher), there are semi-wild areas where bee populations can go through the process of natural selection.

However, this is compounded by the specificity of the US apiary management ways, where a huge part of the bee population is constantly transported to pollinate various crops[142], which creates a huge epizootic threat to bees. In my opinion, this is where we should look for the reasons for the high percentage of colony mortality in North America.

The situation is still different in the British Isles, where amateur beekeeping dominates, and according to official data, the density of the bee population is no higher than 2 colonies/sq. km, while in Poland on average it was 7.5 colonies/sq. km in 2023 – and the numbers are still growing (there are regions where it is even twice as high, for example in the voivodeship where I live it was estimated to be 13,9 colonies/sq. km)[143]. This creates a difficult epidemiological (epizootic) situation in Poland.

I am far from saying, however, that the situation in continental Europe is hopeless. The apiaries of many beekeepers who give up all treatments against varroa are often surrounded by commercial apiaries, where resistance is not selected. Thanks to their consistency and persistence, these bees successfully pass the basic selection process. How is this achieved?

---

142   It is estimated that in California alone, more than two-thirds of all bee colonies from the state (about 2 million out of around 3 million) are transported for the pollination of almond trees, and this number is constantly growing (I sometimes call it: "pathogen redistribution centre"). At the same time, there is no chance that beekeeping will be carried out year-round in the almond growing region, because these are huge monocultures that can only be a source of food during the flowering period of the trees (besides, it is controversial whether the forage from almond trees is good for bees at all, but they have no alternative). Almond growers are therefore forced to import colonies, at least until trees capable of self-pollination are created (work is ongoing). It seems that selecting such almond trees could contribute not only to reducing costs for the orchardist, but also to improving the health of the bee population in the USA, which would not have to be transported *en masse* for almond pollination.

143   https://pasieka24.pl/index.php/pl-pl/aktualnosci/wiadomosci-z-polski/4374-w-polsce-przybywa-pasiek; accessed February 18, 2025.

We should start by answering the questions: what organisms do we consider biologically healthy or effective? We can assume that a wild-bee colony is successful if it ensures future generations by swarming. However, we require a hundred percent survival rate and high productivity from the bees in our apiaries, which of course is unrealistic, because colonies die, some get weaker, and do not produce as much honey as the beekeeper would like. We set the bar higher and higher. However, would we not be satisfied, both from the economic and environmental perspective, if each colony survived for an average of four or five years and remained productive for two or three? This would mean that the average mortality rate would not exceed 20-25%, and we would be able to generate income from most of the hives. For me, this is not only an acceptable situation but also expected. Now I still cannot boast even such statistics, but I still live in hope for the future (Numerous examples cited in this work justify such a conviction) – as maybe more beekeepers would join the breeding effort. It is also worth adding that population mortality (maintained at a certain level) is also necessary from the evolutionary side, to eliminate from the gene pool those individuals that cannot cope[144].

The female varroa mites have camouflage abilities that allow them to very quickly take on the scent of the host, and to hide from the bees. However, due to the considerable size of the mite, it seems that the bee should be able to feel it on its own body, especially while the pest feeds. Meanwhile, the vast majority bees do not notice the mites at all or do not consider it a danger, since the mites are completely ignored by them.

In the hive we can find many organisms, based on observations we know that there can be even over a hundred or so species of small insects and arachnids. They live alongside honey bees and coexist with them in various relationships, most often commensal. Bees tolerate their presence until it becomes a problem for them and then they can remove robbers and various parasites (e.g. wasps, wax mites) from the hive. With propolis they can also cover the cocoon of wax moth larva, or even a dead mouse stung to death. So, they have many mechanisms and tools with which they can fight intruders. There is also nothing to prevent them from using the same tools against mites. Apparently, however, the mite is not considered by bees to

---

144   Such a state (or assumption) is not accepted by most of the beekeeping community. However, let us look at the actions of beekeepers from a biological point of view: is the replacement of queens in an apiary not an action based on a similar assumption? The difference is that the survival of the appropriate genetic lines of bees depends on the economic values of the bees, and not their ability to cope with threats on their own.

be an organism dangerous to them. To neutralize the threat, the bees must implement three subsequent (basic and completely obvious) stages of the strategy to fight the mite. Which ones? First, they must become "aware" of its presence. Second, they must "associate" it with a threat to themselves or at least a nuisance great enough to be worth intervening in. Third, they must take action to minimize the threat.

Bees from many populations can sense the presence of a mite, or rather the specific odour emitted by the infected larva that penetrates the cell cap and intervene. On the Internet, you can watch films showing how bees fight the mite or try to throw it off themselves using their legs. In this context, the question arises whether we really need hundreds or thousands of years of evolution for the desired immune capabilities to develop and become widespread in the population. According to Professor Stephen Martin, in Great Britain this process is already well advanced, and has only been going on for less than three decades[154].

The traits that bees need to develop a survival strategy can be found in the population, including our own apiary population. Unfortunately, many beekeepers select against them, e.g. by eliminating the feature of propolising as inconvenient (they also use hives that do not stimulate the deposition of propolis). They also fight against the natural defence abilities of their bees or use queen bees that produce perfect uninterrupted slabs of brood as breeders. In this way, they exclude from selection the hygienic traits that bees use to detect and remove infected brood from the hive.

One of the main characteristics of bees that are eliminated during selection is swarming. From the perspective of the beekeeper's interests, the emergence of a swarm is disadvantageous. Meanwhile, a natural break in brood production in the middle of the season is exceptionally beneficial to the health of the colony. In addition, the emergence of a swarm is also a way to naturally ensure the continuity of the population.

There are many more such examples, acting against the nature, adaptations and good condition of bees. It seems that to reverse the unfavourable breeding trend, it would be enough take some basic actions: stop selecting against mite-resistance traits and start breeding those colonies that can cope best on their own.

---

154   Stephen Martin talks about this in the podcast Natural varroa-Resistant Honey Bees and Small Hive Beetles, "Two Bees in a Podcast" episode 35, https://open.spotify.com/episode/4d7i8T87lur4wVGgxpblkS, accessed: January 8, 2023.

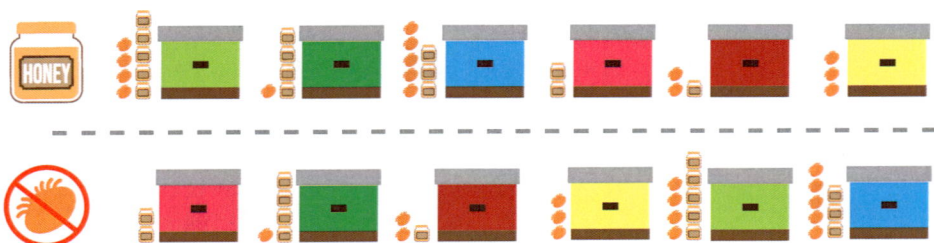

Hypothetical ranking of the same colonies according to productivity (honey production) and mite population growth (indicating the presence of resistance mechanisms). Depending on the priority during the breeding selection, a different breeder will be chosen for reproduction. However, we can find in the population such bees that to a high degree fulfil the criterion of health (resistance), and at the same time are productive and meet other requirements. In this respect, the high position in this ranking is occupied by the bee colony in the dark green hive, figure by M. Uchman.

This could help improve the "genetics" of the bees in the long term, while also allowing them to maintain their adaptations to a changing environment.

Unfortunately, it seems that too many beekeepers still do not understand the importance of a balanced system of bee traits and its significance for maintaining a healthy population. Breeding priorities are increasingly being reversed[145]. First, they choose the most productive, gentle, non-swarming bees, and only from among those that meet these criteria, they choose bees with few mites. Meanwhile, many beekeepers who do not combat the mite promote exactly the opposite approach to the problem: choosing from among those bees that have the lowest number of mites (have the lowest dynamics of parasite population growth) and can survive without treatment. They also pay attention to those that did not require treatment for the longest period and only from among them do they choose for propagation the most productive ones.

At first glance, it may seem that there is not much difference in these approaches, but in practice it turns out that completely different bees pass through the selection sieve.

According to Michael Bush, to conduct healthy beekeeping, one should follow a few simple rules, which the famous practitioner calls "four simple steps to healthier bees"[146]. According to the beekeeper, implementing them would create conditions in which bees would be able to cope with problems themselves. These are:

---

145   The poor condition of bees is the result of the presence of a dangerous parasite, not decades of selection against adaptations, right?

146   https://bushfarms.com/beesfoursimplesteps.htm accessed: January 22, 2025.

1. Clean and natural environment for the bees to live in.

2. Breeding local bees that do not require treatment.

3. Building a natural hive ecosystem.

4. Providing natural food.

Natural beekeepers from all over the world, who have bees that fight mites on their own, agree with these principles. Some add a fifth condition: Peace, following the natural life cycle of the bees plus minimal interference.

Are untreated bees a threat to the surrounding colonies?

For non-resistant bees, varroosis is a merciless disease. By means of horizontal transmission varroa has taken over almost the entire world in just a few decades. So, can giving up treatment in our apiary threaten other apiaries? The answer to this question is yes. Yes, it can. Bee colonies in which the varroa is not controlled can pose a threat to neighbouring apiaries. They can be a source of mite bombs and contribute to the spread of pathogens. However, the question should be asked differently: are untreated colonies really a greater threat to the environment than others? Well, the answer to this question is no longer unequivocal. In my opinion it would even be negative, especially if we consider the total balance of profits and losses and the current situation in apiaries.

Epidemiological threats are common: we should take them into account when organising an apiary. First, for the health of the bees, we must remember to place as few colonies as possible at each apiary. Second, we should move as far away from other apiaries as possible, not only because of competition for food but it also reduces drifting and robbing.

It is hard to deny the fact that in our conditions the percentage of dying bee colonies is higher among untreated bees than among treated ones. In my opinion, however, the cause of the threat should be sought in the way bees are kept, which promotes horizontal transmission and increases the virulence of pathogens. All apiaries can be blamed for this to a similar extent. The least "guilty" are, however, those colonies that live in isolation (including wild/feral ones in forests), and yet they are the ones most often blamed for the poor health condition of bees. The impact of these colonies on the environment (even if they die) is minimal, and often, due to isolation, it is simply zero.

In my opinion, untreated colonies, after the first stage of natural selection, are no greater threat to the environment than others, because the local bees on which they are based are beginning to develop immune mechanisms. The period of their negative impact on the environment passes very quickly, later I would rather consider their influence to be positive (drones can be a source of good genetic material).

The greatest threat is posed by bee colonies dying in summer and early autumn. If the beekeeper does not carry out treatment, non-resistant colonies (with a high degree of infestation) will begin to fail. They can also become "mite-bombs". It is believed that up to 100 mites per day can be transferred to another hive. When I developed my apiary based on bees that survived the first selection period, the late summer and autumn declines in bees stopped completely. In early autumn 2018, at most two colonies died, in 2019 and 2020 none died, and in 2021 and 2022 only one died.

Bees died in winter, most often in January and February. In the autumn of 2019, several colonies suddenly weakened (then died during the winter). Their nests did not look as if they had been robbed, since there was a lot of food in the combs, and I did not find any remains of wax from the capping's on the bottom board, which is characteristic of robbed hives[147].

The initiative to return to the principles of natural selection of bees was pushed to the sidelines for several decades. During this time, most of the beekeeping community (probably 98-99%) performed treatments against mites. In the meantime, the problem of bee losses returned every year in autumn, and it becomes more difficult to solve it with each decade[148].

For the purposes of these considerations Let's assume that individual untreated colonies play a significant role in this. If this were the case, the swarms will continue to escape from the apiaries and look for wild nests. We will always find such colonies around us. So, shouldn't we accept this and try to make changes at a different level? If beekeepers are afraid of (their own!) non-resistant colonies that have escaped from their apiaries, maybe it's high time to change the selection priorities and start breeding resistant

---

147   I am not suggesting at all that the impact of these colonies on the surroundings was zero (even if colonies were not robbed, bees drifting from these colonies could have probably flown into other colonies). In my opinion, this impact would not be greater than that of colonies from treated apiaries, which are also weakening and dying.

148   Scientists and veterinarians often say that the problem is not so much the lack of treatment at all, but the therapies carried out in an inappropriate way. However, I ask a hypothetical question, whether the proper treatment of bees carried out jointly by the entire beekeeping community would solve the problem of varroa for longer than the next season? In that case, however, wouldn't we observe the same negative changes in bee adaptations that we see today?

bees? And this time, our considerations lead us to the same conclusion: we should change the way we think about keeping and breeding bees and start working on their resistance.

Giving up on treating bees and leaving them to natural selection is not only a biological or epidemiological problem, but also in its own way a social one. With a high population density, mite bombs are a serious threat to neighbouring apiaries, which means that they affect not only bees, but indirectly people. Beekeepers who use treatment accuse those who do not of spreading diseases (mites and viruses), and often even blame them for all other failures in their apiaries. Interestingly, at the same time they completely accept the migrations of bees to late nectar flows such as heather and goldenrod, where bees become infected with hundreds of mites. Non-treating beekeepers, on the other hand, accuse beekeepers that treat of not working with the immunity of bees, of importing queens with non-local "genetics", and of maintaining the maladjustment in the population with their drones. Both sides probably have their reasons and motivations. The differences in the approach to the problem results from extremely different views on nature and apiary management. However, both groups are united by their passion and love for bees and beekeeping.

In Poland, however, most beekeepers treat for varroa. Perhaps that is why it is so difficult to keep bees without treatment in Poland. In the county of Gwynedd (Wales), the vast majority are non-treating beekeepers. Those who treat do not blame the non-treating beekeepers of spreading mites around the area, probably because the bees are highly resistant to varroa and so do not spread it at all.

According to estimates, untreated bee colonies in the United States make up as much as 10% of the entire bee population. Based on this data, David Heaf included estimates of the potential impact of 'mite bombs' on local apiaries[149]. Out of approximately 3 million bee colonies in the United States, approximately 300,000 are untreated, and 2.7 million are treated. At the same time, 42% of untreated colonies were reported to die, while 33% of treated colonies died. As a result, over 120,000 colonies that were not treated died, and almost 900,000 that were treated died. If the impact of each dying colony were similar, the impact of treated colonies would be over seven times higher. However, it is impossible to determine the real "flow" (transfer) of mites. It is difficult to estimate whether the treated bee colonies

---

149   D. Heaf, Dealing with Varroa: natural selection or artificial selection? http://www.dheaf.plus.com/warrebeekeeping/dealing_with_varroa.pdf, accessed: January 22, 2025.

died after the treatments (and then the infestation was not so severe), or whether the average dying colony infected others with the same dose of mites, regardless of the group it was in. However, it follows from this, that to speak of a similar level of damage to the environment, the impact of dying untreated colonies would have to be more than seven times higher than that of treated ones. I am convinced that the scale of mite infestation in most of my colonies is smaller than in the case of some beekeepers who treat them (who often claim that their colonies have even several thousand mites), although there are certainly some in which it is not insignificant.

Experienced beekeepers who carry out treatments adjust their dates to the periods when there is the greatest risk of robberies of weakening colonies (late summer and early autumn). One of my acquaintances who treats (who does beekeeping semi-professionally) claims that he is not afraid of 'mite bombs' as he can control them by additional treatments.

Most female mites are found in the sealed brood, many others hide under the bees' tergites, where they feed on the fat body. If mites are visible on the bees' bodies, it can be assumed that the parasite infestation is very high, photo by Ł. Łapka.

It is worth adding that keeping bees in apiaries on a mass scale, especially given their susceptibility to diseases, contrary to popular belief, does not have to mean only positive things for the environment. Too many bees in one place constitute food competition for other insects. There are also ideas that keeping honey bees in apiaries may limit the biodiversity of other pollinators[150], but also plants [151].

In both cases, the phenomenon should be explained by food competition, but in the case of pollinators other than honey bees it is a direct, and in the case of plants, an indirect effect. *Apis mellifera* colonies due to the nature of being a superorganism require huge amounts of food, for example a colony uses about 100 kg of honey and 30 kg of pollen per year. If we add to this the honey collected by the beekeeper, it turns out that each colony must collect over 0.5 tons of nectar during the season. Honey bees therefore mainly benefit from visiting areas where flowers of certain species occur at high densities (e.g. agricultural monocultures). Visiting individual flowers or rarely occurring species is not very effective for them. The presence of one colony (especially a relatively small one, such as those that live in the wild) would probably not constitute much competition for wild pollinators, and it might even have a positive impact on biodiversity, contributing to its growth. At the same time, the presence of 30 or even 40 bee colonies in one location means such great food competition that other pollinators may be

---

150 The problem of the impact of honey bees on other pollinators is controversial and current, mainly due to the fashion for the mass introduction of honey bees to cities. Many scientific studies have been conducted on this topic, some in places to which the honey bee was brought by humans relatively recently (from a biological point of view, i.e. only a few hundred years ago). According to some practitioners, the results of such studies cannot be directly translated to European conditions, in which other wild pollinators co-evolved with the honey bee. There was interesting discussion of many interested parties in a Polish podcast: Pszczoły w mieście a pasieki [Bees in the city and apiaries], https://www.warroza.pl/2021/07/pszczoly-w-miescie-list.html, accessed: November 9, 2022; A. Szaciło et al., Pollination crisis and the honey bee? A cure for all evil or not necessarily? Problems in Biological Sciences 2020, 2.

151 A. Valido, Honey bees disrupt the structure and functionality of plant-pollinator networks, Scientific Reports 2019, 3.

forced out of this ecosystem[152]. Their elimination, in turn, causes that rarely occurring flowers are no longer visited by insects, and as a result they are not pollinated or dispersed, which leads to a reduction in the biodiversity of the local flora.

Studies also show that honey bee pathogens can transfer to other pollinators[153], so transporting honey bees from one location to another, etc., can cause pathogens to mix and mutate, which leads to an increased epidemiological threat. It is difficult to say unequivocally to what extent the described threats are real and to what extent they are just hypothetical. The issue remains controversial and still requires scientific verification[154].

It is time to sum up this part of the discussion. The real problem is not a locally adapted healthy carrier - all bees are carriers of pests and pathogens. The problem is the breeding and raising of bees that are not resistant and are susceptible to infections. They are the real threat. Let us also remember that for bees to properly perform their tasks, they must have contact with the environment. They cannot be locked in a cage. Initiatives to refrain from combating mites in bees are a response to the excessive amount of chemicals used in apiaries, which in turn generates health problems for bees. Blaming beekeepers who do not treat for beekeeping problems is, in my opinion, unfair; it is confusing the effect with the cause. It is like saying that water freezing in a puddle brings winter.

---

152 Let's do some simple calculations. In nature, honey bees most often live in relatively small colonies, which results from the fact that they occupy smaller cavities (e.g. tree hollows); the density is also low (let's assume that it is one colony per square kilometre). At the peak of development, there are probably no more than 30,000 individuals in such a colony, for simplicity let's assume that 10,000 of them are foragers. The situation is completely different in regions with intensive beekeeping, in some areas of Poland there are at least 10 bee colonies per square kilometre. Most often, these are also significantly larger colonies (at least 50,000 bees). Sometimes in one place (in one apiary) there are 30 or even 50 such colonies. With an average density (10 colonies per square kilometre) and the assumed strength (50,000 individuals), we can have half a million bees locally. If we make such calculations for an apiary of 30 hives, it turns out that there can be as many as 1.5 million bees in one place! If we assume that there are 20,000 foragers in each such colony (the number of foragers increases the most with the growth of the colony), it means that there can be 200,000 of them per average square kilometre, and locally in our example apiary there will be as many as 600,000 of them. So, if we relate these proportions to one wild-bee colony, it turns out that on one square kilometre with the same forage base (!) we can have as many as 60 times more bees visiting flowers (10,000 vs. 600,000 forager workers). This represents a huge amount of competition!

153 One such dangerous pathogen is, for example *Nosema ceranae*; R. Manley, Knock-on community impacts of a novel vector: spillover of emerging DWV-B from Varroa infested honey bees to wild bumblebees, Ecology Letters 2019, 8; R.E. Mallinger, Do managed bees have negative effects on wild bees?: A systematic review of the literature, PLOS ONE 2017, 12.

154 In the absence of historical data (well-preserved samples), scientists are unable to clearly determine whether some honey bee pathogens are transferred to other pollinators or *vice versa*. However, there is no doubt that in current studies the same strains of pathogens are found in both honey bees and wild pollinators.

Project COMB

The aim of the COMB initiative (Conventional and Organic Management of Bees), as conceived by Robyn Underwood from Margarita López-Uribe's team at Pennsylvania State University (USA), was to examine the differences in the health of bees kept in three different apiary management regimes: conventional (intensive, in which interventions are often undertaken and artificial chemicals are used to control bee diseases); organic (using so-called natural substances to treat bee diseases)[155] and so-called natural or "chemical-free" (i.e. where no pesticides, whether artificial or organic, are used to treat bee diseases, including varroa)[156].

Underwood's idea, as well as the initial preparation for the experiment, was received with great hopes and expectations, particularly from those beekeepers who do not use any treatments against varroa and felt pushed to the margins of the discussion on healthy apiary management. They often questioned the results of research on certain aspects of bee health, but not because they do not understand the essence of the scientific method or are its opponents or criticize the results of research out of hand[157]. Most often, it is rather about questioning the methodology in specific cases. For example, Prof. Seeley's experiments on the influence of the so-called small cell on the health of bees were criticized, due to the methodologically questionable way in which they were conducted. By creating a research group to function on the so-called small cell, the professor chose "large" bees. To persuade

---

155   This group was deliberately not given the traditional name of "treatment free", but "chemical-free". On the one hand, a certain average management system was adopted, which does not always correspond to the diversity of the so-called natural management, and on the other hand, it was just due to the needs of a scientific experiment. Beekeepers who do not perform treatments also use such methods or hives (e.g. simple frameless structures) that cannot be maintained due to the assumptions of the scientific method. In a scientific experiment, it is necessary to eliminate all variables considered irrelevant. The main variable for the chemical-free group was to be exactly what its name implies, i.e. not using pesticides, chemicals. Although all the bees were kept in ten-frame Langstroth hives, each group had slightly different elements of the nest cavity construction (for example, in accordance with the beekeepers' guidelines, the authors made sure that the chemical-free group had rough walls of the nest cavity used for propolis, or solid bottoms, while the conventional group had mesh bottoms); the method of feeding was also different (in the chemical-free group, the bees were fed only when the colony was at risk of starvation).

156   R.M. Underwood, A longitudinal experiment demonstrates that honey bee colonies managed organically are as healthy and productive as those managed conventionally. Scientific Reports 2023, 4.

157   Beekeepers who do not treat are often compared to the so-called anti-vaccinationists. My observations show that in this group, as in any other, we find a range of views. However, a large part of this group is perfectly versed in the current scientific research (often much more than beekeepers from the "conventional" group), they also have enormous knowledge of beekeeping.

them to build a 4.9 mm cell, he used a plastic foundation, which the bees reluctantly drew out. Then he compared this group to "large" bees building ("their") foundation with a 5.4 mm cell on a wax sheet. The professor's experiment (which lasted only a few months) showed that the bee colonies on the so-called small cell were weaker than on the standard foundation and were not healthier. Most beekeepers who have their own experience with the so-called small cell will admit that the chances of both groups were not equal and even before the experiment they could have stated that in such a situation one could not expect a different result. Various experiments and studies were subjected to such negative reviews by non-treating beekeepers (who questioned the duration of the experiments, the experimental group size or methodological errors).

Robyn Underwood wanted to avoid such criticism. Her experiment was to observe 288 bee colonies (96 in each research group) for three years. The colonies were placed in a total of eight locations (six apiaries in the state of Pennsylvania and two apiaries in West Virginia), with each apiary containing 36 colonies. Each of the eight farms had three apiaries each that were spaced 100 m apart. The practice in each group was to follow the averaged beekeepers' behaviours. Therefore, the researchers did not look for the optimal way of running each group, but decided to ask about how beekeepers would act in their apiaries and adjust it to fit the methodology of the experiment. These assumptions were met with a very favourable reception from the beekeepers (from each of the treatment groups), who were eager to consult it, providing information in the surveys about their own apiary habits and treatment of bees[158].

The experiment began in April 2018. At that time, colonies were created from bee packages of about 1.5 kg. The colonies designated for the chemical-free group were not created from previously untreated bees since the researchers were unable to obtain so many colonies from a single source. Instead, they obtained colonies managed in accordance with the assumptions of organic management (the bees were treated with so-called organic/natural pesticides), and using small-cell comb. In July 2018, in all colonies (from each group) the queens were replaced with daughter queens raised from larvae from a colony that had not been treated for at least seven years. Since then, no other bees have been included in the experiment. The scattered colonies were replaced by splits made from the surviving

---

158   See: Treatment Free Beekeeping podcast, episode 65, "Robyn's Research Project with Robyn Underwood, https://youtu.be/zcw5OZt49g4, accessed: February 18, 2025.

colonies (from each experimental group), and these splits were managed in the same way as the initial colonies in that group.

The following biomarkers were considered to indicate the health of bees: the level of infestation with *N. ceranae*, *N. apis*, deformed wing virus (DWV) and Israeli bee paralysis virus (IAPV), as well as the immune response (expression of certain genes).

The results of the experiment showed that the colonies from the chemical-free group had a reduced ability to survive compared to the other two groups and produced less honey. In the first year of the experiment, no significant differences were observed between the three groups. Differences appeared over time, probably due to the increase in parasite and pathogen infestation in the untreated colonies, which caused their health to deteriorate. It turned out that after three years of the experiment, only one colony (out of 96) from the chemical-free group survived, while 29 colonies from the conventionally managed and 38 from the organic group survived.

The average parasite infestation of all chemical-free colonies was 4.5% (in other groups approx. 1%), and 98% of colonies were infected with *N. ceranae* spores (statistically the lowest was noted in the colonies from the organic group). Colonies from the chemical-free group also had higher expression of genes related to the immune response, which could indicate that the bees were struggling with pathogenic factors more than others.

The conclusions drawn by the scientists from the experiment are to emphasize the importance of the beekeeping system for the survival, health and productivity of honey bees. It was also emphasized that non-treating beekeepers often reject the possibility of conducting chemical treatments due to their potentially harmful effect on the health of bee colonies. Whereas the experiment shows that this effect is positive (the benefits exceed the potential threats). The researchers also found that not using acaricides affects the growth of the mite population, which leads to the deterioration of colony health, and as a result, the risk of death of the bee colony increases fivefold (compared to treated colonies). The scientists do not deny the possibility of survival of bee colonies (populations) without chemical treatments (they refer to numerous scientific studies, some of which are mentioned in this book), but they believe that abandoning treatments is not advisable in apiary management and honey production.

They believe that treatment is necessary when the level of the mite infestation exceeds 1-2%. What they consider important is that it has been proven that apiary management in which bee colonies are treated with organic/natural substances is similarly effective or even better (in terms of bee productivity and colony survival) as that in which artificial/synthetic substances (such as amitraz) are used.

Despite Robyn Underwood's best efforts, the experiment, both the research methodology and the conclusions presented by the author, have been criticized by some beekeepers who do not use varroa treatment. One could say that the results of the experiment (low survival and productivity) of bees from the chemical-free group caused people who expected different results to start looking for errors in the study (according to the principle: if the facts contradict my theory, so much the worse for the facts). Is it justified to question the methodology of the experiment? Well, I consider some of the critical voices that have reached me are irrelevant, but I agree with others. Therefore, I will allow myself to present my own interpretation of some aspects of the experiment.

Each method of keeping bees is a holistic system. If we decide to select only some elements from a certain beekeeping system (whatever it is), we risk not achieving the assumptions. For example (to simplify considerably), if we keep bees that are not resistant to varroa and other diseases, to achieve high survival, we must consider the rigor of performing procedures at specific times of the season (e.g. ensuring that the so-called winter bees are raised in a colony with a low level of mite infestation). We are also rather doomed to frequent disinfection elements of apiary equipment. However, if we decide to limit procedures or completely give them up, then in the assumptions of our apiary management we must include increased mortality of bee colonies (e.g. by developing a system to replenish losses, whether related to catching swarms or increasing colony propagation).

Higher mortality of colonies not subjected to anti-varroa treatment is the norm in most parts of the world (although the experiment in the county of Gwynedd Wales shows that it may be different). In the United States, high bee mortality has been a reality for beekeepers for many years; statistically, about 40% of all bee colonies die there each year. Data on untreated colonies differs only slightly, with about 50% of colonies dying annually. Interestingly, colonies from two treated research groups: conventional and organic showed relatively low mortality, an average of 23%. The authors of the study emphasize that this value is lower than the average for the USA, but they do not venture to hypothesize what could explain this situation. In my opinion, although it is also possible to rule out chance[159], the reason could have been the use in the experiment (also in the treatment groups) of queens from a colony in which no treatment had been applied for at least seven years (in the colonies of the research groups, certain adaptive mechanisms facilitating the survival of the colonies could have been revealed).

I also have doubts about the location of 36 colonies in one apiary, run under three different management regimes. The number of bee colonies in one place may have an impact on the increase in bee drifting and the so-called robbing, because of which varroa and pathogens are transferred between hives. Many experiments show that the accumulation of colonies in one place i.e. in close proximity, may have an impact on the infestation of the varroa, and thus drive higher mortality. Probably none of the beekeepers who do not use the treatment would dare to say, that placing 36 bee colonies in one place is a good idea.

Professor Seeley believes that the distances between wild/feral bee colonies is one of the foundations of maintaining the health and survival of colonies in the Arnot Forest. This factor may be much less harmful to colonies in which pest population is under control by beekeepers. However, any other configuration (e.g. reducing the number of bee colonies in one location) could also be questioned because it would not meet the assumptions of the scientific method.

Other concerns are also raised by the issue of the repeatability of the experiment, depending on the location where it was conducted. Of course, the scientists cannot be accused of a methodological error in this respect: the experiment was conducted in eight locations, in two different states. However, based on many studies and observations, we know that the phenomenon of bee immunity depends on the location and the results may be determined by both the density of bee colonies and other properties of the entire local population. Conducting a similar experiment in the British Isles, South America or in some regions of Norway or Sweden (e.g. the area of the city of Hallsberg) could give completely different results. After all, there, bees have developed immunity mechanisms, naturally or because of artificial selection, which could affect the results of the experiments. There is no doubt, however, that the condition of colonies not subjected to treatment, living in places where bees are not immune, will be much worse;

---

159 This may also be the result of stationary keeping of bees in the experiment. Most bee colonies in the US are transported for pollination over huge distances. This increases the level of stress in colonies, promotes horizontal transmission of pathogens, and thus may cause high mortality of the entire population.

these colonies will also be less productive than those treated (even if we put in them queens from the stock selected for mite-resistance). It's hard to be surprised by this study result.

Critics of the experiment also claim that the young queen bees that were introduced to the experimental colonies in the summer of their first year were inseminated in a place dominated by drones from non-resistant colonies. I am unable to verify the truth of this.

However, if this was the case it could have had an impact on the fact that the colonies from the chemical-free group did not demonstrate the ability to limit the varroa infestation enough to survive. At the same time, most scientists (both those open to the principles of natural selection in apiaries and those critical of it) emphasize that the mechanism of bee resistance to varroa cannot be easily disseminated between colonies using queen bees. It would not be an exaggeration to say that the chemical-free group was not created from bees that non-treatment beekeepers would like to keep (although some would probably like to have such bees to start with in their own practice), but rather a group of non-resistant bees, which at most can reveal some genetic mechanisms of limiting the varroa population. For most beekeepers who do not use treatments, the first few years are just the initial stage of selection, which will allow for the selection of breeder queens (from the population they have). They are the ones to become the basis of a locally adapted population. Perhaps a good idea, regardless of the research groups, would be to create control groups of bees in the apiaries of beekeepers who claim to have a pre-selected population and take samples in the same way as it is done in experimental groups. Although this would not be in accordance with the principles of creating sound scientific methodology (because of introducing too many research variables), it could show the phenomenon in a different light, being a certain point of reference in the assessment of the research methods.

In addition, it should be noted that most beekeepers who do not treat bees do not believe that individual bee colonies in their apiaries are healthier than those treated (in the context of the well-being of individual superorganisms), just because they are not exposed to harmful chemicals. I believe that the vast majority are aware that individual bee colonies may have more health problems than those in which the parasite is killed (and are therefore aware of the short-term positive effect of chemical use).

In this sense, the argument of the authors of the study differs from the views (I think of the majority) of beekeepers who do not use treatment. We rather emphasize that the chemicals used in apiary management have a harmful effect on bees in the long term and perpetuate their maladjustment to environmental conditions. In this sense, the use of chemicals does not solve the problem of varroa but masks the effects of the disease and prolongs the process of co-adaptation of organisms. Therefore, abandoning treatment (gradual or one-off) in apiaries is not a step that will immediately lead to an improvement in the health of bee colonies (in the case of most colonies it will be rather the opposite), but one that in the long term will allow to select from the population those bee colonies that are able to take care of themselves thanks to evolutionarily developed immunity mechanisms.

I am not trying to say that the described scientific study was conducted incorrectly, although as you can see, I have some reservations. The comments regarding the methodology rather concern the imperfection of the experiment resulting from the complexity of reality, which makes it impossible to easily isolate just a few variables for scientific study. After all, even the authors of the study do not deny the ability of many bee populations to survive on their own (the only thing that is still questioned is the possibility of conducting efficient apiary management in such a way). Robyn Underwood also nowhere questions the fact that the breeder from whom all the queens in the experiment came, comes from a colony that had not been treated for at least seven years, and at the same time almost the entire (except for one colony) chemical-free population studied by the author ceased to exist within three years. In fact, some non-treating beekeepers even claim that one must try hard to "kill" almost a hundred colonies so quickly, and they also believe that this is due to management errors. However, this fact does not have to be surprising, considering how difficult it is to keep even a relatively large population of bees alive under unfavourable environmental conditions (large bee colonies, lack of resistant bees in the area, poor nectar yields, etc.).

My criticism (or doubts) would rather concern attempts to apply the conclusions of the experiment to beekeeping without treatment, I would consider such action to be too far-reaching[160]. The experiment rather shows (which we already know from, among other things, scientific research) that transferring immunity of bees by means of queen bees does not bring

---

[160] The author has repeatedly (e.g. in interviews) claimed that her experiment aims to examine the results of treatment-free beekeeping in relation to conventional and organic farming. In my opinion, the conditions in which she conducted the experiment did not meet the requirements that characterize running an apiary without treatment. In a sense, in the study, she only managed to maintain the conditions for establishing such an apiary. Therefore, since the research conclusions referred to a specific experiment conducted in accordance with the methodology described in the work, it is difficult to question them. They also concerned various markers of bee health, which are difficult to question.

the effects we wished to expect. For these reasons, it seems that Robyn Underwood's experiment should not be treated as decisive in the matter of the health of developed (naturally or because of selection) populations of untreated bees. However, it can be treated as a description of what happens in bee colonies in the first years after they are deprived of protection against varroa, even if some of the assumptions of non-treating beekeepers are met (including introducing into the apiary breeder queens from apiaries in which chemical treatments have not been used for years).

NATURAL BEE HIVES

Chapter 2

NATURAL BEE HIVES

Some wag has stated that it is the duty of every beekeeper to invent a new hive before they die. And, looking at the history of beekeeping, something approaching that seems to have been going on.

DAVID HEAF[161]

In the beekeeping community, there has been a discussion for years on the topic of which hive is the best. And so far, there is no indication of an imminent consensus. A large part of my beekeeper friends has hives that are a combination of standard and modified structures. The main areas that undergo modernization and changes are the bottoms, frames and elements that affect the distance between the boxes. I also use modified construction of typical hives (according to my own idea)[162]. All this shows not only the diversity of beekeepers' preferences, but also opinions on the needs of bees.

A hive is a kind of compromise between the needs of bees and the beekeeper. Some believe that bees can cope in any construction. Others do not think about it at all, assuming that if a certain type of hive has been developed, then surely everything has been thought of. Still others try to make the hive like the natural nest cavity of bees, i.e. a tree hollow.

Beekeepers introduce modifications to the basic structure of the hive and as a result, it often happens that they cannot exchange hive elements, even if they work based on the same system. I do not intend to describe the types of hives in detail (I will include a few comments in the chapter on foundation-free management). I will only try to analyse the needs of bees and compare them with my own observations and those of researchers or other beekeepers. I will try to indicate what features a hive should have to meet the requirements of an optimal nest cavity for bees, corresponding to

---

161 D. Heaf, The essence of sustainable beekeeping, http://www.dheaf.plus.com/warrebeekeeping/ essence_sustainable_beekeeping.pdf, accessed: January 23, 2025.

162 Or rather this is largely a compilation of changes introduced by beekeepers.

their species needs (and not only the needs of bees, but also other organisms creating the complex hive ecosystems).

Honey bees, unlike several other species of social bees, try to settle in closed spaces. They rarely build nests in the open air, photo by P. Kasztelewicz.

Bees can adapt to a wide variety of locations, volumes and shapes. Their nests can be found in surprising places. Sometimes they start building combs on a branch, exactly where a swarm has settled[163]. On the internet, we can see bee nests in tires or traffic cones (the ones that are used on the manoeuvring areas where drivers practice their skills). Heaf claims that the strangest place he has seen a bee colony in was a water pumping station in the ground. I am convinced that the bees chose all these places themselves, because no one would think of putting them there. Beekeepers believe that bees do not like drafts in the nest (air flow from the lower to the upper holes in the nest). Meanwhile, they very willingly settle in ventilation shafts,

---

163   Most often, this is due to the swarm not finding a good location.

so apparently either the draft does not bother them, or they know how to deal with it.

The choice of nest sites by bees is often surprising, photo by D. Heaf.

They often settle under the guttering in the soffits or roof spaces of houses. It is completely normal for bees to fly into empty hives, especially if they have already been inhabited and smell of wax or propolis. Sometimes bees inhabit a hive type that we would not expect. It is commonly believed that they prefer nest sites nest sites made of natural materials (wood or straw), and in the meantime they will settle in a Styrofoam hive standing next to a wooden one.

It seems that bees from a swarm, when assessing a potential new place for a colony they do not pay attention to some of its physical properties (thermal insulation, breathability, vapour permeability, etc.), but are guided by completely different premises. So, do these features have no significance for them, or are they unable to assess them? Knowing the physical or mechanical properties of materials, a beekeeper can often decide better than the bees whether a given hive type will be good for their bees (e.g. whether the hive roof will not leak because of temperature changes, whether the inside of the hive would not be too humid in specific time of year, whether the nest site will not be occupied by rodents before winter). While we theorize about the optimal shape (or volume of the hive), bees will surprise us more than once, choosing to live in, for example, an upside-down bucket or cardboard box. Why is this happening? Perhaps because of the lack of other available and convenient nest sites in the area? It is difficult to believe that a traffic cone or a tire would meet their needs better than tree hollows or deserted beehives.

Swarms very willingly to settle in chimneys
(including ventilation chimneys) of old houses,
photo by P. Kasztelewicz.

Bee preferences

The preferences of bees in terms of nest site selection was the subject of research by Thomas Seeley in his doctoral thesis[164]. The research was carried out in places where bees had very limited alternatives, since he wanted a high probability that the bees would choose one of the nest sites he offered.

The swarms most often had a choice of two boxes that differed in one element (a variable used to study bee preferences), e.g. the boxes differed in size (volume) or the location and size of the entrance. It turned out that the bees more often chose boxes with a smaller entrance than the one used in the hives used in apiary management. In this case, the alternatives were boxes with entrances of 12.5 cm^2 and 75 cm^2 [165]. Why? Probably for two reasons. Firstly, it is easier for the guards to protect the hive, and secondly, a smaller opening allows for better thermoregulation of the hive, especially in well-insulated structures.

It also turned out that bees prefer entrances facing south, not north, and rather placed at the bottom of the hive than at the top (i.e. under the nest, not above it). In both cases, it is probably a matter of maintaining the right temperature in the nest. Seeley's research also shows that bees have no preferences regarding the shape of the entrance (round vs. oblong) or the nest (square box vs. oblong box placed vertically in the shape of a tree hole). The researcher also offered boxes of various available volumes: 10, 40 and 100 litres. It turned out that they most often chose the capacity of 40 litres (slightly less than the Langstroth hive body which is 42 litres). This volume is sufficient to store supplies for the winter[166]. Studies of wild bee nest cavities conducted by Seeley in the Arnot Forest led to the conclusion

---

164 In his later scientific work, he also dealt with the decision-making process of the swarm and the natural nest sites of bees, i.e. tree hollows, which allowed him to supplement his research. He shared the knowledge that flowed from this part of his scientific work in an interesting book Honey bee democracy, Princeton University Press, 2010.

165 Which is often used in hives in modern apiary management (height 2 cm at the width of the standard body, i.e. 37.5 cm).

166 Seeley continued his research on the lower limit of the habitat capacity that could be acceptable to bees, offering swarms boxes of 17.5 litres and 25 litres. Given such an alternative, bees were much more likely to choose the larger ones. Other scientists who took up this subject found that the lower limit largely depends on the bees origin. African subspecies will occupy nest sites of just 13 litres, which were not inhabited by European subspecies. According to Seeley, such a profile of preferences results from the evolutionarily shaped need to store supplies for the winter by European subspecies (this requires an appropriate capacity of between 10 to 20 litres just for the food). Bees from African subspecies do not need to store such large supplies, because they evolved in regions where there are no severe winters or correspondingly long periods of no harvest.

that most occupied hollows had a volume of 20-60 litres. Larger volumes (e.g. 100 litres) probably seem too large for bees to easily heat them in winter (or during colder nights in spring and autumn). Nest cavities whose volume corresponds to the volume of hives in apiaries are therefore not attractive to bees. Swarms prefer to occupy those in which there are combs from the previous colony. Probably their instinct tells them that building new ones would have to be done at the cost of additional energy and work, which could be used differently. Or maybe they "think" that the nest cavity must be attractive, since it was previously chosen by another bee colony. On this basis, is it possible to conclude that the risk of pathogens in the vacated nest for the newly settled swarm was low? It seems that if it were high, this preference would be eliminated by natural selection.

Seeley also believes that swarms prefer higher nest sites (e.g. 5 m above the ground than at a height of 1 m), because it is easier to defend the nest from predators or parasites. Rodents or predators would have difficultly

While studying the preferences of bees, Seeley offered the bees two alternative boxes, differing in one detail. The photo shows boxes with different-sized entrances, photo by T. Seeley.

During his cycling trip across Europe, Thomas Gfeller encountered many feral colonies. One of them, in the south of France, settled in a figure of Christ on the cross. According to residents, they lived there for at least four years, photo by T. Gfeller.

accessing a hive located higher[167]. The professor also noted that low-lying hives can cause problems for bees during winter, since before the bees manage to fly up, they fall onto the snow, where they freeze and die.

The scientist also claims that bees are not discouraged by gaps and cervices in the nest, since swarms chose boxes with densely drilled holes (6 mm in diameter) equally as boxes without holes. Similarly, bees didn't mind boxes containing damp sawdust. According to Seeley, it is instinct that tells bees that they will be able to improve the conditions within the nest. They feel that they are not able to increase or decrease the volume or change the location of the entrance, but they can cope with holes and gaps in the walls by sealing them. They can also dry the nest. In my opinion, the lack of a clear preference for a dry nest, without gaps and cracks, may indicate

---

167   There are black bears in the forests of New York State, which often determined the professor's actions (for example, he was forced to suspend hives on ropes under tree branches). He admits that when he was examining wild bee nests in the 1970s, he drew incorrect conclusions about their height above the ground (he gave a smaller distance). To collect as much data as possible on naturally occupied nest sites, he placed advertisements in newspapers asking for information about wild bees. The response was considerable (within a few weeks, it was learned that there were about 30 wild bee colonies in the area), but it turned out that people knew about the existence of nests all located relatively low (sometimes directly above the ground with an entrance in the tree roots). Whereas those located higher escaped their attention, which means that bees are good at hiding in the treetops.

DIFFERENT CHOICES OF THE BEES

ENTRANCE SIZE
Function: Colony defense and thermoregulation

12,5 cm > 75 cm

CAVITY VOLUME
Function: Storage space for honey; colony thermoregulation

10 L < 40 L > 100 L

ENTRANCE DIRECTION
Function: Colony thermoregulation

south > north

CAVITY SHAPE
No preference was found

cubical = tall

ENTRANCE HEIGHT
Function: Colony defense

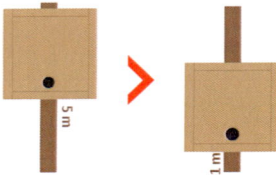

5 m > 1 m

CAVITY DRYNESS
No preference was found (Bees can waterproof a leaky cavity)

wet = dry

ENTRANCE LOCATION
Function: Colony thermoregulation

bottom > top of cavity

CAVITY DRAFTINESS
No preference was found (Bees can caulk cracks and holes)

drafty = tight

ENTRANCE SHAPE
No preference was found

circle = slit

COMBS IN CAVITY
Function: Economy of comb construction

with > without

Seeley studied the criteria that swarms use to select nest sites. This allows us to indicate the characteristics of a nest cavity that meets the preferences of bee colonies (the needs of *Apis mellifera*), figure by M. Uchman.

that bees are not fully able to correctly assess some of their properties[168]. This may be because over millions of years of evolution, the nest sites were of a similar nature: most often tree holes, and in warmer climates, rock crevices. Perhaps for this reason, bees have not developed the ability to choose a place for a colony that would be characterized by good thermal insulation properties during the wintering period. Therefore, they do not have the appropriate tools that they could use to "assess" non-breathable nest cavities made of plastic, or those with very narrow walls (without a guarantee of proper insulation). When choosing a place for the colony, bees were guided by preferences that, according to Seeley, indicated the need to secure appropriate thermal conditions for the nest (i.e. the size and location of the entrance and the volume). At the same time, it turned out that the bees did not dislike the box with the holes drilled in it.

Although the workers could quickly seal all the holes in the walls, the perforated surface, even if patched, would not provide thermal insulation to the same extent as the solid and thick wall of a natural hollow. Perhaps, therefore, the bees' preferences did not result from the need to have good thermal insulation of the nest at all, but for a reason that we do not know. Recently, Seeley began experimenting with well-insulated polystyrene hives. As a practicing beekeeper, he claims that his bees fare well in them[169].

Properties of the natural nesting site

Torben Schiffer was one of those who dealt with comparing the physical properties of wild, natural bee nest sites, i.e. tree hollows, and hives used in apiaries. The conclusions he draws seem somewhat exaggerated, however. According to Schiffer, modern hives do not meet the needs of the *Apis mellifera* and even classifies them as '*pessimal*' (lethal) niches. I do not intend to defend the view that modern beehives are optimal nest sites for bees, they certainly are not. The results of Schiffer's research suggest that

---

168   The lack of preferences for the discussed nest cavity features, which from our perspective may seem unfavourable, may result from various reasons: a) bees are not always able to assess this, it may depend on unknown circumstances (e.g. the same swarm will behave differently under different conditions); b) only some colonies are able to assess potentially unfavourable conditions, while others cannot (there is individual variability); c) for all colonies these factors are irrelevant (other factors may decide on the choice of a given nest cavity).

169   A team led by Seeley is conducting research on the properties of isolated natural nest sites (tree hollows) and insulated artificial ones (hives). The professor claims that the conditions inside insulated and uninsulated hives are like day and night!

we should agree with him when he claims that tree hollows or structures modelled on them (not only in terms of shape, but above all physical properties) can provide bees with better living conditions. However, I do not believe that hives (even those made of one board, with a mesh bottom) are a lethal niche. Since in such a niche organisms are unable to develop and reproduce, and some bees manage to do so perfectly, we have no grounds to claim that. A healthy colony can cope in any nest site, even one far less friendly than a modern hive.

The experience of many beekeepers (including tree-beekeepers from Germany and Switzerland) shows, however, that colonies that are not adapted to local conditions and infested with varroa may not survive the winter even if their nest sites conditions are close to optimal[170]. It is difficult to say to what extent a good nest site can help bees survive (e.g. by reducing environmental stress). The problem of the health of a bee colony is not limited to whether the hive wall is built of a single board, thick solid wood or insulated with a 15cm layer of straw.

Let us move on to Schiffer's research, because its results will undoubtedly allow us to look at many beekeeping problems from a new perspective[171].

The German beekeeper believes that the further we distance the bee from the environment in which it evolved, the more it becomes susceptible to threats.

By disrupting the process of natural selection, by bringing bees that are not adapted locally to apiaries, by keeping them in barren hives, by using the methods of modern intensive beekeeping, and finally by poisoning the bees' living environment with so-called drugs (and their metabolites), we have caused a significant decrease in bees' tolerance to external factors over the last few decades. Schiffer assumes that to be able to deal with the ecology of the honey bee, one must first learn the features and properties of the tree-hollow, the natural bee nesting site. This is where bees evolved and developed their many ecological associations. Studies of the micro-climate of hollows differ significantly from those found in most modern hives. Hollows are more stable biotopes, so both temperature and humidity fluctuations (daily and seasonal) are much smaller there (e.g. air humidity is lower than in

---

170   "Arboreal Apiculture Salon", interview with Andre Wermelinger from Switzerland (episode 12) and Sabine Bergmann from Germany (episode 16), https://arboreal-apiculture-salon.libsyn.com/, accessed: November 13, 2022.

171   They are described, among others, in the Guide to species appropriate beekeeping with pseudoscorpions, 2017; but also in many other publications by the author in the beekeeping magazines - e.g. T. Schiffer, Beekeeping (r)evolution, Natural Bee Husbandry 2019, 8.

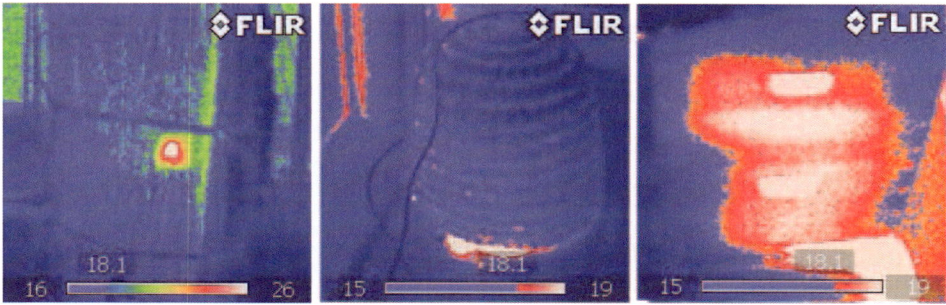

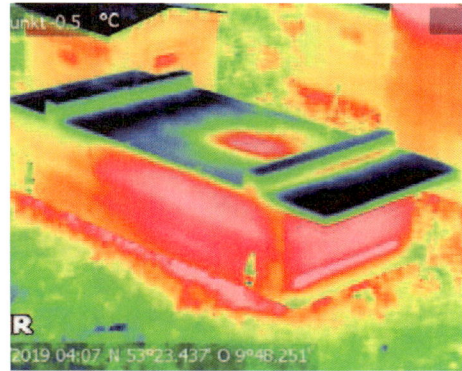

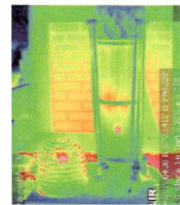

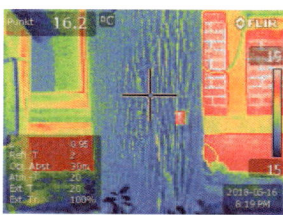

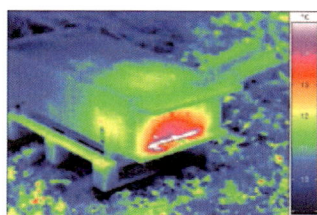

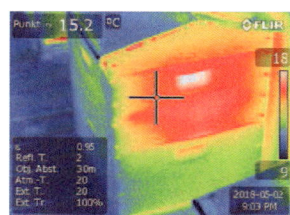

The analysis of hive images using a thermal imaging camera shows that the greatest heat losses occur in small uninsulated hives, photo by T. Schiffer.

TREE HOLLOW IN WINTER: MOISTURE ABSORBTION AND REGULATION OF THE CLIMATE

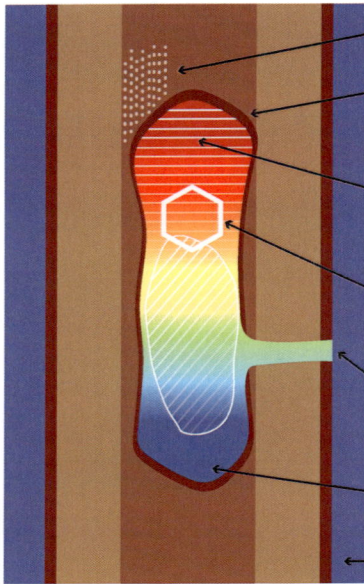

The wood absorbs moisture out of tree hollow air until the fibers become saturated with water.

The layer of propolis which covers the walls and the ceiling of the nest acts as a barrier that prevents condensated water from seeping back into the nest. It also limits the air circulation within the cavity.

4,87 g/m³ of the absolute humidity at 24ºC of the bee cluster temperature is only 22,4% of relative humidity. The additional humidity secreted by the clustering bees will be systematically absorbed by the wood.

The bee colony produces heat and vapour. The wood of the hollow is a very good insulator and the heat will stay up in the area around the cluster, simultaneously warming the honey stores and preventing crystalization.

The entrance will allow for the exchange of warm, moist air for colder and drier air. The height of the cavity can affect heat loss - the higher it is, the slower it cools down.

At 4ºC inside the tree hollow and 80% relative humidity, the absolute humidity is 5,12 g/m³.

At 0ºC outside and 100% relative humidity, the absolute humidity is 4,87g/m³.

Physical phenomena occurring in the tree hollow differ from those in a hive (especially a non-insulated one), figure by M. Uchman, based on T. Schiffer.

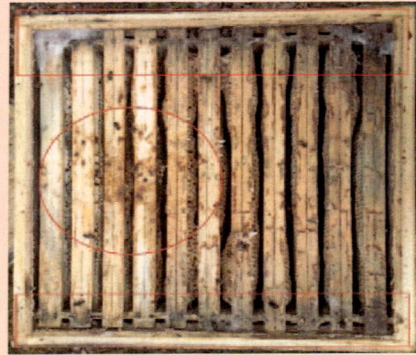

In the corners of an uninsulated hive, water vapor condenses (thermal bridges) and mould appears over time, photo by T. Schiffer.

a standard modern hive, although a lot also depends on the shape, volume, tree species or location of the entrance). Research also shows that bees live mainly in hollows, where the level of relative humidity throughout the season oscillates between 40-70%, while in most hives the level of relative humidity is periodically higher, which favours the development of many organisms

The difference between the cross-section of a standard hive body and a tree hollow is immediately visible, photo by T. Schiffer.

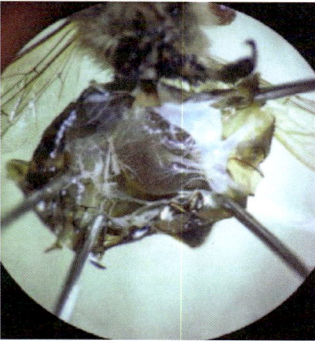

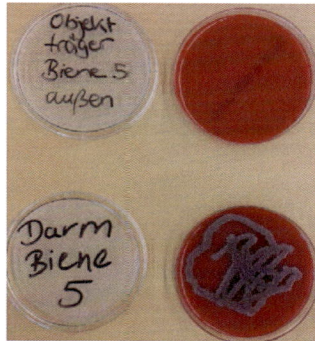

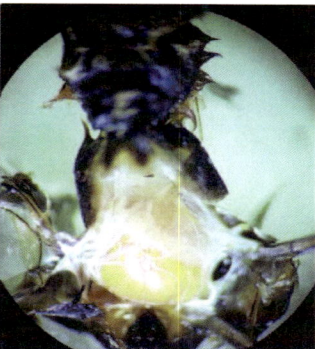

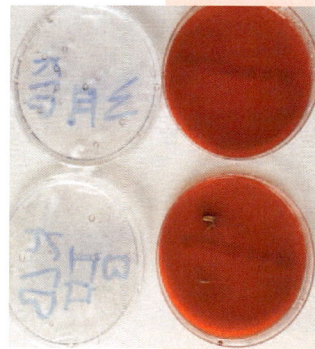

The intestine of a dead worker obtained from a colony in a tree is light and bright in colour, indicating no infection. Whereas the intestine of a dead worker from a non-insulated hive is darker and greyish, indicating infection with mould spores. Pathogenic flora was found in samples taken from the intestines of dead bees obtained from the bottom of a non-isolated hive, which was not observed in samples of bees from a beehive log, photo by T. Schiffer.

that are hostile to bees or even pathogenic. Moulds harmful to bees develop, for example, when humidity reaches 70-80%. This means that in the case of most hollows there is no risk of mould, but in hives it is much higher. This process is also facilitated by other factors such as the location of the nest (low above the ground, e.g. on a pallet, especially with mesh bottoms); the materials it is made of (non-breathable or plastic); poor thermal insulation (cold promotes water vapour condensation); shape. Square or rectangular cross-sections, i.e. shapes most characteristic of hives, are particularly unfavourable.

The coldest places in hives are the corners, where thermal bridges are formed, moisture accumulates, and condensation occurs first[172]. It is precisely the differences between the physical conditions of natural bee nest sites and hives that cause the emergence of completely different ecological relationships.

Varroa is not the only threat to bees[173], pathogens causing specific diseases are causing havoc. As it turns out, we can have some influence on them by creating appropriate living conditions for bees. Studies conducted by Schiffer on the contents of the intestines of bees wintering in damp standard hives and tree hollows show that pathogenic microflora flourished on samples prepared from dead bees from damp hives but were not found on the ones from dead bees from tree hollows.

Hollows also differ from hives in their diameter, which are small in natural nest sites. Meanwhile, the cross-sections of the types of hives that are very popular today are large[174]. A smaller diameter makes it easier for bees to heat the cluster in winter, especially when heat losses are small. The larger the volume of the nest cavity, the more difficult it is to maintain the appropriate temperature. Although beekeepers emphasize that bees heat themselves, i.e. the cluster, and not the volume of the nest cavity, the cluster also emits (loses) heat, and with a small volume of the nest cavity, it remains in the surroundings (completely different than in a large and uninsulated hive). For this reason, energy expenditure is much lower. According to

---

172　Relative humidity increases with decreasing temperature, so it is highest in the coldest places, the so-called thermal bridges, i.e. primarily in the corners of the nest cavity. There, water vapour precipitates from the air and condenses the fastest.

173　Or rather, it is not the only immediate problem.

174　Many beekeeping practitioners believe that large hives are better for bees. They also believe that if we want to take care of the health of bees, we should start by providing them with a large space. Research shows that this belief is completely wrong (large hives are better primarily for honey production as they stimulate the growth of the bee colony).

Schiffer, bees in a hollow do not have to tighten the cluster as much (except during periods of long-lasting severe frosts), which allows them to move easily towards the honey stores.

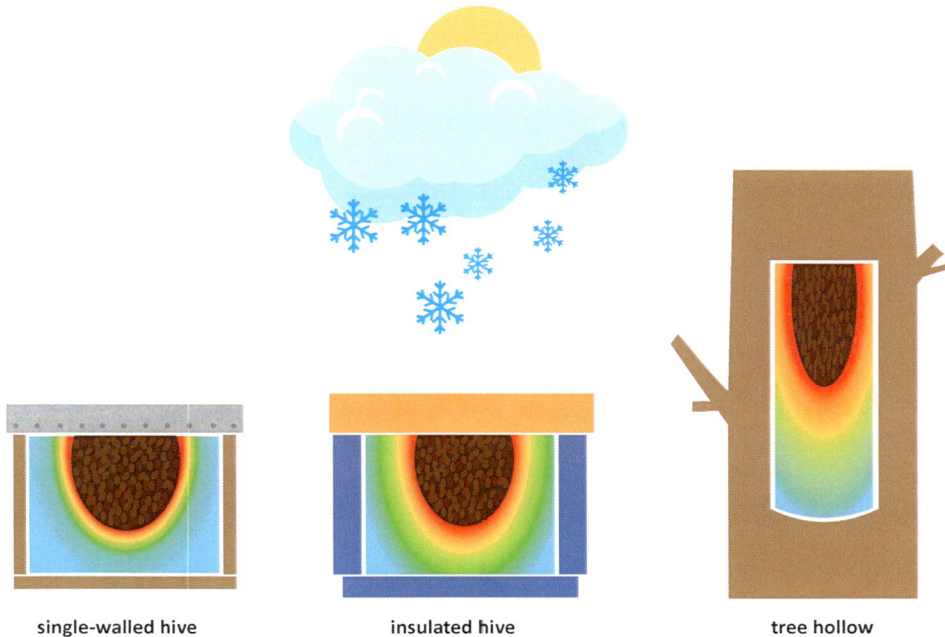

single-walled hive   insulated hive   tree hollow

Temperatures inside the nest depend on insulation. In winter, bees try to maintain a constant temperature around the queen, inside the cluster. During severe frosts, the temperature in the centre of the cluster is close to 20° C, while in its outer layers it can drop to about 0° C (or below). Although the bee cluster provides excellent insulation, it also gives off heat to the surroundings. If the walls of the nest are well insulated (like in a hollow), the heat will remain in the bees' surroundings, which means that the bees will use less energy to heat the cluster, figure by M. Uchman.

A large hive means the need for greater effort (work), which in turn means the consumption of a larger amount of food (honey). During periods of severe frost, the bees tighten the cluster and are only able to move on the combs they occupy since the neighbouring ones become inaccessible to them. Therefore, it happens that they die from hunger, even though there is food on the comb next to the cluster. To counteract this, some beekeepers arrange and tighten the bee nests for the winter and control the availability of food. Consuming more food causes the bees to speed up their metabolism, which causes them to work faster and produce more water vapour.

In a well-sealed hive with a foil or polystyrene cover, water vapour condenses on the walls in the coldest places of the hive (corners, furthest from the cluster, unheated combs). Excess water vapour promotes the growth of pathogenic micro-flora, primarily mould (higher humidity creates optimum conditions for it). Meanwhile, in a tree hollow, the situation is different. The construction of the "ceiling" plays a major role here, as does the layer of propolis, which most often covers the inside of the hollow, especially in the upper part. Propolis allows water vapour to penetrate but stops the drops forming (water in liquid state). Water vapour therefore penetrates the tree through the propolised surfaces and there, after cooling, condenses, but thanks to the layer of propolis it cannot return to the nest, so the bees' environment remains dry and warm. This is of course most important in winter, when the bees remain clustered (this phenomenon occurs without their interference). However, this process does not proceed in the same way in every type of tree. For example, the environment in oak hollows is much more humid than that of trees with a lighter structure, because the dense structure of oak prevents moisture from penetrating (draining) and the water flows down the walls of the nest cavity. Thanks to the propolis covering the walls of the hollow, the natural nest cavity is also free from pathogenic flora[175]. Moulds and pathogens do not develop on propolis, and the excess water can be drunk by bees (when they need it). It turns out that water washes out some components of propolis and thanks to this it has antibiotic properties[176]. However, if the only source of water is that secreted by the bees, then as it is devoid of microelements it can lead to the washing out of necessary electrolytes[177]. In hives, the processes take place differently than in tree hollows, even very humid such as oak ones.

Most of the wooden inner covers cannot help with excessive moisture in the hive. The structure of the tree is of fundamental importance, especially the arrangement of channels (tubes, fibres) through which water, sap and nutritional components are supplied. These channels are like narrow pipes, and are arranged vertically, along the trunk, so the holes are only in the ceiling of the nest. Because they are like tubes, moisture can only penetrate through the holes, and the walls of the fibres are closed to water. A wooden roof, where these channels run across the nest, is not able to effectively drain moisture away from the hive.

---

175   A similar situation could theoretically occur in the hive, but bees are very reluctant to propolise smooth walls, and beekeepers often sterilise the surfaces by burning them or destroy the propolis layer in other ways.

176   Honey stored in cells with a propolis layer may also wash out some substances, which may additionally increase its antibiotic (and therefore health-promoting) properties.

177   Therefore, bees must replenish their water supplies, taking it in, for example, ponds or other such reservoirs. In winter, they can replenish the deficit of microelements by eating honey.

Propolis acts like a Gore-Tex membrane, allowing water vapor to pass through it but retains water droplets, photo by T. Schiffer.

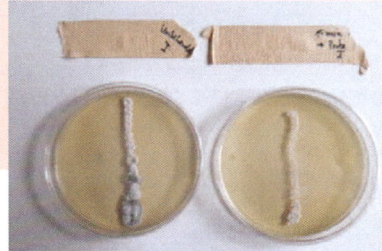

Two Petri dishes with the same medium and mould spores. The one on the right was sprayed with water that had contact with the propolis for only 15 minutes. After 24 hours, the two samples were significantly different, which means that the water can leach substances from the propolis that inhibit the growth of harmful micro-organisms, photo by T. Schiffer.

Many natural beekeepers use so-called eco-boxes (eco-roofs). These are boxes filled with organic matter: straw, shavings, forest litter, twigs etc.. They are not only good for removing moisture from the nest but also act as thermal insulation (in winter they protect against heat loss, in summer against excessive heating. At the bottom (from the side of the bees' nest) they can have a metal mesh, stiff and thick fabric (one that the bees will not chew through) or rough boards, preferably thin, separated by narrow gaps through, which, although they will be propolised by the bees, moisture from the hive will penetrate to the top faster than in the case of a uniform covering (e.g. plywood). Eco-roofs are very popular among beekeepers who keep bees in Warre hives. Some (including me) also use eco-floors. These are boxes that are placed on the bottom of the hive. Both eco-roofs and eco-floors not only isolate the nest, but also provide good nest sites for various creatures that accompany bees in the hives, which contributes to the development of complex ecosystems of organisms co-existing with honey bees.

Moisture is not the only problem of modern hives. Beekeepers deal with its excess in various ways. Some use additional upper outlets, ventilation in the covers or mesh bottoms (that go hand in hand with a tight cover over the nest, e.g. foil). All these solutions reduce the humidity level in the hives, but they also contribute to large heat losses in the nest and more temperature fluctuations, which significantly increases the demand for energy in winter, and thus reduces bees' honey stores. It can therefore be said that these methods help to solve the problems that we ourselves cause by building hives without insulation, from materials that do not absorb moisture (or do not drain it outside, etc.).

Currently, queen bees lay eggs longer in autumn and start laying eggs earlier in spring. While the early start of the colony development is considered a valuable feature of bees (because colonies are better prepared for earlier flows), laying eggs in autumn is considered a problem. Many beekeepers see the need to cool down the hives, believing that this will stop the queen from laying eggs. However, is this not an attempt to solve the problem that we have caused by decades of selection (to increase the strength of bee colonies); artificially accelerating development of bees (e.g. by using heaters in the hive); maintaining their strength until late harvests through various procedures; and introducing foreign queens to apiaries that are not adapted to local weather and climate conditions?

In the past, bees themselves finished laying eggs in late summer (or early autumn at the latest), even though they occupied tree hollows, and therefore nest sites that were much better insulated and warmer than hives. I therefore believe that it is an oversimplification to attribute the cause of extended egg laying periods by the queen bee to be caused by a hive that is too warm or by climate change[178].

A modern hive, regardless of whether it is made of wood, polystyrene or other materials, has smooth walls, without cracks, gaps, holes or crevices. This is what beekeepers value, because the bees do not stick propolis on them, and even if they do, it can be easily scraped off and the hive disinfected by different substances or, in the case of wood by burning. Very interesting research that sheds new light on this problem was conducted by the team of Prof. Marla Spivak from the University of Minnesota (USA)[179]. The positive effect of propolis on the health of bees is a well-known issue[180], but not everyone is aware that in a natural tree hollow unlike in a hive, bees live surrounded by propolis. The irregular and rough walls of the hollow (and sometimes also the floor) stimulate bees to propolise these surfaces. Meanwhile, in hives, bees usually only fill the gaps between the hive boxes or frames, corners and sometimes the ventilation mesh in the ceilings, smoothing or rounding them.

---

178 For decades bees have been selected to be strong from early spring to late autumn; today we are convinced that the extension of autumn brood rearing is the result of, as beekeepers say, a hive that is too warm. In my opinion, beekeepers too often fail to see the consequences of their own actions and blame them on external factors.

179 R.S. Borba et al., Seasonal benefits of a natural propolis envelope to honey bee immunity and colony health, Journal of Experimental Biology 2015, 11.

180 New facts are still being discovered. Seeley believes that our knowledge of the importance of propolis for bee health has been revolutionized in the last few years; see also: M. Simone-Finstrom et al., Propolis Counteracts Some Threats to Honey Bee Health, Insects 2017, 4.

The natural bee nest cavity (or one close to it) has walls that stimulate propolis production, photo by Treebeekeeping Brotherhood.

With the help of propolis, bees also reduce the size of their hive entrances (usually before winter), which minimizes heat loss and allows for better protection of the interior of the hive against invaders, photo by T. Gfeller.

American scientists installed special grids on the walls of the hives, used usually to harvest propolis, thus motivating the bees to coat the walls, and then conducted comparative genetic studies of bees living in a standard hive and in one surrounded by a propolis layer. What did they find out? The expression of genes important for the immune response of bees was significantly lower in the propolised hive than in the hive without it. This means that bees living in the standard hive invest more in maintaining health, because their immune system is constantly forced to fight pathogens. Meanwhile, the propolis coating acts as a shield against pathogenic micro-organisms. It seems, therefore, that we should make sure that the surface of the hive walls stimulates the bees to propolis. Some beekeepers believe that this is burdening the bees with additional work (claiming, that since they do not propolis smooth walls, they clearly do not need it). Studies conducted at the United States Department of Agriculture in Baton Rouge[181] show that stimulating bee propolis production does indeed affect honey production in the first year. In the second season, however, not only were there no differences in honey production, but it was found that bee colonies that propolised a lot were on average one frame stronger during almond pollination[182]. So even if we add work to the bees, it is an investment that can bring many benefits to the bees in the long run.

It is worth adding that Torben Schiffer, based on his own research and experience, concluded that the life cycle of a bee colony is different in a tree hollow than in a hive adapted to apiary management, regardless of the type of hive construction we are dealing with (e.g. whether the hive is insulated or not). He claims that the ability of the colony to independently shape its environment is even limited by the use of frames. For this reason, he developed an artificial nest cavity for bees, modelled on a natural tree hollow, calling it a 'Schiffer Tree'. The structure, which was by no means created for collecting bee products, has a round cross-section, (so we will not find the thermal bridges characteristic of modern hives), is frameless, with a volume of 40 litres. It is rather supposed to be an unmanaged nest cavity, in which bees can organize their lives and live in accordance with the natural cycle. The 'Schiffer Tree' is rather supposed to make it difficult to

181   Honey Bee Breeding, Genetics and Physiology Research Unit, known as the Honey Bee Lab.

182   From: M. Simone-Finstrom, conversation: "Two Bees in a Podcast" episode 39.

The construction of the Schiffer Tree was to resemble a tree hollow, the natural nest cavity of bees, as closely as possible, photo by T. Schiffer.

obtain the products of the bees' work, not easier[183]. Further additions (such as honey supers) could cause the creation of new thermal bridges, which would change the properties of the nest cavity, and that could disturb the internal micro-climate (after all, its parameters were the basic premise of the project). The walls of the artificial tree are thick and therefore insulate the interior properly. The covering is also thick, which absorbs moisture well. It is difficult to say to what extent this structure is better than, for example, a beehive log, but heat losses are certainly smaller here, because there is no side opening that is used by tree beekeepers to collect honey but only a single entrance for the bees.

From Schiffer's research, as well as the observations of many natural beekeepers, it follows that among the structures used by men to keep bees, the most like natural bee nest sites seem to be the skeps. Straw hives are dry, warm, have rough walls stimulating propolis, and an oval cross-section, which eliminates the formation of thermal bridges.

Also, straw hives (skeps) are usually relatively small, which favours bee swarming and in conclusion their health. It is worth bee enthusiasts considering such a solution for their amateur apiary.

Symbiotic organisms of bees - pseudoscorpions

As a species, the book scorpion (also known by the name 'house pseudoscorpion', Latin: *Chelifer cancroides*) was first described in 1790 by the Czech naturalist Jan Daniel Preysler. In 1891, it was noticed by the German beekeeper Alois Alfonsus[184], who claimed that it hunts the bee louse (*Braula coleca*) and other mites, as well as insects, so it can be very useful in the hive.

---

183 So it is exactly the opposite of the assumption of those beekeepers who believe that a hive should facilitate apiary management and access to the bees. 'Schiffer's Tree' is also a response to the legal problem related to the activity of natural beekeepers who do not want to administer pesticides to bees in their hives. Beekeepers from Germany claim that there is an obligation to report apiaries and treat bees for varroa. Meanwhile, the researcher defends his structure as a nest cavity for the protection of the honey bee species, and not a hive for the management of apiaries. He claims that if a beekeeper hangs such a structure (or e.g. a beehive log) on a tree and then a swarm of bees settles in it, he does not have to report it as an apiary (he would have to do this if the bees were placed there by humans). According to Schiffer, this also exempts him from the legal obligation to combat varroa mites in the bee colony.

184 Alois Alfonsus (1871-1927); Schiffer points out that the symbiosis Alfonsus writes about was not initiated by humans but occurred naturally.

Recently, Torben Schiffer has been involved in research on book scorpions[185]. He also drew my attention to this species, indicating its potentially positive impact on the health of bees. He came across pseudoscorpions while searching for a natural enemy of varroa, which was the subject of his research (along with the issue of the physical properties of bee nest sites).

I mention the book scorpion because the topic of symbiotic organisms is indirectly connected to the nest sites the bees occupy. While we can keep bees in hives by force (even if such nest sites are not optimal for them), we cannot do the same with symbiotic organisms (if the nest cavity is not favourable to them, they will abandon it).

Chelifer cancroides, like *Varroa destructor*, belongs to the class of arachnids. The book scorpion species is classified in the order of the pseudoscorpions, whose name in both Latin (*pseudoscorpiones*) and English refers to another order of arachnids, the scorpions. Both groups have characteristically shaped pedipalps, ending in pincers. Pseudoscorpions differ from scorpions in that they do not have a venomous stinger at the end of their abdomen.

Pseudoscorpions originate from tropical and subtropical zones. There are around 3,000 species of them in the world, several hundred species live in Europe, and about 100 in Poland. Some like high humidity, others prefer a dry environment. They can be found under stones, tree bark and in various crevices. Due to their small size (4-5 mm), individual species are difficult to distinguish from each other. They are predators, hunting primarily for mites or small insects.

The book scorpions which occur in dry and warm environments, can be found on bookshelves, in libraries, since they feed on mites that eat paper (cellulose). They are also common in barns and stables. In our climate, these arachnids have difficulty surviving outside their nest sites and residential buildings, as they do not tolerate temperatures that are too low. Pseudoscorpions have very efficient pincers, which they use to orient themselves in space and to hunt (the pincers have venom glands). They lie in wait for their prey, attack quickly, injecting venom, and keep the victim at a safe distance until it is dead. Thanks to this, they can hunt down prey that is significantly larger and stronger than themselves (even wax moth larvae).

---

185   I met Torben Schiffer in 2018 during the conference "Beekeeping Without Treatment" in Neusiedl, Austria and that was when I first heard about his discoveries.

We can also find pseudoscorpions in hives, among bees, if the nest cavity meets their species requirements. Apparently, even in a state of extreme starvation, the book scorpion does not eat bee larvae, and even less so adult workers (in such a situation cannibalism becomes an option). If it does not find another individual of its own species near it, it will starve to death (this is a factor regulating the size of the population). Perhaps this is why bees have evolved a complete tolerance of book scorpions in their environment. Otherwise, they might fight them off, treating them as a potential threat.

In honey bee nests, these arachnids originated in a completely natural way. They can cling to a bee flying out of the hive, and thus disperse to new places. Possibly up to several dozen pseudoscorpions can move with the swarm to a new nest. With the help of other insects, including honey bees, the book scorpions spread between tree hollows, where they lived alongside the bees. Later, together with them, they found their way to the beehive logs and finally into modern hives. Old wooden hives, insulated with straw or shavings, were very good nest sites for them. In hives, used in modern apiary management, there is no place for book scorpions. These nest sites are too cold and damp with solid surfaces devoid of gaps are not a suitable place for them. Mesh bottoms are equally unfriendly to them. Not only because of the greater fluctuations in temperature and humidity within the hive, but particles of hive matter fall out through the mesh, which are the food for both the nymphs (immature stages) of the book scorpions and other organisms that also are food for them in various stages of development. Adults leave the hive through holes in the bottom mesh and rarely return.

Pseudoscorpions disappeared from hives also because of the toxic and biocidal compounds used by beekeepers, since the composition of chemical preparations intended for use in apiaries is supposed to be the least harmful to bees but most deadly to arachnids, their victims are not only mites, but also the book scorpions.

Formic acid is exceptionally toxic to pseudoscorpions, they die even at small doses. Thymol, in turn, causes chronic poisoning in them, which begins with a reduced appetite, then causes a state of apathy and ends with death. Synthetic pesticides used to combat varroa mite also have a toxic effect. Oxalic acid is considered the least toxic to pseudoscorpions. It seems that it does not kill adult pseudoscorpions but can harm nymphs, although the results are not unambiguous.

Pseudoscorpions like to build their nests from straw, which is why beehives where this material is used for insulation provide an environment favouring pseudoscorpions, photo by T. Schiffer.

Under laboratory conditions, the pseudoscorpions were able to hunt up to several females varroa mites per day, but we do not know how they will behave in the conditions of the hive, photo by T. Schiffer.

It is also important to remember that all substances used to combat the varroa have a destructive effect on the micro-fauna of the hive, destroying the complex ecological relationships. Even if the book scorpions survive, these agents may cause the severing of other links in the hive ecosystem, e.g. limit the population of organisms that are food for them. A hive is by no means a container made of boards, polystyrene or polyurethane, in which one species of insect lives. It is a place where life should be vibrant.

Laboratory studies have shown that the pseudoscorpion could hunt about nine female varroa mites per day. However, it is difficult to say how many mites could fall victim to the arachnid in the hive, in the uncontrolled conditions of its natural nest cavity. It seems that the pseudoscorpion population is not able to stop the excessive growth of the varroa mite population. Therefore, it is difficult to consider them as a fully effective "biological weapon". However, it seems that they can limit the rate of reproduction of the mite population. Schiffer estimates that if each pseudoscorpion hunted two female mites per day, then 20 individuals could stop the growth of a 1,000-strong varroa population. This means that they would be able to hunt as many individuals as the daily growth of the infestation. However, this is a hypothetical model, and the researcher emphasizes that to achieve such an effect, the population of book scorpions would have to be significantly larger than occurs naturally.

So-called eco-floors and/or eco-roofs (boxes filled with organic matter: straw, humus, forest litter) can support the survival of pseudoscorpions even in single-walled hives. In the photo a pseudoscorpion in the debris of one of my (single wall) hives, photo by author.

Book scorpions are also considered an excellent bioindicator of a healthy bee nest cavity and a verifier of proper beekeeping methods. Their presence in the hive indicates that the management is favourable to bees. However, it may happen that the pseudoscorpions noticed in the hives turn out to be another species of the same order. Some of them, such as *Lasiochernes pilosus*, prefer high humidity. According to Schiffer, its presence in the hive is in fact evidence of the use of beekeeping methods that are harmful to bees.

Is it worth catching pseudoscorpion yourself and introducing them into the hive as a weapon to fight varroa[186]? It's hard to say. It's not a hunter efficient enough to change the fate of bees. However, if someone decides to

---

186   In barns or stables they can be caught in trap boards in sheltered places (e.g. ceilings). Under the board there should be a crevice: there is a chance that a book scorpion will use it as a nesting nest cavity or shelter.

do this, they should first take care to ensure the right nest cavity, i.e. adapt the hive to the presence of these arachnids. Otherwise, it will be a wasted effort, and even condemning the organisms to death.

Stratiolaelaps scimitus mites (previously called *Hyposeismites*) are also used for biological control of various pests including varroa. They are used in horticulture, as they are quite effective in dealing with crop pests. However, these mites must be bred outside the hives, as they require moist soil to reproduce. Once introduced under the roof of the hive, they begin their journey downwards (towards the soil) and, in the process, hunt for female varroa mites quite effectively. Some even believe that several such treatments during the season would allow for controlling the entire varroa population in the hive and so forgo other treatments. However, this method has not become widespread, because breeding *Stratiolaelaps* is quite laborious and purchasing it is by no means cheap. These organisms are not able to reproduce in the hive, so it is not enough to introduce them to the hive once.

Natural bee hives - Conclusions

Considering Seeley and Schiffer's research, the obvious conclusion is that healthy bee nest site/cavity should be dry and warm. This is especially important during the wintering period. In the summer, bees can actively control the micro-climate of their nest cavity. They can control the temperature and humidity levels. In the winter, however, they remain more passive to the processes taking place in the hive.

A natural and healthy hive should be conducive to creating an appropriate micro-climate, beneficial not only for bees, but also for organisms that have been companions of honey bees for centuries, co-creating the complex ecosystem of honey bee nests. It should therefore be an optimum niche for them, and a worsening or the worst niche for pathogenic micro-flora[187]. Hives with additional insulation will work much better here, because they provide more stable thermal conditions (smaller temperature amplitudes, both daily as well as annually).

---

[187] For bees, the hive is an important, but only a certain part of the environment, or biological niche. Bees obtain food and water from the outside, and thanks to evolutionarily developed abilities, they can also compensate for certain nest cavity deficiencies (ventilate, heat, cool, plug holes with propolis, etc.). For micro-flora, the hive is the only living environment, and therefore the conditions that prevail in it must allow for the development and maintenance of complex relationships in which organisms live.

It seems that beehives made of natural and breathable materials: wood, straw or reed[188], will be healthier for bees and their symbionts. Straw, not necessarily as the main building material, but as insulation, is very friendly to various organisms, including bees.

Bees prefer smaller volumes than the capacity of standard hives. To satisfy their biological needs, a nest cavity with a volume of 30-40 litres is enough for them. According to many natural beekeepers, it should be constant, because by increasing the volume (e.g. adding another box as honey super), we disturb the harmony of bees' work.

Smooth surfaces should be avoided inside the hive, and a good solution are walls made of unplanned and unsanded wood. Straw or reed hives or such division boards can also be used for this purpose. Porous, uneven surfaces stimulate propolis production. The hives should be equipped with full bottoms, these can be so-called eco-floors. The ceiling should allow (or absorb) moisture, and at the same time act as a thermal insulator. A good solution, for example, are straw or reed mats. Eco-roofs/inner covers can also be used.

The hive should have a lower exit: these are the preferences of the bees. This type of solution also helps the bees to optimally heat the cluster.

My hives (I will return to this topic in another chapter) lack many of those things. But bees that are healthy, locally adapted and disease-resistant will cope in any structure, even if it does not meet all the criteria for an optimal bee house.

When building a bee nest site, we should consider whether we want to run an apiary or create a shelter which we do not intend to look after and/or manage. If we only want to enjoy the view of working bees, we can create a much better nest cavity for them, better than a regular hive. It can be, for example, a log, a Schiffer Tree hive or a natural tree hollow. On the other hand, a much cheaper and easiest way to make a nest site for bees is

---

188 Many practicing beekeepers, as well as scientists, including those who are close to natural
beekeeping, and even beekeeping without chemicals, also appreciate plastic structures
(e.g. polystyrene hives). They claim that they have very good thermal insulation properties, which
meet the needs of bees much better than a hive built from a single board. Currently, polystyrene
hives are a kind of discovery in the USA, where traditionally wooden hives were built, so-called single
wall (from a board about 2.5 cm thick, without any additional insulation). In terms of optimal nest
cavity properties, the basic disadvantages of these hives are smooth walls (not stimulating propolis
production), which are also impermeable to air. If we do not provide good ventilation (which will
dissipate a lot of heat to the outside), then such a hive will have high humidity, and that may promote
the development of mould on the combs.

A reed hive (covered with a mixture of cow manure, clay, sifted wood ash and whey) and a so-called wild hive, in which the covering was made like a skep from woven straw, from **David Heaf's apiary,** photo by D. Heaf.

Bee-Skeps are nest sites that are most similar in physical properties to tree hollows: they have a similar volume (favourable for swarming), very good insulating properties, rough walls that stimulate bees to propolis, photos: T. Gfeller (top), J. Rüther (right), author (skep with a roof), .

Although most studies confirm that modern single-wall hives are not optimal for bees, healthy bees with immune mechanisms will also cope perfectly well in such structures. Erik Österlund keeps bees in non-insulated Langstroth hives with a large cross-section, photo by E. Österlund.

to simply build a box built of regular rough, unsanded boards. If we do not run an apiary, the hive should have a small volume (about 40 litres), and it should also be placed higher, for example on a house wall, tree, roof or terrace.

BEEKEEPING WITHOUT FOUNDATION

Chapter 3

BEEKEEPING WITHOUT FOUNDATION

> *He must be a dull man who can examine the exquisite structure of a comb, so beautifully adapted to its end, without enthusiastic admiration. We hear from mathematicians that bees have practically solved a recondite problem, and have made their cells of the proper shape to hold the greatest possible amount of honey, with the least possible consumption of precious wax in their construction. It has been remarked that a skilful workman with fitting tools and measures, would find it difficult to make cells of wax of the true form, though this is effected by a crowd of bees working in a dark hive.*

CHARLES DARWIN[189]

Advantages and disadvantages

Importance of combs for bees and in apiary management

The comb plays a much more important role in the life of bees than a house does in the life of a human. We should think of wax structures as a kind of organ of the bee superorganism. They are not just a place where bees stay. They also raise new generations here and store food.

---

189   Ch. Darwin, The Origin of Species, PF Collier & Son, New York, 1909., page 279-280.

The comb allows them to regulate the right nest temperature and protect it from unfavourable external conditions[190]. So, it seems to me that we should try to change the way of thinking about the bee nest, let the bees shape it according to their own needs[191].

The abandonment of foundation allows for the creation of new alternatives to apiary management and natural beekeeping. These methods have many advantages, but they are not without disadvantages, which should not be significant for an amateur beekeeper running a stationary apiary. For this reason, it is worthwhile for hobbyist beekeepers to at least try this type of management.

To practice beekeeping without using foundation, you need to change your way of thinking about several elements of apiary management. Which one depends on the method of practice you choose. Beekeeping without foundation can be practiced in several ways. You can let the bees build their combs completely freely (so called wild nest), but you can also practice top-bar beekeeping or completely standard management in a modern frame hive, on any type of frame. I will briefly discuss the advantages and disadvantages of beekeeping without using foundation.

Pure wax

The most important advantage of not using foundation is the bees' living environment is free from chemical pollutants[192].

It is worth mentioning that dishonest manufacturers add other cheaper substances to foundation, mixing them with beeswax. Most often, these are paraffin wax, a product that is foreign to the natural environment of bees. Bees are reluctant to draw this type of foundation. They bite it off and carry it out of the hive. It also happens that queen bees do not lay eggs in combs that contain some of these substances. By deciding on a foundation-free management, we interrupt the closed wax cycle that leads to the accumulation

---

190   I once came across information about a colony that managed to overwinter in a nest built under a branch (without external cover), but I am unable to find the source of this information. Such cases are, however, extremely rare. It happens that a swarm that fails to find a suitable nest cavity after a few days starts building a nest in the open air, where it has just tied itself up. Most often, however, it does not manage to survive the winter.

191   Many hobbyists, even those who have only a few hives and declare that they do not care about honey, only about contact with bees, believe that the best thing that can be done for the bees is to provide them with foundation done by themselves from their own wax. So, they do not even consider the obvious alternative, which is to give up providing foundation at all.

192   This has been largely discussed in the chapter on chemicals.

of an increasing number of various impurities. On the one hand, we eliminate substances that are potentially harmful to bees. On the other hand, we cut the circular chain of fraud associated with wax adulteration. Additives are often very difficult to remove and require special treatment. It should be added that detergents are used when pressing foundation in the production process, which prevents the foundation from sticking to the surface of the rollers. This means that each sheet of foundation has already been doused on both sides with detergent-water solution before it is placed in the hives[193].

Wax obtained from frames with foundation that came to my apiary together with the purchased bees (on the left). Wax from my foundation-free frames (on the right). Beekeepers can obtain comparable wax quality only by melting frames consisting of drone cells (cut out as a mite removing treatment) or by uncapping's, photo by the author.

By waiving to use the foundation, we have a guarantee that the bees will live in a cleaner environment. Removing the comb is tantamount to eliminating contaminants from the hive that could have penetrated its structure. It must be added that our product in the form of beeswax will also gain the highest quality, which is unattainable by normal beekeeping.

---

193   The detergent is rinsed off, but not very thoroughly; I have observed this process and I am convinced that some of the chemicals remain on the surface of wax (or are dissolved in its structure).

At the peak of the season, a strong bee colony can quickly build a wax comb, photo by Ł. Łapka.

Natural worker-cells

At this point of the discussion, it is worth paying attention to natural comb (built by bees without any guidelines) and the so-called small cell (i.e. a cell 4.9 mm in size created based on the foundation). These issues are connected to some extent. As a rule, the size of the cells of each foundation available on the market is not natural, because the cells built by bees on one comb vary in size[194]. Some are treated as a warehouse (the bees store honey there), others are incubators (the queen lays eggs in them), although they can be used interchangeably.

For this reason, cells located in the upper part of the natural bee comb and on its sides are generally larger than those located centrally and at the bottom. Therefore, a comb in which all the cells have the same size is unnatural, regardless of the size of the cells. It is worth adding that the use of the so-called small cells is treated by many beekeepers as a specific method of combating varroa. However, practice has not supported this hypothesis. Hopes for a long-term solution to the problem just by using foundation with a 4.9 mm cell turned out to be vain. Despite this, a small cell or

---

194   Even if the bees build the cells that we suggest to them with the foundation, it will be a bit like the proverb: "Even a broken clock shows the right hour twice a day".

a natural comb can to some extent help bees in minimising some stresses or environmental pressures (including mite pressure)[195].

In central Europe, in the pre-foundation times, the average cell size was estimated at 5-5.1 mm. François Huber, one of the greatest Swiss naturalists and beekeepers at the turn of the 18th and 19th centuries, a period when no one had thought about foundation yet, estimated the cell size at 5.08 mm, Teofil Ciesielski mentioned "almost 5 mm", and Leonard Weber gave the size as 5.115 mm[196]. If I am thinking correctly, the individual numbers mean only an average, obtained based on measuring cells of different sizes located on the comb.

Some subspecies of bees, mostly African ecotypes, build cells of 4.4 mm, others 4.6-4.7 mm. The cells of northern subspecies are larger. European ecotypes raised on a foundation with a 5.4 mm cell, after being placed in a hive without foundation, will build cells of 5.3-5.4 mm in size. Sometimes even 5.2-5.6 mm, but with each subsequent generation, raised in a cell of increasingly smaller size, more and more cells approach about 5 mm and their average width should also approach this value.

Some beekeepers believe that the size of a bee cell is constant, because it is written in the bees' genome. They therefore believe that the size of the cell that is created immediately after the bees are taken away from the foundation will be preserved. They therefore deny the process of bees systematically reducing the size of the cell. Those beekeepers usually use foundation in their apiaries, so their experience in the discussed issue is limited. In turn, beekeepers not using foundation claim that for the bee cell to return to its natural size for a given subspecies, ecotype or "genetics", bees need 6-8 generations, successively dropped onto empty frames. Only after this period will they start building combs with the natural size of the

---

195 K. Olszewski points out: "Our research also shows that untreated colonies kept on small-cell combs usually entered overwintering in a much weaker condition than treated colonies kept on standard-cell combs. However, they were much stronger than untreated colonies kept on standard-cell combs. Colonies kept on standard-cell combs, created from splits and untreated against varroa, did not survive overwintering following the season in which they were created. Meanwhile 30% of colonies created while were kept on small-cell combs and were untreated against varroa survived. It also turned out that when the treatment of the colonies was discontinued, colonies kept on combs with standard cells survived the first overwintering and died during the next one. The colonies treated in the same way on small-cell combs survived one season more after discontinuing the treatment", K. Olszewski, Wpływ plastrów o małych komórkach na cechy pszczół, rodzin pszczelich oraz odporność na *Varroa destructor* [The influence of small-cell combs on the characteristics of bees, bee colonies and resistance to *Varroa destructor*], Pszczelarstwo 2022, 1.

196 Ł. Łapka, O tym, jak chciano powiększyć pszczołę [About how they wanted to enlarge the bee], Pszczelarstwo 2016, 3.

cell[197]. This is because while drawing comb in a way they "refer" to their own size, as well as the size of the cells in the nest. During the first generation, they will not return to a cell size of 5.0-5.1mm if they are hatched from a larger cell. A "large" bee will not build a small cell, at most a little smaller than the one it hatched from. This process will continue until the size of the cells in the nest and the genetically recorded size match, because genetic mechanisms are regulated by external factors.

Similarly to the construction of the comb, perhaps the size of the bee's body will never be reduced to such dimensions as in the pre-foundation times. After all, for decades there was a selection of bees that better tolerated the enlarged cell of 5.4 mm, so the genetics could have changed in this manner.

When I put away the foundation in my own apiary, the bees started building mainly cells of 5.3-5.4 mm, but after a few years I managed to find more and more combs with smaller cells (5.1-5.2 mm). I found that the cell sizes in my natural combs fluctuates between 4.9-5.5 mm. The largest number of cells are 5.2-5.3 mm in size. However, this is not a statistically calculated value. I measure cells only sporadically on some combs withdrawn from the hives. So far, I have not tried to measure cells on all combs in a specific bee nest built by bees from scratch. after placing the swarm on empty frames. It is certain, however, that they have different sizes, and that smaller ones dominate among them than those found on standard foundation.

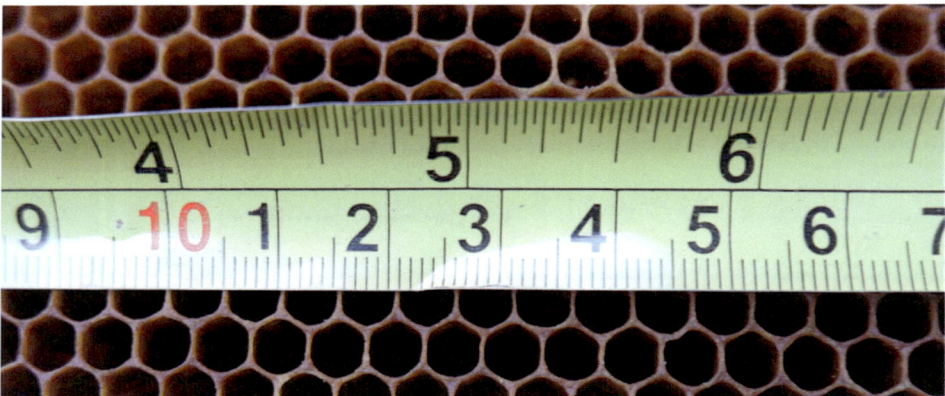

One of my foundation-free combs: cell size - 5.15 mm, photo by author.

---

197   This process may take years, especially if we left many of the drawn comb in the nest and withdraw only some of them yearly. Bees as a rule build larger cells in which they store supplies. Therefore, if we put empty frames in the honey supers for them to draw comb and then move them to the nest where the bees propagate, we could have the situation in which we would have always larger cell sizes than those that bees build in nature.

The size of the cell depends on many factors. These include the current size of the workers, the size of the cell they emerged from, the time of the season, the cell's location on the comb, and the location of the comb in the nest.

On one comb, cells with a very large range of diameters can be formed (sometimes the differences in cell size can reach up to about 1 mm), photo by D. Murrell.

Moreover, there is also a specific genetic trait that dominates a given subspecies of bees. Bees also build larger cells during times of abundant forage. It can be assumed that this is for purely economic reasons because they need a place to store nectar. Larger cells save building material: the ratio of the amount of wax used to the volume of the built storage is then more favourable.

But is the size of the comb cell so important for the health of the bees and the functioning of the entire superorganism? Studies of bees raised in small-cell combs suggest that they exhibit heightened hygiene instincts. We can therefore go a little further and hypothesise that the size of the comb cells could serve to enhance other characteristics or behaviours of the bees[198].

Beekeeping textbooks often contain information suggesting that individual stages of a bee's life are associated with specific hive tasks. Seeley's research has proven that this is an oversimplification. It turns out that the division of tasks in the hive based on the age of the workers is quite fluid. One could therefore consider the hypothesis that predispositions to perform tasks may be determined by the size of the cell from which the worker emerged.

In the 19th century, scientists claimed that a larger bee has a longer tongue, as well as a larger honey stomach, and can bring more nectar to the hive at one time. Their observations led them to the conclusion that a colony with larger bees can produce more honey. Today, we can consider this thesis to be a kind of simplification, and perhaps even false, resulting from a methodological error.

Many beekeepers claim that bee colonies have become stronger and honey yields have increased since they started using comb foundation with

---

198 K. Olszewski's research found that bees hatching from smaller cells not only have a different size of their whole body, than bees from cells drawn on standard foundation, but they also differed in the proportion of the sizes of various organs, K. Olszewski, Wpływ wychowu w plastrach o małych komórkach na cechy morfometryczne i masę ciała robotnic [Effect of rearing in small-cell combs on morphometric traits and body weight of worker bees], Pszczelarstwo 2022, 1. J. Tautz's research proves additionally that bees that developed at a higher temperature (36 degrees °C) have better cognitive abilities than those that developed at a lower temperature (32-34 degrees °C), J. Tautz et al., Behavioural performance in adult honey bees is influenced by the temperature experienced during their pupal development, Proceedings of the National Academy of Sciences 2003, 5. A small cell (or more precisely, the density of "packing" larvae and pupae, which means less space in the cell next to the developing pupa) is conducive to maintaining a higher temperature during brood rearing, and therefore may indirectly have an impact on improving the cognitive abilities of workers in the colony. I do not know of any research results that would directly indicate the relationship between the behaviour (or potential) of honey bees and the size of the cell from which they emerged. Although there were such studies on bumblebees. This may suggest that the size of the bee cell from which the bee hatched may have some influence on its later behaviour (e.g. through gene expression). What would be that influence? We do not know. According to suggestions from representatives of the world of natural beekeeping, bees from cells of different sizes may perform some tasks better. K. Olszewski in the work cited before notes: "Some believe that maintaining colonies on combs with small cells is a method of increasing the effectiveness of selecting bees resistant to *V. destructor*. This provides a basis for concluding that the width of the comb cells as a factor of the nest environment has an impact on the modification of the biology of bees and entire colonies to a greater extent than previously assumed, creating additional opportunities for breeders". Many of these assumptions and hypotheses were confirmed already at the editorial stage of the book. See also: P. Dziechciarz, K. Olszewski, Naddominowanie behawioralne [Behavioural overdominance], "Pszczelarstwo" 2022, 12. The authors of the study suggest that providing the hives with combs with cells of different sizes (4.9 mm and 5.4 mm) caused an effect similar to hybrid vigour in the colony: "The effect of maintaining colonies on two types of combs exceeded our expectations, because the colonies with two types of combs in the nests significantly exceeded in terms of strength (in autumn), cold hardiness, spring development and honey yield, the colonies maintained on combs of one type, both those living only on combs with standard cells and those living only on combs with small cells".

a cell size of 4.9 mm. Not everyone agrees with this. It is possible that the reasons for these dependencies should be sought not directly in the size of the comb cells, but in other variables, e.g. bee subspecies, hive type or even the type of or availability of nectar[199].

Recent studies have also led to ambiguous conclusions. Some scientists suggest a positive influence of a small cell (e.g. the research carried out by Krzysztof Olszewski), which is not always confirmed by others. However, I do not know of any studies that would prove that a smaller size of a bee cell has a negative effect on the health of the bee colony[200].

Some observations indicate that the small bee shows more natural immune behaviours in relation to varroa, which means greater tolerance to varroosis. Beekeepers often claim that the same bees placed on a foundation with a cell size of 4.9 mm show an increase in the intensity of hygienic behaviours or so-called grooming. According to beekeepers, the small bee is long-lived and more resistant to diseases. Observations also allow us to conclude that the small cell can serve to shorten the time of pupal development (sealed brood) by several hours and in exceptional cases even a dozen or so hours.

The issue of natural comb is related to a certain hypothesis, named after the researcher, Michael Housel, 'Housel Positioning'. The honeycomb has a specific structure, the cells on one side of the honeycomb are shifted in relation to the cells on the other side. If we lift the honeycomb up and look at it against the light, we will see a structure in the shape of the letter "Y" in the bottoms of the individual cells, while on the other side the letter "Y" will be upside down (it resembles a symbol like the Mercedes logo). According to Housel, wild bees build honeycombs in a specific way. When we look in both directions at the honeycombs against the light from the middle of the

---

199 It has been confirmed that the size of cells can affect the size of bees and the proportion of the sizes of their various organs, which in turn can be related to the quality of using some of the blossoming flows (e.g. because of the size of flowers, their nectarines etc.). The length of the tongue or even the mass of the bee can be important here, particularly in the case of small and very delicate flowers. The condition and size of the bee colony have an impact on the use of blossoming. The size of the harvest can indirectly result from how the colony copes with the pressure of mites. The reaction to the foundation with a small cell may depend on the subspecies or ecotype of the bees. Perhaps this is a partial explanation of why in some experiments and research a small cell helps to limit varroa populations, while in others it has no effect. However, some of the experiments might not have been carried out without certain errors, because we know that in the first-generation large bees can have problems functioning on combs with smaller cells. Therefore the procedure, if carried out improperly, may affect later results.

200 For more on the small cell, see Ł. Łapka, Mała komórka, a pszczelarskie kłopoty [Small Cell size and Beekeeping Troubles] "Pszczelarstwo" 2019, 12; 2020, 1-2 and K. Olszewski, Wpływ plastrów o małych komórkach... [Influence of small-cell combs...] op. cit.

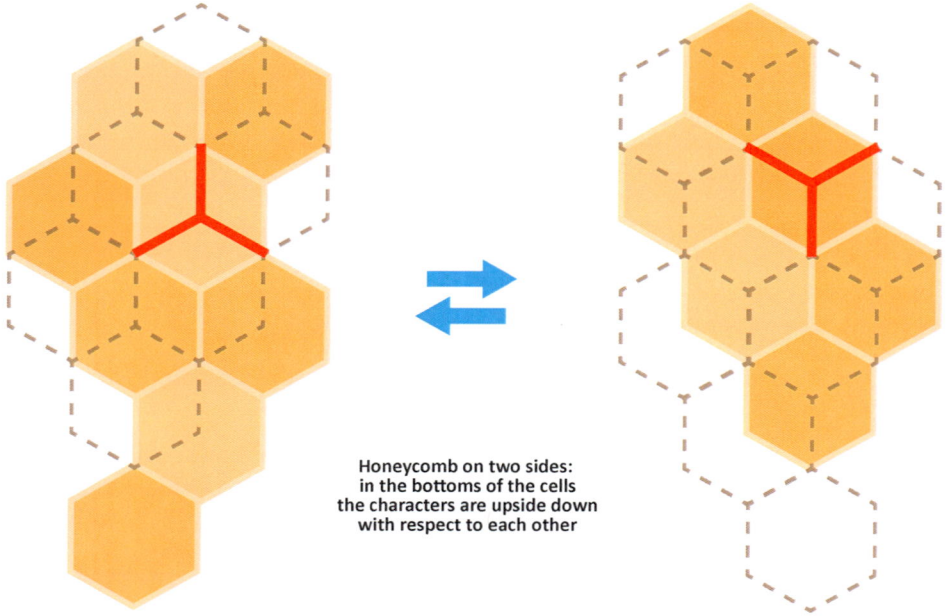

Honeycomb on two sides:
in the bottoms of the cells
the characters are upside down
with respect to each other

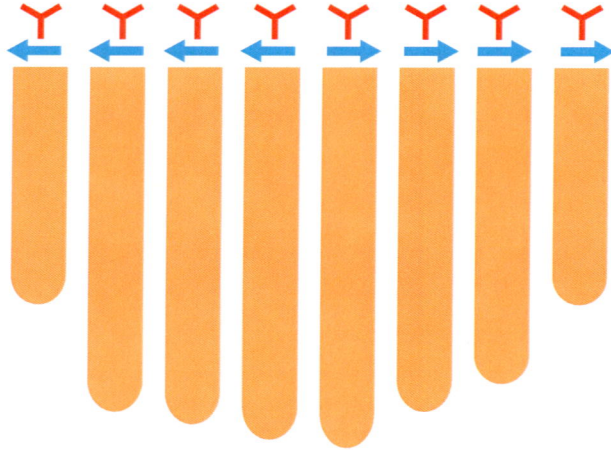

The "Y" side faces outwards in the bees' nest

Housel's hypothesis concerns the arrangement of "Y's" in wild bee nests. He suggests that beekeepers should place the "Y" in foundation in managed colonies in the same direction in which wild-living colonies build their combs, figure by M. Uchman.

nest we should see the letter "Y". The researcher therefore suggests that when placing honeycombs in hives, we should follow the example of wild bee nests, i.e. direct them with the "Y" outwards. Some beekeepers believe that this direction is good for the bees. It calms them down and introduces harmony in the nest and thus in the work and communication of workers and as a result increases productivity[201]. Others, however, claim that they do not notice any difference resulting from the comb direction/positioning.

It is difficult for me to unequivocally deny Housel's hypothesis, especially since my bees do not live in wild nests, and I also happen to rotate the combs, which generally disrupts their natural arrangement. However, my personal experiments using a natural comb allow me to conclude that bees, when building comb, do not follow the direction suggested by Housel. It therefore happens that on the same comb, the "Y" has a vector in one direction, the other in the opposite direction, and sometimes it is in a horizontal position rotated 90 degrees to the left or right. It sometimes happens that even on one comb, the "Y's" are turned in every possible direction. Therefore, it seems that even if Housel's hypothesis is based on a statistical regularity, it is not a rule.

Freedom to raise drones

One reason for using foundation is to try to limit the rearing of drones. The foundation suggests to the bees the size of the cell they should build. However, the colony instinct is so strong that if the bees do not find a place in the nest to build drone comb, they will start rebuilding the foundation and create drone cells.

Some claim that our knowledge of how honey bees function as a species is at an early stage. Regardless of whether they are right or not, it seems that we really know very little about the role of drones in the bee colony. It is assumed that they only fulfil a reproductive function, which is hard to accept. Drone brood develops longer than worker brood (24 days from the time the queen lays an unfertilized egg). they also consume more food, so raising drones requires more effort from the bee colony and exploits its resources more. Raising thousands of drones just for the purpose of having a few of them fertilise queens during mating flights seems simply unimaginable extravagance. However, it must be admitted that in male-

---

201   One proponent of this type of nest arrangement is a professional beekeeper from Arizona, Dee Lusby, who is also a great promoter of small cell beekeeping.

female relationships, nature often shows evidence of an almost Byzantine mismanagement of resources. This is due to so-called sexual selection. Perhaps it is similar in the case of bees? Reproductive success will be achieved by those colonies whose genes can afford such great waste, such as several thousand of drones doing nothing except consuming enormous amounts of resources. Beekeepers do not appreciate the importance of the role of drones. Most of them believe that queen bees "will somehow fertilise themselves".

Drones are treated almost as parasites, freeloaders, that only contribute to the impoverishment of honey harvests. The mating areas are therefore dominated by drones from colonies from which beekeepers have not removed them, and not from those that are doing well. Perhaps only queen breeders are interested in raising as many drones as possible.

In my apiary the bees have complete freedom to build drone combs and raise drones, photo by author.

However, in breeding apiaries drones will come from genetic lines of bees that produce the most honey but not from those that cope with the varroa that well. Meanwhile, many scientific studies provide information indicating that high genetic diversity within a bee colony benefits the health of the bees. A correlation was found between the genetic diversity within a colony and its ability to cope with various pathogens, including those associated with varroa or the *Paenibacillus larvae* bacteria[202].

We should therefore appreciate the need to raise drones, especially those coming from the healthiest colonies that do not require treatment.

Maybe drones play some important roles in the bee colony. For example, it is believed that they produce much more heat than workers. Maybe they are underestimated in that matter, and their presence allows more workers to forage? Perhaps they help ventilate the nest?

We must not forget about their impact on the health of the bee colony. They can be a kind of magnet for brood parasites, since drone brood is more attractive than worker brood. Because drone pupae develop longer than worker brood, varroa prefers drone brood, which also has various consequences. Firstly, drones are an important factor in the selection of untreated populations. Large infestations of non-resistant colonies result in the fact that they cannot afford to mass-rear drones and therefore do not pass on their own genes. Drones originating from cells in which varroa has reproduced are usually weakened to such an extent that they are either unable to fly or are unable to effectively compete with others in the air. Therefore, thanks to drones, only healthy and resistant colonies can achieve reproductive success.

Secondly, the concentration of varroa mites in drone brood may relieve the pressure put on workers to some extent, as their brood is less infested. This means that they will be healthier for most of the season. A similar mechanism of drone mite burden is observed in the colonies of the eastern honey bee (*A. cerana*). In both species drones are not necessary to secure the basic life functions of the colony, other than mating. However, it is difficult to compare both species of bees directly, because although many of the immune mechanisms work quite similarly, there are also those that function differently, which results from differences in their biology.

In the colonies of the eastern honey bee, the mite generally does not develop in the worker brood. Most often the offspring of varroa developed in the worker brood remain infertile. In turn, the drone brood has such a thick

---

202   K.A. Palmer, B.P. Oldroyd, Evidence for intra-colonial genetic variance in resistance to American foulbrood of honey bees (*Apis mellifera*): further support for the parasite/pathogen hypothesis for the evolution of polyandry, Naturwissenschaften 2003, 5; S. Desai, R.W. Curie, Genetic diversity within honey bee colonies affects pathogen load and relative virus levels in honey bees, *Apis mellifera*. Behavioural Ecology and Sociobiology 2015, 7; M. Lopez-Uribe, Higher immunocompetence is associated with higher genetic diversity in feral honey bee colonies (*Apis mellifera*), Conservation Genetics 2017, 2.

capping that the bees are unable to chew through it. Mites are therefore relatively safe in the drone brood of the eastern honey bee. In the case of our honey bees, there is no difference between the capping of the drone brood and that of the workers. Bees can therefore remove infected drone brood with complete impunity. Erik Österlund even claims that this is a specific, original "drone removing treatment" of resistant bees and notes that thanks to this feature, *A. mellifera* could be better adapted evolutionarily in the long run to the presence of varroa than *A. cerana*. Of course, our bee populations still must learn to detect varroa. In this respect, they are still far behind the Eastern honey bees, but detection of varroa can be strengthened during selection. Additionally, during the so-called drone expulsion at the end of the season, some mites are completely naturally removed from the hive together with the drones.

However, raising many drones by a bee colony that has not developed mite resistance mechanisms may lead to a rapid increase in the mite population to an incomparably greater extent than in a colony that has not raised many drones. This makes it easier for an epidemic to develop there, caused by pathogens spread by the mite. It can therefore be assumed that many drones are a threat to those bee colonies that have not yet developed varroa resistance or tolerance. In resistant colonies, this threat is probably much smaller. Some supporters of the small cell believe that the use of drone comb removal treatments, especially in combination with unnaturally large standard foundation cells, results in long-term selection of mites for development in the worker brood. This is therefore a kind of artificial direction of evolution towards the development of a mechanism precisely opposite to that observed in the Eastern honey bee. The role of drones in the aspect of the proper functioning of the colony cannot be overestimated. It is thanks to them that it is possible to satisfy the bee's instinct to reproduce, which regulates the functioning of the entire colony.

This can be observed every time a beekeeper removes the drone brood[203], since the bees rebuild it repeatedly until the reproductive instinct is satisfied or the need for rebuilding ceases, which usually occurs in late summer.

Although research does not confirm this, some beekeepers claim that drone rearing does not always occur at the expense of honey harvests. It is true that drone brood eats huge amounts food, but observations also show that the colony, satisfying the reproductive instinct, works more calmly

---

203   Some people call this castrating the bee colony.

and efficiently, thus balancing the greater consumption of food. Of course, such a situation probably does not occur in every bee colony, it depends on many conditions. Undoubtedly, in regions with poor nectar flows, or during unfavourable weather conditions, rearing a larger amount of brood or worker bees can negatively affect honey harvests. This is why to increase honey harvests in regions with poorer nectar flows, beekeepers decide to use queen isolators. However, if the nectar flows are good, the presence of drones does not have to significantly burden the production balance of the bee colony. Moreover, if the drones took over some of their sisters' duties, even the simplest ones, more workers could go out and forage.

Regardless of the typical economic and management issues many drones can be very beneficial in many respects. Drones from varroa-resistant bee colonies can be exceptionally valuable for the entire area and contribute to improving the genetic characteristics of the surrounding apiaries. I therefore believe that investing in drones should be important for every beekeeper who applies the principles of natural beekeeping and selects for bee resistance. I do not remove drone frames from nests and most often leave them exactly where the bees build them.

They are used when the bee colony feels the need to raise drones, but not all season long, although the queens have constant access to them. If the bees do not feel the need to raise drones, the frames will be left empty, filled with honey or sometimes even bee bread.

Saving time and money

If we have two or three hives in the garden, wiring frames, buying and installing foundation is neither a big effort nor an expense. However, with each subsequent hive, the costs increase. With 10 hives, the costs become noticeable, and with 30 or 50 hives the costs start to become a problem. Someone will say that the more hives, the greater the income from the apiary, and therefore the unit costs do not increase and with good work and organisation they can even decrease. It is hard to disagree with this. Some costs of running an apiary are necessary. You must buy hives or material to make them, a chisel, etc. However, in my opinion, foundation is not one of them. In some situations, it can make the work easier, but in most hobbyist apiaries it means unnecessary effort. Here is a simple calculation. Each hive usually requires 0.5-1.0 kg of foundation. With 30-40 hives, the expenses can already reach hundreds of Euros or dollars, but it does not end there. Wiring each frame requires at least a few minutes of work, although specialists who are skilled at using a pneumatic stapler can do it in a few seconds.

If we assume that we replace a dozen or so frames per year in one hive, this means at least 30 minutes of work. Whether this is a lot or not, everyone will judge for themselves. If we have two or three hives, we can prepare for the entire season in one afternoon. However, even with 30-40 hives, the effort becomes quite large.

Disadvantages of foundation-free beekeeping

Colony management without using foundation also has several weak points. First, combs built without foundation are more fragile and delicate than those made based on a wax or a plastic sheet. In the case of small frames, it is not a problem. The problem begins with larger frames. In my own apiary, I use a Polish "Wielkopolski" standard hive that has frames 36x26 cm, although I also use some frames lowered to 18 cm. I also "Warsaw" hives which are the same size as Dadant frames, but with the frames put vertically. While in the first case, especially with lower frames, I do not see any serious inconveniences, in Dadant-vertical hive frames you must be careful with new combs loaded with brood or honey, because with too sudden movements or excessive leaning of the frame from the vertical during inspection may cause the comb to break. I have experienced this several times. Large-frame hives also create more difficulties for migratory beekeeping[204]. Of course, the more times a comb is re-used, the stronger it is; a dark-brown comb can be treated like any other drawn on a foundation and wired. Larger frames must therefore be reinforced. Special attention is required when extracting honey from "virgin" combs without foundation. In the case of an unwired Warsaw or Dadant frame, extracting from such a comb without the risk of damaging it is rather impossible, even if we significantly reduce the speed of the honey extractor. However, I have successfully extracted unwired "white" combs from the smaller frames, although I did it with caution.

A separate problem is the bending of the surface of subsequent combs and their deformation. Bees build new combs parallel to existing combs. Therefore, if we place an empty frame between two drawn ones, we can expect that the comb that the bees build will be straight. However, if we wanted to settle a swarm in an empty hive or simply add a box to the colony, we can expect that during the next inspection we will find combs built in every possible direction and not necessarily along the frames.

---

204   I believe that combs, in frames up to 20 or so centimetres high, should not suffer any damage during transport of the hives, even if the combs are freshly drawn. The lower the frame is the less danger of damage there is.

Bees simply do not know the concept of a frame hive; they build according to their own needs and current conditions.

While we can save time and work by not adding foundation, when increasing the volume of the hive, adding an additional box with empty frames is not enough, so the inspection or maintenance may take us more time than if we used foundation. In the management of a frame hive, we must consider that here and there we will have to cut a piece of the comb and straighten it. The bees will then attach it in the frame and strengthen it. When adding frames, we should also be prepared for the fact that the bees may build a slightly bent comb on the side of the nest and the same fate will befall each subsequent one, unless we intervene in time.

Sometimes everything goes wrong: bees build crooked combs, and some of them can break off due to the beekeeper's carelessness, photo by author.

Cells in existing combs, especially during high nectar flow, will be elongated, because bees do not have time to build new ones. All those issues can be avoided or their negative effects can be easily minimised. However, we need to change the way of thinking and some aspects of apiary management, the comb, and the hive.

Ways of keeping bees without foundation

Natural comb

In many respects, beekeeping based on so called natural comb (i.e. comb built without any frames or guidelines) is undoubtedly the most primary and simplest form of keeping bees. This is where it all began many thousands of years ago. The oldest known cave painting depicting bees and a silhouette of a man sneaking for honey probably comes from the 6th millennium BC. It is in a complex of caves near Valencia in Spain. At least since then, people have been using the work of bees, creating 'an economic relationship' with the insects. Until modern times, this relationship consisted primarily of searching for wild nests and collecting honey, brood and bee bread. Today, we encounter this form of beekeeping less often, because modern beekeeping should be compared to cultivation rather than gathering.

When it comes to the construction of bee nests, it seems that the imagination of beekeepers knows no bounds since a great many of them have been created. I will only discuss the most popular solutions. Ways to benefit from the management based on nests with natural development depends on the construction of the nest cavity used - 'the hive'. A nest that is completely impossible to disassemble to parts, such as a fragment of a tree trunk with a natural hollow, significantly limits the options of obtaining bee products but still allows the colony to perform their most valuable work of pollination.

Each of us can put a log or a suitably constructed box hive in the garden, practically without interfering with the lives of bees. However, it is difficult to call it beekeeping, which is rather associated with taking away the products of bees' work. That is why beekeepers create such structures that allow them access to the nest and control the life cycle of bees. In logs-hives or tree-hives ("barć" for the tree hive and "bartnik" for a person who took care of the bees or similar terms are used in most of the Slavic languages, in Germany term "Zeidler" is often used), access was provided by cutting a high hole through which the hollow in the log or trunk was widened from the inside to create an appropriate space for bees and was then closed with a 'plug'. If a log rotten inside served as the nest, access to the interior was possible from below. In many regions this type of keeping bees has fallen into oblivion[205] and one must rely solely on historical accounts, which are fragmentary. Some claim that original tree-hives were nests carved in trunk of living trees, while others say that they were rather logs (a cut fragment of a tree trunk) placed on a tree. For sure there are historical records of the laws proving that the cavities were made in the trunks of living trees. We also have some remains of such trees with carved "hives". If a tree with a man-made cavity has fallen the keeper of the bees ("bartnik") could cut the part with the hollow out and use it as a log-hive. There are also some claims that beekeepers most often used logs rotten inside, because this meant keeping their work of making the hollow to the necessary minimum. Among supporters of returning to such type of beekeeping, there is a lively discussion on which nest cavity is most beneficial for bees: a living or dead tree? Some believe that a living tree creates a more favourable micro-climate, while others claim that a rotten tree has better thermal insulation or hygroscopic properties. A tree beehive hole limits access to bees but allows for basic inspections. Bees were visited rarely, usually a few times a year. During the spring inspection, early beekeepers checked if the colonies were alive, cleaned the hive of dead bees, cut the combs for new construction and to obtain wax and prepared a place to receive a new swarm if the colony had run its course. During the autumn inspection, honey was obtained by cutting out part of the combs if there was enough of it.

Most of the oldest combs were cut from the side of the nest, only to a certain height. At least 30-40 cm were left in the upper part of the nest. The undoubted value of this way of keeping bees was low density of colonies. Thanks to this, not only did the so-called over-stocking of an area not occur (i.e. less competition for food). Also the risk of drifting was eliminated and the risk of robberies was significantly reduced. This then greatly limited the horizontal transmission of pathogens. Beehive nests and nest holes were usually located at least 4-6 m above the ground, which forced tree beekeepers to make a lot of effort when looking after them. However, the advantage was the relative peace of the bees, both on the part of

---

205 Bee enthusiasts are trying to revive old traditions, but this type of beekeeping remains a marginal activity. However, in some parts of Belarus, Ukraine and Lithuania it is believed that this is not a part of only historical re-enactment, but a notable part of local ways of keeping bees, since the tradition was kept up for generations.

beekeepers, who minimised the number of inspections, and predators or pests, for whom the bees' homes were less accessible.

The bees of our local Central European subspecies (*Apis mellifera mellifera*) have survived the last thousands of years of evolution in tree hollows, in the forested regions of Central and Northern Europe. It should therefore be assumed that beekeeping, as a practice that most fully recreates the historical direction of evolution of honey bees in our geofigureal region, would be most beneficial for them, by fully corresponding to their adaptations and thus their population health.

Keeping bees in logs was very similar to tree beekeeping. In principle, the structure of a log does not differ from the structure of a tree-hive. However, they were most often placed in apiaries where access to honey and wax was undoubtedly easier than in a hollow up in the trees, so the density of bees in one spot would be like the one in apiaries today.

Beekeeping based on wild bee combs was also carried out in skeps, i.e. baskets woven from wicker or straw[206]. Sometimes, to improve insulation, they were covered with clay or a mixture of clay and manure. Since the volume of a basket is usually relatively small, like hollows in logs and tree beehives, bee colonies often swarmed. In this way, beekeepers propagate naturally without making splits or nucs.

In a world dominated by varroa, returning to this concept of conducting a natural extensive type of management of bees would have some justification[207].

Skeps are handy, they allow for making artificial swarms and their construction also facilitates the settlement of bees inside the construction or even migrations with bees for pollination. Obtaining bee products from such constructions involves cutting out honeycombs. Often, colonies were combined in autumn, thus obtaining all the collected honey from some of the skeps. Thanks to this procedure, wintered colonies were stronger. In the following year, new swarms were caught and the apiary was very quickly replenished in numbers.

---

206 Nowadays, other straw structures are also used rather than the classic skeps, for example, the sun-hive structure is becoming popular among natural beekeepers; it is a basket woven from straw that does not lie on the ground but is suspended from ceilings (usually on verandas or under shelters).

207 Compare: the Darwinian concept of beekeeping.

Modern tree beekeeping is gaining popularity, but most often it is a purely recreational activity, photo by Treebeekeeping Brotherhood.

On the Iberian Peninsula, beehives made from the bark of the cork oak (*Quercus suber*) were popular, a very durable material with good insulating properties. Some people still use them today. Sheets of cork bark were tied with wire, creating most often cylindrical structures. However, sometimes they were squarer shaped, like the one in the museum in Portugal, photo by T. Gfeller.

Even beekeepers who run their apiaries in a standard way sometimes have logs in their garden, often decorated: carved and/or painted (sometimes they are adapted to hold frames), however, these are not classic beehive logs, but rather decorated hives. Below, the collection in the private open-air museum of Mr. Wacław Ratyński (Lower Silesia) photos by E. Österlund and the author.

Another way to obtain products from skeps was to sulphur (i.e. kill) bees in autumn. The colonies that had accumulated the most supplies were killed and then the honey was taken away. Some claim that in this way the bees were selected for lower efficiency, those that had accumulated less honey were left alive. Honey harvesting, which involves killing bees, is obviously ethically questionable. However, even the merging of colonies may raise doubts among supporters of natural beekeeping, because it is undoubtedly associated with stress for the bees, the possible reduction of fighting workers and death of one of the queens[208].

Keeping bees with wild comb can also be carried out in structures resembling ordinary hives. However, the beekeeper must be provided with adequate access to the nest. For this reason, hives often have a box structure, with a side opening for inspections like tree beehives or logs. There are also no constraints to keep bees in box-type hives without any frames. Smaller cross-section hives work better here with relatively low height and volumes i.e. the nest layout is then like a natural tree hollow.

Such a construction is characteristic of traditional Japanese hives, with a square cross-section, with the length of the side inside no more than 20-plus cm, the classic dimension being 22 cm). Each box has a small round entrance, which allows the colony to fly from any storey and ensures adequate ventilation. Crossed sticks are placed in the hive, which serve to support the comb. The advantage of small boxes is the ease of obtaining honey, even during a season poor in nectar, or in regions poor in nectar. In addition, taking one of several boxes with bees and brood can be treated as making a kind of nucleus colony, i.e. the beginning of a new bee colony. The boxes are separated using string or wire. The combs are cut by it. The inspections are performed by looking from the bottom through the combs. Like the case of skeps or logs, it is impossible to examine each comb separately. We obtain honey by removing the upper body. Then we fill the hive's volume from the bottom with an empty box.

We can conduct this type of management in almost every type of hive. In large boxes, e.g. similar in size to Dadant, Langstroth or Nationals). However, there is a risk of wild construction combs breaking off, either during inspection or transport of the hive. Large boxes are also very heavy, which makes them difficult to handle, so small boxes work best.

---

208   In the case of so-called colony combining, two (or several) completely independent superorganisms are merged into one. In this case, with the death of one of the queens, the organism ceases to exist. Colony combining therefore leads to the killing of one of the superorganisms.

Traditional cylindrical reed hive in the Spanish Pyrenees, photo by T. Gfeller.

Wild comb beekeeping is not efficient. Obtaining honey is always connected with limiting the amount of storage space by cutting out combs, which are then destroyed, crushed and strained. But this type of beekeeping is quite cheap and does not require a lot of work. In addition, assuming that it is conducted in a rational way, it will prove to be beneficial for the bees because interference from the tree beekeeper in both the nest and in the activities of the bee colony is very limited.

Bees have not yet been fully domesticated, and any interference is stressful for them. This can be apparent in their behaviour even a few days later. Due to the advanced breeding and the ongoing domestication of honey bees, they should tolerate human interference better[209]. However, the beekeeper's manipulations do have an impact on the functioning of bee colonies. The assessment of their health and the increasing mortality rate indicate that we are approaching the limits of the bees' tolerance. The management based on wild comb, which makes human interference difficult, may therefore be the answer to at least some of the problems caused by intensive/industrial beekeeping. It cannot be omitted that this type of management provides a completely different type of satisfaction from interactions with the bees.

---

209   The pressure of intensive apiary management, based on the laws of economy, is reflected in the genetics of bees. For decades, those bees that react badly to the stress of constant inspections and attempts to control their lives by beekeepers have been eliminated. Although we cannot rule out that they would cope perfectly well in natural conditions. Those colonies that, despite all these pressures, maintain vitality, dynamic development and productivity are selected for further reproduction. In this way, the process of domestication is constantly deepening.

It allows one to observe the construction of natural comb, in comparison to which the drawn-out foundation frames seem uninteresting and artificial. The methods of primary beekeeping described above are perceived today by some beekeepers as folklore or the passion of a handful of hobbyists. Yet many "natural" beekeepers run larger apiaries, even several dozens of such natural comb constructions, obtaining a reasonable income[210]. Although it is difficult to treat this type of beekeeping as the primary source of income, one cannot deny its recreational potential, which is probably its greatest strength.

Top-bar hives

Top-bars are the wooden strips from which bees start building combs. In top-bar hives, upper bars are therefore used similarly to frames, so to maintain a bee nest with combs that can be pulled out independently of each other by grabbing the bars. This requires more attention during inspections, because the comb is attached only to one bar from above, and not to 3-4 walls as it is in case of a surrounding frame. Such a construction, like any other solution, has advantages and disadvantages. Using top bars is easier than using a full wooden frame, cheaper to use, and allows bees better management of the whole hive's space. Bees have a bit more freedom and the beekeeper can perform some tasks more easily than in wild beekeeping, e.g. creating nucs, searching for the queen, feeding the colony, introducing brood, etc. Due to their simple construction, top-bar hives are easier to make than standard ones. Beekeepers often use unprocessed wood for this purpose. For the beekeeper, all you need is a box and some slats. Some people even use bamboo, or skewer sticks as the bars.

Similarly to the case of management in a regular frame hive, it is also worth having one standard size of comb, which helps manipulations between hives and facilitates logistics.

Top-bars can be used in almost every type of hive, but similarly to the case of wild comb construction in box-hives, this solution will be more practical with a smaller comb surface, because then the risk of it breaking off when loaded with brood or honey is smaller. Bees can attach a comb to the hive wall to strengthen it and they often do. So if we want to remove it, we must cut it off and thus weaken it. Therefore, a sharp hive tool, a knife, or other similar instrument is needed to remove the top-bar frames. Awkward

---

210   Sometimes the income is the result of educational activities, not the sale of apiary products.

handling can end in tearing and damaging the comb structure. Despite certain disadvantages, top-bar hives have quite a large group of supporters.

One of the most popular top-bar hives is the 'Warré hive', in the apiary of David Heaf, photo by D. Heaf.

The two most popular designs are the Warré hive, also called the people's hive (French: la ruche populaire, English: people's hive), and the so-called Kenyan top-bar hive, sometimes just called the Top-Bar Hive. The first one is a simplified construction of the box hive, created by the French Catholic priest, Father Emile Warré, to whom the hive also owes its name. The body is small, with dimensions just a little bigger to those of a Japanese hive box.

The difference is that instead of crossed support-sticks to strengthen the wild comb construction, top bars are used to hold the combs. Although a properly made construction allows each comb to be pulled out separately, in practice many users use the hive practically in the same way as a Japanese hive by not paying attention to the shape of the combs and their even distribution. However, top bars give the combs significantly greater stability than crossed stick structures.

Bees, to have their own passages, often do not build each comb to the top-bars from the lower box, which in effect makes the work of the beekeeper easier. Practitioners also point out that once the combs are cut off when the box is removed during inspection, they are not always attached back by the bees. Some beekeepers use various substances, mainly oils, to lubricate the bars from the top. This in turn makes the work of the beekeeper easier, since it discourages bees to stick the comb to the upper side of the top-bars from the lower box.

However, it is worth remembering that such a procedure can also discourage bees from moving to the lower level. Honey is collected from the upper body, and the empty box is added from the bottom. Sometimes round entrances are made in each of the boxes, although the classic design has one entrance in the bottom of the hive. Additionally, in the Warré hive, so-called eco-roofs are often used.

Keeping bees foundation-free it is worth having a tool that will allow the beekeeper to reach deeper into the hive and cut the combs, both vertically and horizontally, photo by author.

The management of the Warré hive has been given much attention[211], which is evidence of the popularity of the French beekeeper's solution with the supporters of simple and natural beekeeping. This idea was also widely popularised in the USA, after David Heaf translated Emile Warré's book into English and after the Welsh beekeeper's monofigure on the management of Warré hives was published[212]. According to practitioners, the Warré hive is one of the least demanding types of hive for home amateur beekeepers, although it is also used by professional beekeepers. One of its big supporters is Sam Comfort, a professional beekeeper from the United States, who manages about a thousand hives, of which hive constructions based on

---

211   The bibliofigurey on this subject is very rich.

212   E. Warré, L' Apiculture Pour Tous, 12th ed., 1948; E. Warré, Beekeeping for All, translated by Patricia and David Heaf, published by Northern Bee Books, 2010; D. Heaf, Natural Beekeeping with a Warre Hive. A manual, Northern Bee Books, 2013.

Warré hives constitute a large part[213]. Their design has been slightly modified and even more simplified. At the same time, it works perfectly in a decidedly extensive model of management.

Kenyan top-bar hive, photo by M. Zarek.

For me, the Warré concept is not so much a type of hive as an idea or philosophy of working with bees, in accordance with their natural life cycle. The bee colony grows by expanding the nest from the top to the bottom, as in nature[214]. As a rule, the beekeeper does not interfere with the comb construction and his only role is to add boxes at the bottom to increase the

---

213   To distinguish them from original Warre hives, they are sometimes called Comfort hives; the differences are primarily that round sticks (skewers) are used instead of top-bars. The bottom is meshed and there are small round outlets in every other box. The roof is simple with a regular cover. There is no insulated eco-cover here. Comfort makes sure that the combs in these hives are removable, to comply with legal regulations. The law in the USA does not allow the use of wild comb construction. The aim is to enable beekeeping inspectors to take out each comb individually, primarily to determine possible brood diseases, particularly American foulbrood.

214   Contrary to the principles of modern management, where additional boxes with honey supers are added above the nest to speed up occupation, the bees tolerate the empty space above them less well than when it is under the nest.

volume of the hive and to take away the upper body that contains the honey. In this way, the oldest combs in the nest are also withdrawn. The idea of this type of work is undoubtedly appealing. Although in this case we cannot talk about a very efficient type of management, one should appreciate the simple construction of the hive, the principles of management, the small amount of work and costs, as well as the minimal interference in the life of the bee colony. Beekeepers who use the Warré hives very often do not carry out any swarm control procedures. Management, including inspections is limited to a minimum.

In the so-called Kenyan top-bar hive, the side walls form a 'V' (trapezoid) shape downwards, which determines the shape of the comb. Because of that, it often happens that bees do not attach the comb to the sides of the hive, which makes it easier for the beekeeper to take it out[215]. Unlike in the previously discussed structures, in this type of hive the development of the bee colony does not proceed vertically by occupying the next, lower or higher box, but horizontally by building another comb behind the last one. Such a structure might be appreciated especially by people who have spine problems as there is no need to lift boxes. The Top Bar Hive slats are most often arranged in such a way that each of them is directly connected to the neighbouring ones, creating a kind of tight ceiling. In this way it is possible to significantly reduce the interference with the bee colony, because it is not necessary to open the entire nest. It is enough just to remove a single comb at a time. Most often the inspection is performed from the back of the hive up to the first frames with brood.

The greatest strength of top-bar hives lies in their simple construction. All you need is a regular box with access from the top and slats of appropriate length and you can start your beekeeping adventure. This type of management can also be practiced in any type of hive, regardless of the construction and dimensions. Frames in a standard hive can also be replaced with top bars.

Frame hives

Frame hives dominate our apiaries. This is primarily because they allow the beekeeper to have the most complete control over the development of the bee colony. Manipulation is made easier by surrounding the comb with a wooden frame. The comb is usually reinforced with a stretched wire which

---

215   Thanks to this shape, the comb is also stronger - at the top, it is wider and narrows towards the bottom, which makes the weight there smaller.

minimises the risk of damaging the comb. This solution makes the combs stable enough to use in a centrifuge honey extractor and after centrifuging the frame can just be put back into the hive. The possibility of reusing the comb is the greatest advantage of using frames, so it is no wonder that the method quickly gained popularity around the world, allowing for much larger honey harvests than in the case of more primitive beekeeping methods.

What is an advantage on the one hand, can be a disadvantage on the other. Hives with removable frames contributed to the transformation of the previous 'bee and nature' friendly branch of agriculture into an industrial practice. Also in small apiaries, which could have been the stronghold of bee-friendly beekeeping, they instead began using the methods of modern commercial beekeeping to increase the amount of honey extracted from the hive. The apiary management has therefore changed from extensive to intensive and at the same time more efficient and profitable. Transferring combs between hives has significantly increased the risk of epidemiological threats, intensifying the horizontal transmission of pathogens. Over the decades, this behaviour could have caused the death of a huge number of bee colonies.

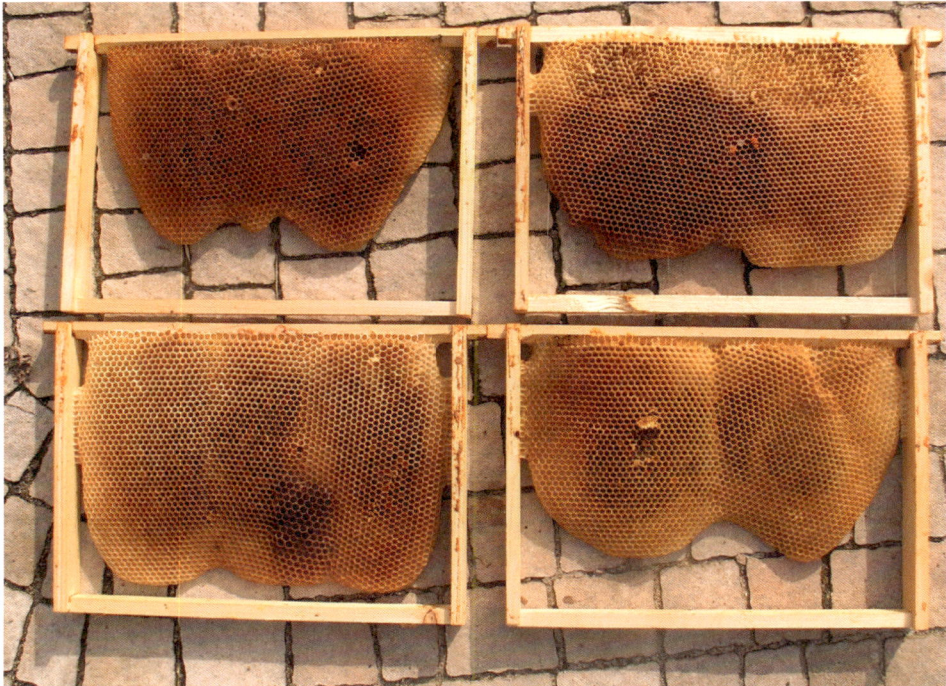

Keeping the combs straight and removable in a frame system is not a problem if we pay a minimum of attention, photo by author.

The newly built comb most often appears in several places at the same time, then the bees connect them, creating a larger structure, photo by P. Słotwiński.

Removable frames and modern beekeeping methods have also made it easier to breed bees to increase their efficiency. What was once difficult has now become easier, but it has also indirectly contributed to the deterioration of both the individual health of the bees and the diversity of the population.

Should we therefore assume that using a frame hive is harmful to bees? I certainly wouldn't go that far in my assessment. Of course, looking at it on a global scale, it must be said that using this structure for decades has indirectly caused a deterioration in the health of the bees. However, the cause of the damage should be sought not in the tool, but in the way it was and still is used. After all, it is not about the movable frames themselves, but rather the fact that the use of them has led to the intensification of breeding and apiary management methods, which together with the appearance of varroa mites has resulted in the violation of the biological integrity of entire hive ecosystems. The presence of varroa has prompted beekeepers to decide to use pesticides in their hives. The risk of rapid transmission of pathogens has required disinfection of combs and their storage in fumes of various biocidal chemical compounds, which has caused another wave of disturbances in hive ecosystems and an increase in pathogen virulence. This has led to increased chemicalisation of apiaries. So the vicious circle continues; the effect becomes the cause, and the cause becomes the effect. Beekeepers have fallen into a trap from which it is impossible to escape while maintaining the current methods of apiary management and expectations regarding honey production. Due to the methods used in modern industrial beekeeping, it is difficult for bees to find biological balance. Natural bee-friendly ways of keeping bees and rational use of frame management methods are, in turn, a simple way to minimise their stress. It is up to us how we use the tool that we inherited from 19th century beekeepers.

Frame hives, like top-bar constructions, can be divided into horizontal and vertical. Due to the proportions of the sides of the frames, we divide them into two main groups: narrow-high, where the height of the frame is greater than the width (e.g. Polish Warsaw type, i.e. Dadant vertical) and wide-low, in which the height is less than the width (e.g. Dadant, Langstroth, National).

In addition, hives with square or trapezoidal frames are also in use. Some hives that combine different solutions are also in use, e.g. a nest built on narrow-high frames, top-bars or on wild comb construction, and honey supers on wide-low frames. All of them have advantages and disadvantages. This is not the time and place to enter considerations on these various structures and their significance for bees and apiary management, because I intend to focus on matters around the issue of foundation-free management.

I would also like to mention a rather unusual solution, which is a frame without a lower bar, combining some of the advantages, but also some of the disadvantages, of frame and top-bar hives. Charles Martin Simon[216], a deceased American beekeeper, drew attention to this and claimed that he "invented" them in a landfill with decaying beekeeping equipment. These were frames from which the lower cross-bar had fallen off due to rotting. The advantage of this solution is that it reduces the space inaccessible to bees between the hive boxes. In a standard hive with a box or super, looking from below, the bees have above them the upper cross-bar of the frame, the space between the frames, the lower beam and then often an empty space under the comb, which they do not fill with wax. For this reason, a space 3-4 cm high is created between the combs, which the bees cannot use directly. This usually does not bother strong colonies. However, the weaker ones reluctantly pass between the boxes, which can stop their development during the season.

---

216 Although my beekeeping was shaped by a more practical model, Simon was something of a discovery for me. He helped me look at bees in a different way; https://beesource.com/point-of-view/charles-martin-simon/, accessed: January 27, 2025.

Frames without a lower cross-bar allow bees to extend the developing combs, photo by M. Adamczak.

It also happens that if we do not manage to persuade bees to move between the hive boxes in the summer, they will rather swarm from one box than occupy another. In the winter, this can even lead to the death of the colony[217]. If the food is not properly distributed, in unfavourable conditions, they will not move to the upper box. Then even if there is honey in the combs they may starve to death. This is of course a worst-case scenario.

In most cases the bees deal with this problem themselves, or the beekeeper helps them. Nevertheless, the solution I am mentioning minimises the negative effects of this type of situation. The side slats of the "frame" prevent the comb from attaching to the hive wall and therefore makes it easier to get the frames out than in the case of management with top-bars.

---

217   This is sometimes referred to as the artificial bottom or artificial ceiling effect, depending on which way the colony is developing. Some small colonies, having a gap of a few centimetres between the hive boxes above them or below them are reluctant to occupy another hive box. Heating the brood also requires heating the space between the boxes, which may need some effort. Bees limit their development to one box and often, if the beekeeper does not move the frame with the brood to the next one, they will not occupy it. Larger colonies, occupying at least two low boxes, usually have no problem moving to the next.

Management without foundation in a frame hive

Adding empty frames

I do not intend to teach anyone the methods of apiary management. I would only like to suggest how to induce bees to build a straight comb in a frame. Taking into consideration some specific differences, this advice applies either to the top-bar hive or the frame hive, because a straight and mobile comb is one of the foundations of both types of apiary management. I would also like to point out that for several years I have exclusively been using foundation-free frames of various sizes and types, as well as frames without a lower beam, with most of them being unwired.

If the bees live in already constructed nests with drawn comb, changing to a foundation-free system can be done easily and painlessly for the beekeeper. If we move the combs apart to the right distance and place an empty frame in between them, the probability that the bees will draw a straight comb is high, although it won't be 100%. A similar effect is when we put an empty frame next to a hive wall, because bees most often build parallel to the existing "barriers", while maintaining the appropriate distance between the combs, which is usually 8-9 mm. Although this distance may be different in the areas where honey is stored.

This gap which is called the bee space[218], allows the bees to move freely and perform their hive duties such as feeding brood, constructing cells, etc. Unlike the nest for egg laying and brood, an empty space left in any part of the hive that bees by themselves recognise as being for honey storage will probably be constructed by them with the rules of bee-economy and colony

---

218   It is commonly believed that 'bee space' was first discovered by Lorenzo Langstroth, often called 'the father of modern beekeeping'. He realised that bees would build comb in any space larger than three-quarters of an inch (about 22 mm), but will usually fill with propolis, or propolis mixed with wax, any space smaller than one-quarter of an inch (about 6 mm). Therefore, he recognised that space between these values will be perceived by bees as free space and will be used as a communication route or a place for work (e.g. handling brood, drying honey, etc.).
This discovery, among others, was the basis for the concept of the modern frame hive in 1852. Independently, similar observations at the same time were made Polish priest Jan Dzierżon (1811-1906). He said he was a Pole by birth and German by upbringing and education, as he lived in the then German city Breslau, presently Wrocław, Lower Salisia, South-Western Poland. He came up with a construction like a frame hive even as early as 1835, on which as it is believed the modern construction of Langstroth's hive was based. What is also interesting is that Jan Dzierżon discovered that drones are produced from unfertilized eggs, which was then a revolutionary thesis.
See: https://en.wikipedia.org/wiki/Johann_Dzierzon, accessed: February 19, 2025.

efficiency, but not according to what beekeeper would hope for. For example, during the honey flow, bees strive to quickly increase the storage volume and can therefore extend the cells of neighbouring combs. In this way, using less wax, they very quickly increase the possibilities of collecting more nectar or honey. It may also turn out that some empty spaces are filled with new comb, but other places, most often in the upper corners of the frame, will be taken by the prolonged/extended cells of neighbouring frames. So, if an empty frame is placed in an area that has already been designated by bees for storage, even with limited nectar flow, it can be drawn not according to the expectations of the beekeeper.

When moving the combs apart and placing an empty frame inside, the beekeeper assumes that the new comb will be straight. However, we should not be surprised if it happens differently. Bees function according to their own logic and they can attach the comb anywhere, for example by connecting it to other combs with bridge-comb. The frame bar is therefore not necessary for them to uphold a new structure. By appropriately adding empty frames in the nest, rather than in the honey super, the problem will most likely cease to exist. Using this method, there is also no need to prepare the frames so that the bees build their constructions as we expect. Having built combs, we can add another box to the hive under the existing nest with frames arranged alternately empty and drawn. The bees will occupy it when the colony develops and they will probably not use it for storage.

In the same way, we can expand in horizontal hives, by adding on to the side of the nest several frames in an alternating arrangement, one drawn, one empty. However, it would be best to add frames closer to the exit, e.g. moving the entire nest away from it. This is because adding empty frames at the back of the nest can lead to the elongation of the cells in the existing combs.

A similar method of enlarging the nest can also be used to prevent bees from swarm preparation. For example, when there is a high probability of a swarming, we can move every two or three frames with the brood apart, placing an empty frame in the gaps. In this way we can kill several birds with one stone. Firstly, the combs will be straight and will be drawn very quickly, because the bees do not tolerate empty spaces in the nest. So the queen will lay eggs in the newly built cells in no time. Secondly, the nest expands and the bees, having a lot of free space, enter or remain in a 'working mood', building new cells and rearing brood. For natural beekeepers, however, the method is controversial. Some, as a rule, very rarely, or never, separate the brood with an empty frame, treating the brood nest part as sacrosanct. It must also be remembered that such a procedure may threaten the brood with cooling during a break in the weather. Therefore, if using this method it is worth considering this in advance, choosing the right time of the season and adjusting the number of empty frames and their location to the external conditions and the size of the colony.

Placing the colony on empty frames, expanding the nest

In a situation where we want to settle a swarm in an empty hive or create a small nuc, and we do not have drawn combs so obtaining straight and even combs may not be easy. So we must prepare the top bars of the frames in such a way that they will encourage the bees to follow our ideas of producing parallel combs built in a straight line.

If we provide the bees with a new hive body with empty frames, without any guidelines, the first parts of the comb will be built attached to the wooden slats, but then they will start to bridge and distort the combs. If we leave such a structure without intervention, we will lose the possibility of easily removing individual frames from the hive, photo by author.

Frames for foundation-free management with "guidelines" for bees; from the left: a strip of wax left, a strip of foundation, a wooden slat glued to the upper beam, the crossbeam planed into a point (V-shaped), photo by the author.

I must admit that it is not always possible. Sometimes if we want the combs to be straight, then during the inspection we would have to cut them off on some part and bend them appropriately, or even cut off some pieces of wax. In any case, the key to success is either the use of division boards or preparing frames appropriately and have more precise upper bars.

Frames with strips of foundation

Instead of a whole sheet, we can attach a strip of foundation to the upper beam (1.5-2.0 cm is completely sufficient). Thanks to this, the comb should be drawn in a straight line. However, this method is laborious. We must place the strip of foundation in a wired frame or glue it with wax. However, as a result, we will get what we want, i.e. combs evenly placed in frames, built of natural-sized cells and pure wax, the kind that is best for bees.

Slats glued to the upper bar of the frame

This method is quite time and labour-intensive and consists of gluing, preferably with wax, 1.0 mm thick and 0.5-1.0 cm high strips to the underside of the upper beam of the frame, along its entire length. I have only tested this method once and I am not convinced. Although the combs built by the bees were nice and even, I am not sure whether this was the "merit" of this method. It cannot be ruled out that it happened due to other favourable circumstances.

The upper bars are planned to a point

I use this method in most of my frames. The bees start building from the tip, which is the lowest line in the underside of the upper frame bar. In fact, the bees almost always start building the combs from the tip of the upper bar. However, if the nectar flow is good, each subsequent comb will be more and more bent and distant from the frame axis due to the elongation of the cells of the previous frame. Preparing such bars will not be a problem if we have a hand-held electric planer, but it requires work. The upper beams must also be thicker than the standard ones. Minimum dimensions should be 1.3-1.5 cm and it is best if the cross-section is square with a side of about 1.5 cm).

The upper bar of the frame is set vertically

For this method the height of the upper beam should be greater than its width. Most frame slats are 2.0-2.5 cm wide and about 1.0 cm thick. If these proportions are reversed, the distances between successive beams will automatically increase. Bees are more likely to start building along the frames and not across them. The comb will also be stronger if more of it is attached to the solid upper frame bar. However, in this method, bees will most likely build comb on the sides of the frame bars creating wax bridges above the frames, which increases the risk of bees being crushed during inspection. On the other hand, such "bridges" make it easier for bees to move between the boxes.

Leaving a few rows of old comb cells

When we have rebuilt combs, when we want to remove the old wax from the hive and let the bees build new wax, we leave a few rows of already constructed cells under the upper beam. To be on the safe side, we can also leave a few rows of such cells at the side beams of the frame, where the combs most often built by the bees have curves. Then we can be most certain that the comb will be straight and even and will fit in the frame. This method is the easiest, the least laborious and prone to producing irregular comb.

Attaching the comb strips to the frame

A variant of the previous solution is to attach strips of comb to the upper beam of the frame. Then we only need one drawn comb for several new frames. So, we cut the comb out of the frame leaving a few centimetres at the top and then cut it lengthwise into two 3.0 cm strips, which we "glue" under the frame beams, e.g. with hot wax. The bees will quickly re-attach everything and construct the comb in accordance with the direction of the strip.

Other advice

First, it is important to realise that bees build the combs they need, in the way they think is right. It is amazing how many beekeepers think that a hundred years of providing wax foundation to bees contributed to the fact that they lost the ability to build combs in accordance with the needs of the swarm. They claim that if we do not add foundation, the bees will create strange structures or will only build drone cells. In my opinion, such a view rather indicates that the beekeeper has lost the ability to understand the natural needs of bees.

Bees will fill any empty space left for them in the hive, photo by author.

A common practice when transferring bees from a wild nest to a hive is to cut out combs and place them in frames, photo by P. Kasztelewicz.

If bees used combs only for brood rearing, it would be enough to place one comb in the middle of the nest. The rest would most often be drawn parallel to it, i.e. straight in the frames. Most curves in the combs are created in the storage area, where the bees store honey. My experience shows that the problem of bent combs is significantly larger in smaller colonies which do not have their nest cavity naturally divided into a brood chamber and storage space. Although, it cannot be ruled out that this is because a smaller colony, no matter whether it is a nuc, a packet or a swarm, suffers from a shortage of combs. If the volume of the nest is limited to a few drawn combs, the queen lays eggs in the middle of the comb and the bees store honey on the edges and in the upper part of the comb. When the honey flow appears, they need to rapidly increase the surface area of the larder to ensure they have enough food for the colony growth. A small colony, for purely economic reasons, begins to lengthen the cells in which it stores honey. When constructing another comb, they build it at the appropriate distance from the previous one. However, because the line of last comb runs in an arc, the curvature is repeated on subsequent combs. During rich flow, the queen bee will start to lay eggs in the middle of a new comb, and its sides will be increasingly distant from the expected frame axis. These remarks apply mostly to management in the vertical box hives with supers with the frame in a wide-low arrangement. A mature colony has a larger nest with many cells at the outset. Due to the number of bees, it also has a significantly greater potential for rapid construction of new combs in empty spaces in the hive.

Maintaining a straight line of combs in large frames with a narrow-high arrangement, e.g. vertical Dadant frame, is probably even easier. Bees always have the possibility to expand the nest downwards. When adding cells down the comb, they designate them for brood, not storage space. However, this forces the beekeeper to level the hive so the comb will be built vertically downwards, regardless of whether the frames are arranged in this way or not. This is particularly important with high frames.

Honey is deposited in the upper part of combs, most often on previously built, suitably arranged combs. Moreover, it is easier for a beekeeper to use a division board if he uses a large and high comb than if he were to use a small frames. In the latter case, limiting the volume of the colony could easily end in swarming. On the other hand, a comb in a small wide-low frame is less susceptible to damage due to its shape and size. To effectively manage without foundation, one must learn the basic principles of comb construction and adapt them to a specific type of hive.

In the case of larger frames, to be able to work freely, we will probably have to strengthen them by wiring. In my experience, a comb with a surface area like National or Langstroth hive requires reinforcement to eliminate the risk of damage. In most cases, when working on unwired vertical Dadant frames, I had no problem, but freshly drawn combs heavily loaded with brood or honey broke off several times. A quick balance of advantages and disadvantages of wiring frames shows that I cannot give up this activity, which I particularly dislike in the case of large frames. However, I found more friendly solutions such as reinforcing the combs with bamboo sticks which can be used for frames without a lower beam or dividing the frame with a wooden strip (approx. 1 x 1 cm) about halfway up. This allows me work faster and much more pleasantly.

Each bee colony has its own style of building combs. Some build combs that will always be slightly wavy or uneven, while others build them evenly, regardless of many factors, including the season, the size of the honey harvest, or the size of the colony. Michael Bush, thanks to whom I discovered the advantages of working without foundation, believes that it depends on the genetics of the bees. In my opinion, it is worth making the effort to learn how to work without foundation, especially in small, home hobbyist apiaries that are not focused on migrating with bees to increase their honey production.

It is worth reinforcing large frames: on the left a frame with bamboo sticks, on the right a wired one, photo by author.

Honeycombs in the Einrambeute hive in the apiary of David Heaf reinforced with horizontal sticks, photo by D. Heaf.

The weak points of natural combs can be accepted, and besides, they are easy to eliminate. If we are late with the inspection and the bees have constructed some unusual structure that cannot be corrected, we can move it next to the hive wall so that the curves are not repeated. However, if the curves really bother us, we can make a nuc from the comb and use it to collect honey or move it to a place where the queen bee is reluctant to lay eggs. After the bees have hatched we can withdraw it from the hive, obtaining clean wax. Of course, we can simply accept this shape of the combs. If they do not bother the bees, why should they bother us?

Honey extraction from foundation-free combs

Honey extraction by spinning

Many beekeepers claim that combs constructed without foundation or reinforcement with wires cannot be centrifuged in a honey extractor. This opinion is mostly expressed by those who have never tried it. It is true that comb without foundation and not wired, is more fragile and susceptible to damage, especially if the comb is loaded with honey. However, this does not mean that it cannot be centrifuged safely, especially if the comb is attached to at least three frame slats, the top and side ones. The comb is also strengthened with each subsequent brood rearing cycle. A comb that has already been used several times is not easy to damage. Of course, it will be slightly weaker than the same comb drawn on foundation and reinforced with wire, but still strong enough to thoroughly empty the cells at full speed after initial centrifugation at low speed. Today, most honey extractors are equipped with a mechanism for precise control of the rotation speed. It is sufficient to increase the speed slowly and control how much the frame can withstand. With careful spinning, the combs should not be damaged. I have many experiences of doing this. For example, I was spinning newly built combs in a four-frame diagonal honey extractor, constructed in 'Wielkopolski' frames (standard and 18 cm high). I increased the speed gradually. I collected the remaining honey from almost empty cells at maximum speed. It simply took a bit longer. I realise that in a professional apiary this would affect the efficiency of work. I do not think that such methods will be a problem for a hobby beekeeper. However, they will be troublesome for a beekeeper who is accustomed to handling frames without taking excessive care.

Section honey is gaining popularity, photo by Ł. Łapka.

Honey squeezing and section honey

Obtaining honey by squeezing and draining has several advantages. First, a beekeeper running a small apiary does not have to invest in a honey extractor. To obtain honey, all you need is large pot or a bucket, a tool such as a potato masher and a large sieve or colander. Cut slices of comb into a container, then finely chop or crush them and pour the whole thing into a sieve. The honey will drain into the container within a few hours, and the sieve will contain wax with honey residues.

Place the crushed honeycomb on a sieve and wait until the honey flows into the container, photo by D. Heaf.

Thanks to the use of presses, less honey remains in the wax and the process of obtaining it is faster, photo by D. Heaf.

Some emphasise that squeezed honey wins in terms of quality over spun honey. During the spinning process, many valuable chemical compounds evaporate from the honey[219], especially aromatic ones, which determine the quality of the honey. Many valuable qualities such as taste, smell and a wealth of health-beneficial chemical compounds may be lost or reduced. During squeezing, particles of bee bread contained in the combs also get into the honey, which increases the nutritional value of the product and its taste. However, this method has its opponents, who prefer clear honeys, often filtered on fine sieves.

Honey can also be obtained by cutting out fragments of the sealed honeycomb, then we are dealing with section honey. This is a way to obtain the purest and most valuable product, because we do not lose any valuable chemical substances that are enclosed in wax cells. For this reason, comb sections have a large group of supporters.

You chew the honey together with the wax and possibly the bee bread. You can also spread the comb sections on bread. The wax, although not fully digested by the human body, is not harmful. However, you should approach this with caution if you do not know the source of the wax as there is a risk that it contains pesticides.

The disadvantage of this type of honey extraction method is that it means destroying the work of the bees: to collect more honey the bees must build new combs. Removing the combs reduces the storage volume of the hive and can overall affect the amount of honey harvested.

Some beekeepers limit the bees' comb-building to the bare minimum of no more than 2-3 combs per year per bee colony. They claim that bees use several kilograms of honey to produce one kilogram of wax, so multiple use of combs should increase the harvest by reducing losses. This may be true, but it must also be remembered that regardless of whether the bees have space to build combs or not, they will produce wax scales anyway. Although this would be to a much smaller quantity than if they need to draw comb. Some are concerned about keeping older combs in the hive due to the risk of pathogens. They believe that systematic recycling of wax in the nest can have a beneficial effect on hive hygiene. Thus, it turns out that crushing and

---

219   Centrifugal force throws honey out of the cells of the combs; it hits the walls of the centrifuge and breaks into droplets. The less liquid there is in a droplet, the greater its surface area to volume ratio. The droplet size decreases with the increasing force of liquid hitting the wall. Small droplets therefore have a relatively large surface area, which affects the faster evaporation of various volatile chemical compounds that constitute the exceptionally valuable properties of honey and its rich aroma. For this reason, it is better to obtain honey by crushing the combs, because then the evaporation surface is smaller, and many valuable chemical compounds remain in the honey.

straining honey has another advantage, if the combs are not put back to the hives, we reduce the risk of horizontal transmission of pathogens.

Efficiency of an apiary run without foundation

Opponents of the foundation-free beekeeping often give the argument that it is not very efficient. They also claim that obtaining honey requires a lot of effort. In my opinion, these accusations are unfounded. It is true that my apiary is far from efficient, but I am not convinced that a lack of foundation affects my honey harvests.

Undoubtedly, management based on natural comb or in top-bar hives, where we remove and destroy the combs, is less efficient than that in which the combs are centrifuged and then placed back in the hives. In addition, it is difficult to conduct migratory beekeeping when using foundation-free big surface frames. However, the lack of foundation has no direct effect on this, only indirectly. It completely does not matter for the bees into which kind of combs they collect nectar. By providing foundation, we provide the bees with building material. Theoretically, they can build more combs at the same time. It seems, however, that if we have pure wax and do not use any chemicals or pesticides in the apiary, there is nothing to prevent us from using such combs for longer than those drawn on foundation[220].

---

220   In my opinion, fears about pathogens present on/in combs should be considered largely unjustified. Every year, this is proven by long-lived wild colonies, whose combs are never replaced. M. Lopez-Uribe's research also shows that the level of infection with viruses is not as important for the survival of colonies, as the level of immune response. Certainly, viruses associated with bee mites (including DWV) present in combs should not pose a real threat. Since colonies free from mites are rare, it should be assumed that, in principle, there are very few colonies free from DWV either. The most dangerous are primarily viruses "injected" by mites, so it is more reasonable to fear excessive growth in the mite population, not viruses on older combs.

MY APIARY

Chapter 4

MY APIARY

99 *My intention is not to return beekeeping to the dark ages, but to take stock of what works and what doesn't and to mix and match methods toward the goal of maximum bee health rather than maximum production. Bees are incredibly powerful creatures. Given half a chance, they are unstoppable*

CHARLES MARTIN SIMON[221]

History of selection

I do not believe that the results of my work should be treated as an argument for giving up chemical treatments (the examples I present in other chapters are more convincing). I also know beekeepers – who are also from Poland - whose results are more promising; they record lower bee mortality and significantly higher honey production than my own. However, I decided to devote some considerations to my situation, to show my choices and perspectives.

I bought my first two bee nucleus colonies in 2013. Before that, I read a book about apiary management and attended a meeting of the local beekeeping association. At that time, I had no idea what beekeeping was like. Like most people, I didn't connect it with an industrial branch of agriculture but rather a hobby activity in nature, e.g. in a home garden.

From that time on, I began to take a serious interest in beekeeping reding more and more about the subject. One of my first major concerns, amongst others, was *Varroa destructor*. All sources I came across at that time presented a clear idea that it is impossible to keep bees without controlling the mite. I accepted this and in the first year I bought Apiwarol

---

[221] Ch.M. Simon, Bottomless Beekeeping, https://www.beesource.com/threads/bottomless-beekeeping.365777/,accessed: January 28, 2025.

tablets (with Amitraz). I burned at least two or three in each hive, according to the instructions on the labels and advice of more experienced beekeepers. Despite this, it was hard for me to get rid of doubts whether what I was doing was right and consistent with my approach to nature. So, I started searching on my own. Quite quickly I found a lecture on YouTube about a small cell (4.9 mm), where it was suggested that using foundation with this size of cell would allow keeping bees without chemicals. Even though it turned out not to be true, this was a turning point in my thinking. I became hungry for knowledge about natural beekeeping and began searching in earnest. At first, I came across Michael Bush's lectures, then his website. I was amazed to find that there was a whole field of knowledge about beekeeping, which for some reason is not talked about in my country - Poland. After some time, it became clear to me that people who openly share their experiences in this matter are united by their point of view on the issue. The accounts differed only in the details, such as the location of the apiaries, the local subspecies of bees and their views on nature and beekeeping were manifested in specific practical choices. Some ran a completely normal apiary, close to intensive, while others allowed the bees to live freely and only rarely or never opened the hives. So, I realized that it was enough to follow a few basic rules, and the rest was just perseverance and determination.

The year 2014 started very badly for me. In the first days of January, while working with a saw (which I bought when I started working with bees), I cut off three fingers of my left hand. The doctors (one of them was my brother, actually) managed to sew two back on, however, months of rehabilitation awaited me (and in this case, the saying that every cloud has a silver lining applies). I used this time, among other things, to gain knowledge about natural beekeeping. I entered the 2014 season with many more thoughts in my mind. I decided to stop using synthetic pesticides to treat bees for varroa. I was already aware then that sooner or later I would base my beekeeping solely on natural selection. It happened even faster than I had expected. The package of oxalic acid that I bought at the end of 2013 is still lying untouched in the depths of my garage. I also bought; I am ashamed to admit it a bottle of RAPICIDE ™. At that time, this product had a good reputation among some practitioners and beekeepers assured that it was much less harmful than other synthetic pesticides, while still being very effective. I never used it. Fortunately, I managed to dispose of it.

Contrary to the optimistic news from the west, I knew perfectly well that the mite pressure was too great to simply stop treating the bees and run the apiary as beekeeping textbooks teach. I knew that I had to reach for some kind of initial selection method (the so-called transition period) to reduce the mite pressure, without killing all the mites or my colonies. Pressure is essential to achieve progress in selection, as it allows us to weed out (through natural or breeding selection) those bees that cannot cope with it. Enriched with this knowledge, I decided, first, to use a small cell comb in the apiary; second, to make numerous splits (which would cause the mite population to divide, and thus temporarily weaken the parasite pressure); third, to introduce a break in brood production, due to the beneficial effects of this procedure. In addition, in the first year (or within two years) I planned a treatment with powdered sugar[222].

I also quickly gave up on using 4.9 mm cell foundation instead opting for a naturally built combs, which I am faithful to today.

I was aware that two or even five bee colonies were not enough to be optimistic about the selection results. This thought motivated me to increase my number of colonies. In early spring, I bought a few more colonies, and in May, a few more nuclei, by the end of summer, I was preparing 23 bee colonies for winter on two bee yards. However, the end of summer did not inspire optimism. Despite using the method with powdered sugar (I admit that I did it without any conviction), I noted the collapse of the first colonies already in August. Each time I visited the apiary, I found more empty hives, there were also some surrounded by wasps trying to break in. Most of the colonies collapsed in the autumn. I noticed the death of the last, my 23rd colony at the beginning of January 2015. It was a huge blow to me.

I started the next season with new hope by buying several new bee colonies that had overwintered. Previous experiences, contrary to what one might think, confirmed that I had made the right choice. The collapse of my apiary was not the only case in the region. Beekeepers who had been meticulously carrying out the recommended treatments also lost all (or almost all) of their colonies. The average loss rate that year in my association was 70%, and in the surrounding one was said to be 90%. This was almost unbelievable, but we all experienced it then, and as far as I know this was locally the worst year ever. Since the first colonies had already died by the end of August and most of them were lost by the end of October, I had reason to believe that treatments applied in the autumn would probably not have helped much. The alternative to completely abandoning the idea of treatments was therefore not a single treatment in the autumn but controlling the mite population throughout the entire season. I did not consider this

---

222 Sprinkling bees with powdered sugar is supposed to irritate them so that they start cleaning themselves (groom), throwing off mites, this stimulates hygienic behaviour.

a good solution, so I really had no alternative.

The 2014 season brought many new thoughts and equipped me with several important ideas. Firstly, I had kept rather strong colonies that year (in the following years it will be the other way around). In addition, used some of them in the production of herbal honeys.

It also turned out that many smaller colonies (nuclei) that I did not feed lived longer than those that had looked quite impressive during the summer. I was enriched by further conclusions: bee colonies die regardless of whether they are strong or weak, and whether they are fed during periods of no nectar flow or not. I do not deny that diet is one of the key elements for maintaining the health of bee colonies (in fact I believe that the problems with forage are one of the most important reasons why I have poor results here), but feeding with artificial food does not meet these assumptions. Conversations with beekeepers confirmed my belief that it is very difficult for them to go beyond the stereotype popularized in apiary management textbooks or resulting from many years of experience in the struggle to increase honey production. Situations that occur in untreated apiaries are difficult to understand until we ourselves stop treating the bees. Only then can we be convinced that regardless of what we do, nature has its own key to selection (perhaps bees do not read textbooks on apiary management?).

In the next years I acted more boldly, I split colonies, creating more smaller nuclei than in the first year (one, two or three-frame nucs). I was also less willing to supplement their food supplies during the periods of nectar dearth. I reacted only when the colony was on the verge of starvation. From mid-July I also fed the nuclei, as part of preparation for wintering[223].

Year 2015 was also a time of taking up beekeeping initiatives. Together with several people, we established the Natural Beekeeping Association "Wolne Pszczoły" ("Free Bees").

We started exchanging experiences and promising each other help in case of bee colonies falling apart. This is how the "Fort Knox" project (https://bees-fortknox.pl/) was created, consisting of providing each other with guarantees in case of losses of bees given over to natural selection.

In 2015 I wintered 34 treatment-free colonies and nucleus hives and about 70% of them survived. With about two dozen overwintered colonies, I was convinced that I would easily be able to split them and stabilize the

---

223 I am an advocate of slowly feeding colonies in preparation for wintering and I do not practice supplementary feeding "for development" with later flooding them with large amounts of food in the fall. I believe that bees should "feel" that they are building a stable supply before wintering.

size of the apiary before the most difficult period entered, to minimize the risk of inevitable losses. I expected that sooner or later my apiary would face a crisis, when the mite population will grow to a level that would break the bees' immune barrier. However, the season surprised me and was another lesson. Nothing worked: individual colonies would collapse even during the season, queens died in the queen cells, those that hatched did not mate, colonies were robbed, the season was poor, and the mite situation proved to me that both I and my bees still had a very long way to go. By the end of the year, I had not even managed to double the number of colonies and overwintered 41 of them in five bee yards. My pessimism deepened day by day. As it quickly turned out my intuition was right.

In the spring of 2017, I had only one colony left[224]. Although I had lost almost my entire apiary, my situation was not as bad as it had been two years earlier, for five reasons. Firstly, I had gained new insights and experiences. Secondly, I had collected quite a lot of beekeeping equipment that I could use to compensate for the losses. Thirdly, the one surviving colony began to develop relatively well offering (as I had hoped) valuable genetics. Fourthly, I was a member of the Fort Knox project, so I could count on help. Fifthly, I finally received the gift of an overwintered colony from Łukasz Łapka (www.llapka.blogspot.com), which had already survived 2 full years without treatment.

Additionally, I bought some overwintered colonies, which I placed on my foundation-free, chemical-free combs. From that time on for a few years I noticed the growing improvement of the bees health and vigour, even though the honey crop for me was most often poor or no existing and the losses were greater than I hoped for and expected.

Each year, even if colonies didn't overwinter in much strength and were too weak to build up for spring harvests, they developed dynamically in relation to their size and conditions later in the year. The colonies did not die off during the summer season, and for many years each subsequent autumn there have been fewer weaker colonies.

In the years 2018-2021, I prepared about 40 colonies (mainly nucs) for overwintering, and the mortality rate in my colonies oscillated around 50-60%[225].

---

224 The 2016/2017 winter was difficult, also for other beekeepers, the losses of bees in the region were very large then (estimated around 50%), although not as tragic as the winter of 2014/2015.

225 See: https://pantruten.blogspot.com/p/zestawia-podsumowania.html, where I document the selection process and share my experiences and thoughts on beekeeping.

In the autumn of 2020, I moved from the Kraków district to the Limanowa district (about 100 km apart), taking some of the colonies with me. I left the rest in their previous locations. I hoped that will change my beekeeping situation for the better. Where I live now it's a mountainous region with a lusher environment, with less farming and no large monocultures and more diversified forests. However, each year I've noticed a worse survival rate in the new places compared with the previous one. Whereas in the previous location it has not been satisfactory, in the new one it almost always has been close to complete die off during the winter months. I cannot tell for sure what is the reason for that, but of course I have my hypothesis.

Firstly, I observe that year by year the weather is the reason for rather bad nectar flows and poor forage opportunities for the bees. This is often linked to the climate change which is making very poor conditions in Poland in some regions (or some years). Although some of the beekeepers have quite a reasonable harvest, usually it comes from short and very intensive flows (followed later by prolonging dearth). Colonies that are kept in an intensive way – so very strong for all the season – may use such flow and give good surplus for beekeepers. This however means that the rest of the season beekeepers must feed bees with sugar, prevent robbing and constantly fight bee diseases that spread quickly in such conditions. Of course, this also means high loses for some of the beekeepers – treating or not.

Secondly, in Poland – and especially in southern regions – populations of bees grew rapidly for more than a decade. According to official data bee density in Poland more than doubled since 2006 (then the average was on the level of 3,5 and now it exceeds 7,5 colony per km^2) and it grows constantly up until now. I've found data, that in 2023 in the United States of America there was between 2,5 and 4 million western honey bee colonies, whereas at the same time in Poland there was 2,35 million. The territory of the USA is 30 times bigger than Poland (even if we agree that many regions in USA are not good enough for keeping bees, this is still a great disproportion).

The population of bees is not evenly distributed throughout the country. According to official data in the region of Małopolska, where I've lived before and still do after I have moved, in 2023 honey bee density reached 13,9 colonies per km2[226]! Knowing the surrounding areas of where I used to live that had around 10 colonies per km^2, to where I live now, which appears to have up to twice that number! That is an enormous epidemic

---

226   https://pasieka24.pl/index.php/pl-pl/aktualnosci/wiadomosci-z-polski/4374-w-polsce-przybywa-pasiek, accessed January 28, 2025.

threat, especially combining this with the poor foraging opportunities which weaken the bees' immunity.

Since 2024 it is more and more spoken of a crisis in beekeeping in Poland, and many beekeepers declare that they will stop keeping bees or at least decrease their apiaries. Maybe in the long run this crisis will adjust the bee population and density to more realistic levels in line with the forage opportunities.

Thirdly, Poland is said to be one of those countries that imports the most foreign bee-queens from within the European Union. Although I don't know specific data, I can clearly see it in the queens-for-sale offers. For example, one of the apiaries a few kilometres from my house offers 12 (!) different lines of bees including: artificially created race Buckfast, subspecies from Africa *A.m. sahariensis*, Iranian *A.m. meda*, Turkish *A.m. anatolica*, Primorsky (Russian), *A.m. ligustica* (Italian), Elgon and few other lines of *A.m. carnica* bees from different countries in Europe. I don't know the reason for that phenomenon (other than business and marketing – so maybe this is the reason enough for some?). Beekeepers seem to believe that if they import queens their situation should improve – although it may be helpful for some and may induce the heterosis effect in the first generation, in the long run for us all, it's the other way round. Although this practice is disastrous for the local adaptation of bees, it flourishes usually with the tacit consent of bee authorities or at least understanding the need for improved production for some at the expense of others and bees' adaptations.

Of course, the direct cause of my untreated bees dying can be deduced from those. For example, in the year 2023 was an enormous dearth and hunger for bees in my region. For the first time I was forced to feed bees throughout the entire season even in May, which usually brings an abundance of nectar and pollen. I heard from beekeepers that such a nectar dearth occurred across the entire southern belt of the Polish Carpathian region (beekeepers in the Czech Republic and Slovakia also complained), while in the central and northern part of the country the season was very good and for some even record-breaking. However, my colonies located in Limanowa County developed very poorly due to the year-round pollen and nectar shortage, with my colonies reaching only half or two thirds of the strength of colonies located in locations near Krakow. After that year the overwintering rate was exceptionally bad, and – again, as in the first years after skipping the treatments – my losses were almost 100% (in fact only one extremely poor "colony" survived the winter, but in spring what was left of it was just a queen with few workers – I saved it by adding some of the worker bees from my friend). This, however, was again not just my situation –

in my local association the losses were 50% even when treating the bees for varroa – many of the beekeepers lost all their colonies. At the same time, I have many friends and acquaintances who have reasonably good survival rates while doing more-less the same that I did in my apiary. As a rule I can tell that most of them live in a region with better forage opportunities (without long dearth periods during the summer) and with a lower density of bee colonies.

I went to a bee conference in Austria by bike, photo by author.

A lot has happened during those years. In 2018, I participated in two international beekeeping conferences, presenting our "Fort Knox" project. In the spring, a small conference entirely devoted to treatment-free beekeeping took place in the Austrian town of Neusiedl am See. The meeting was organized by a local beekeeper, Norbert Dorn. At the end of the summer, I had the chance to participate in a conference organized by the Natural Beekeeping Trust in Doorn in the Netherlands[227]. These were opportunities to talk to people whose experiences and breeding work I had followed with interest, but whom I knew only from their publications.

---

227   See the reports on the conferences: https://bractwopszczele.pl/art/art3.html; https://bractwopszczele. pl/art/art4.html; accessed February 19, 2025

In 2018, during the "Learning from the Bees" conference in the Netherlands, I met Prof. Seeley, with whom I managed to talk about Darwinian beekeeping, photo by author.

Later because of differences in the views on the direction of the association's development I decided to leave the 'Wolne Pszczoły' association and together with several beekeepers, we founded an informal group 'Bee Brotherhood' [Bractwo Pszczele] (https://bractwopszczele.pl/), adopting as our goal the education of beekeepers in the field of natural beekeeping and convincing them to give up treating colonies for varroa and subjecting them to natural selection or varroa resistance breeding. During several years of activity, I translated many texts of non-treating beekeepers from English into Polish, wrote many articles for the monthly "Pszczelarstwo" (Polish Beekeeping Magazine), in which I tried to convince people about the validity of selecting bees for mite-resistance.

As already mentioned, I'm still struggling with problems in my own beekeeping practice. By no means can I tell that I've solved the problem locally in my apiary. I can assume that this is partially due to overpopulation of bees and unfavourable nectar flow pattern (as the bees' health issues are strictly linked to their diet) - normally we can count on good early nectar flows (May, early June), but then, until the end of August, there is a long period without forage opportunities, after which goldenrod appears. Such a system causes problems because most of the nucleus colonies are not able to gather enough natural food for healthy development. Most often I do not winter strong colonies but use smaller or larger splits. Therefore, in spring they are smaller than colonies of other beekeepers, and they can be additionally weakened by varroa. When the bees finally achieve a reasonably high strength, the late spring and early summer nectar flows usually end. Then I split the colonies, providing each colony with a brood break. It is therefore very difficult for me to coordinate the growth of my bee colonies with local nectar flows. However, I've noticed that this does not have to be directly connected to the size of the colony. Smaller colonies or nucleus hives – if they are in good shape – sometimes manage to gather nectar for themselves while the stronger colonies must be fed or will suffer hunger.

Summers now are exceptionally hot and dry, and the plants do not produce nectar or it dries quickly in the flowers. There were times when the field margins were blue with cornflowers, and yet there was not a single bee on them[228]. Hunger not only affects the health and condition of individual bees but also has a severe impact on the development of young colonies.

Considering the statistics of my apiary, it is difficult to speak of any successes, especially economic ones. The apiary production is negligible, and the losses of bees are at a level that in the eyes of most beekeepers disqualifies the method.

However, even with my personal poor results I want to believe that from a biological point of view we, as cooperating beekeepers, can talk about a certain breakthrough. Several of my friends who trusted natural selection, have proved, although this is only anecdotal evidence, that from year-to-year bees are able to ensure the replacement of generations without the need for treatment against varroa[229]. Many colonies are now developing very well, and we hope that over time we will spread these genes into the local population[243]. This proves that the problem can be social, economic, for some maybe ethical, but not biological. This conviction gives me satisfaction.

I must emphasise, that the general situation is difficult here. Although the official statistics for Poland say that year-to-year bee mortality rate is 10-20%, I do not believe this. Mostly it is made in voluntary surveys with only a minor sample (few hundred of beekeepers while there are almost 100 000 of them in Poland), and thus it may be subject to error and bias. Beekeepers rarely admit their losses, because in that case they are seen as bad beekeepers. One of my colleagues who is the president of the local association told me once that many of the members often declare very high survival rate, and then the same people order new nucleus colonies

---

228 One of the beekeeper friends who runs an apiary and treats bees in the neighbourhood (in the Krakow district) claims that since 2014 he has had to feed his colonies in the summer each season, which he did not do before. There has clearly been a decrease in the nectar flows, especially in the summer.

229 Some of them have a significantly better beekeeping situation, with seasons where they obtain several hundred kilograms of honey from several dozen colonies. This may not be a satisfactory result for professional beekeepers, but it is sometimes better than for neighbouring hobby beekeepers who run apiaries in a standard manner.

almost every year, while their apiaries don't grow... Some of the beekeepers combine a large proportion of their weakening colonies in autumn and they don't include that in the mortality, either. Most beekeepers I talk to admit that their losses are 20-70% depending on the year. Almost every year I hear about some total die-offs of the apiaries – sometimes even quite big ones from a few dozen hives up to one hundred. I don't know what the real mortality rate in Poland is, but I would assume it is on average not less than 25-30% with local bigger problems each year in different regions. For the last ten years my local association had losses that averaged at least 50% and once 70%. Each time the local losses are big, my losses are big too. However, in the good years for the others, my losses are smaller too. It seems, that some local factor (whatever it is in detail, but mostly directly or indirectly connected to varroa) influences each apiary – treating or not. But the latter are hit stronger.

As for myself, I believe that the only way out for us all is to take up a joint breeding effort based on local stock. When I started keeping bees, I believed that locally I could have reasonable losses and at least enough honey production for myself. I probably could (as some my friends and colleagues do), if the bee density was no higher than 4, maybe 6 colonies per km2, and the bees did not suffer 2-month or even longer dearth periods every year. Right now, I see that with a constantly growing bee population and weakening forage opportunities we have to keep together and share the effort of selection and bettering our landscapes. This way we could create the situation in which bee pathogens become less virulent, and our bees more resilient and locally adapted. I see that there is no other way for a sustainable future.

Bee hives

Decisions regarding future apiary management, including the type of hive, I made (like many beekeepers) when my awareness of the subject was very insufficient. When choosing a hive, I was guided by its popularity, so I chose typical for Poland 'Wielkopolski' hive (vertical hive type with square box and frame size 36 cm wide, 26 cm high). I introduced modifications largely under the influence of American beekeeping. In the United States, single-wall hives, made of so-called inch boards, are exceptionally popular, and they are also becoming popular in Poland. The experiences of beekeepers that I considered then gave me reason to believe that it does not matter much to bees whether we keep them in warm or single-wall hives. Many people even believe that the latter will be better, especially when using mesh bottoms, the tendency is therefore more noticeable to cool the nests than to keep the bees in warm hives.

At that time, I also found myself thinking along similar lines. For practical and economic reasons, I very quickly decided on hive boxes made of one board. I also quickly switched to a more modern version of the 'Wielkopolski' hive with a frame lowered to 18 cm, thinking that bees are so flexible that they can cope with any type of hive construction. The reason for that was quite simple: it is much easier to buy a board that is 19 cm wide rather than a 27 cm wide one, and make boxes out of them.

Currently, the trend in apiary management is to prefer low frames in large-section bodies (e.g. Dadant or Langstroth) and to use one type of frame in all colonies, most often low: 16-20 cm. It is often suggested to move away from traditional structures, in which deeper frames were used in the brood nest and shallow for the honey super. Modern solutions not only simplify the entire logistics in the apiary but also make it easier to work with the bees.

For several years I ran an apiary using 'Wielkopolski' hives (frame height 18 cm), using only upper entrances, photo by author.

For several years now, most of my bee colonies have had a nest the size of two "eighteen" frames arranged only in the upper box. Large combs facilitate the development of the colony during the season and overwintering. Although I do not use queen excluders, they rarely go up to lay eggs in the upper hive-bodies, photo by author.

Today, based on my knowledge and experience, I believe that a hive with shallow frames (and boxes) and a large cross-section is a construction created more with beekeepers in mind than bees. Why? A beekeeper simplifies his work to a stage where the method does not interfere with the bees too much. However, when it begins to bother them, either other bees or management methods must be sought that will allow the problem to be solved.

For example, a large and "cold hive" (not insulated) can lead to situations in which bees will not properly distribute food. This can lead to starving to death in winter, even though they had a large (and certainly sufficient) amount of honey supplies in the hive. The problem can be solved in many ways: by arranging nests for winter; by pouring so much syrup into the nest that the bees have enough of it everywhere; by monthly checks of the bees' condition throughout winter or by other interventions when needed (it seems that this has been one of the standards of modern beekeeping for several years, although a few decades ago people were warned not to disturb their hives in winter). Of course, some bees will manage without help from the beekeeper, but others will die without it.

Considering the convenience of management and control over the colony development, I cannot find much to fault with the "Wielkopolski" hive with the "eighteen" frame system. The only thing is that with each season, the concepts of natural beekeeping grow to be even more and more attractive to me, where I do not have to inspect the hive often. After the season, for half a year (winter) I can think about the bees, but I do not want to disturb them and worry about whether they will survive or not because of the hive construction. So, I think that after some time I will slowly start to switch to a hive in which the concept of work is like the Warré method. For now, however, the hive with a shallow frame dominates in my apiary and to a certain extent I have become its hostage. What I mean is that it forces me to intervene when I do not feel like it.

Shallow frames are not optimal for wintering bees within just one box (although some colonies successfully winter). Developing nucs in one box throughout the season and then setting up a colony on two boxes before wintering is not the best solution for those who want to leave the development of the bees, to them alone, and limit their interference in the nest to a minimum. This is certainly not good for the bees either. In addition, smaller colonies may be affected by the gap between the boxes (the effect of an artificial ceiling). A hive with a low frame therefore requires a lot of interference or forces the maintenance of very strong colonies, that do not notice the disadvantages of those constructions.

Instead, my apiary is dominated by rather smaller colonies (which results from the adopted selection for mite-resistance and multiplication of colonies) and at the same time I am reluctant to arrange frames in the nest for bees.

Over time I started to think about using a deeper frame, at least such as in the classic Wielkopolski hive, which I abandoned for the reasons described above. Over many years, however, I have collected a lot of apiary equipment, which for some time determined my beekeeping. I decided to keep the bees on two "eighteen" boxes, but with frames set only in the upper one. In this type of solution, the bees can freely build a comb down the nest, i.e. develop as if on one large frame measuring 36 x 36 cm. Of course, this causes many different problems, such as enlarging the nest with another box or transporting the hives (e.g. after creating nucs), because a large comb can break more easily. However, it is a more beneficial solution for smaller colonies, as it provides them with more freedom in the nest and does not require me to interfere or change the nest layout with the development of the nuc. I also do not have to worry about arranging the nest for the winter: it is basically organised the moment the nuc is created. For now, I manage my colonies in this way, because it provides a kind of compromise: it is not a perfect method for the bees, and it is not very convenient for me (as an amateur I can afford it), but it allows me to use the equipment I have without having to completely rebuild the apiary.

I will also add that, following the advice of beekeepers using a small cell, I have narrowed the side slats of the frames. The standard body of a Wielkopolski hive is 37.5 cm and holds 10 frames. The axes of the combs are spaced 3.5 cm apart. My frames have a side slat 3.2 cm wide. So, I place 11 frames in a 10-frame body. David Heaf claims that in the nests of wild bees he found combs whose axes were only 2.9 cm apart. In turn, Roger Patterson from the British Isles, also an experienced beekeeper, claims that in many nests of wild bees he found combs spaced the standard distance of 3.5 cm apart. Since I do not provide foundation and thus do not mark the axes of their combs for the bees, I believe that under the conditions I provide, they can build comb according to their own needs.

A few years ago, I received a few old Warsaw horizontal hives from a beekeeper friend (vertical Dadant frame). When working with them, I noticed that they allowed the bees to develop much better than in the Wielkopolska hive with eighteen cm high frames, at least when it comes to smaller colonies, nucs or swarms. So, I decided to build a 10-frame (11-frame in my case) Warsaw boxes, which I also tried to use in my apiaries. Their volume was about 53 1iters and provides the colony with more than sufficient space in their first season. If necessary, in subsequent seasons, I can increase the volume by using the Wielkopolski honey supers as an extension.

Inspired by Michael Bush's practice, for several years I managed mainly hives with upper entrances. Over the years I have greatly appreciated the advantages of this solution as it is quite a convenient method of beekeeping. After the winter, there was practically no moisture in my hives (in fact, there was no mould in the corners of the hive or on the combs). In the winter I did not have to worry about the build-up of debris on the hive bottom, which could block the bees' exit. Rodents do not enter the hives through upper entrances, so they colony does not need to be protected from them[230].

---

230   For several years I wintered about 40 colonies, the vast majority of which had small upper entrances, unprotected against mice; I encountered a rodent only once in one hive (it was stung to death near the entrance).

The entrance is higher, so it is not covered by overgrowing grass or other plants.

Few years ago, however, I decided to return to the lower entrances. The main (and probably the only serious) disadvantage of using the upper entrances is that the nest cools down more in the winter. The bees must produce more energy to maintain the right temperature in the cluster. The knowledge I gained after 2018, made me look at this issue from a slightly different perspective.

I still have doubts whether this is really such an important issue, especially when using a single-wall hive, which, by necessity, is a cold hive anyway. Every year I wintered at least a few colonies in hives with lower entrances and I did not notice any obvious relationship between the location of the entrance and the survival of the colonies or their condition. My experiences did not confirm the assumption that the location of the entrance is of fundamental importance for bees in terms of survival and maintaining health, so other factors must be much more important. Despite this, the upper entrances started to bother me - I still don't know if they bother me or the bees more.

The bee

The bee you need to be able to practice natural beekeeping is the one that has lived in your apiary for several seasons. Every non-treating beekeeper uses their own locally adapted bees, which are subjected to a natural selection process every year. Yes, some breeders do bring in queens for testing or in the name of genetic diversity. However, these are bees with the rights of a local experiment, not those that are supposed to replace the current population.

When starting to select bees, we must decide which ones to start with. If we already have an apiary, there is basically no dilemma: we start with our own. The experience of many natural beekeepers shows that regardless of whether we start with pre-selected material but imported from outside or our relatively local, but treated bees, the difference in survival will not be large (resistance to varroa is highly connected to local conditions).

The results of some experiments lead to the conclusion that bees selected for varroa resistance have a higher survival rate. Thomas Seeley in one of his studies confirms that bees from Kirk Webster's apiary (in the state of Vermont which is in a similar climate zone to New York) coped with Varroa better than Italian bees bred for the VSH trait, and similarly to bees from the local population of feral living bees from the Arnot forest[231]. The COLOSS studies, which covered many European countries, lead to the conclusion that local colonies were more likely to deal with varroa better than non-local (foreign) colonies.

In Poland, we can hardly count on the wild-living/feral bee colonies that are resistant to varroa. Even if there are colonies somewhere that have been living without the beekeeper's interference for years, catching a swarm from such is unlikely. The Polish bee population is dominated by colonies that are treated, and, unfortunately, are not selected for varroa resistance.

When starting selection in my own apiary, I considered two criteria for bees to buy. Firstly, I looked for queens from apiaries where bees are selected for varroa-resistance. Secondly, I looked for relatively local bees, from apiaries where beekeepers do not replace queens, but breed their own. In 2015, I acquired bees that were very likely local crossbreeds of the Central European subspecies (we called them "pre-war")[232]. From a Polish breeder, I managed to buy queen bees imported from Juhani Lunden's breeding stock. In addition, from a beekeeper from Lower Silesia, I bought Elgon bee nuclei and queens. These were the next generation from the stock that he had imported from Sweden, from Erik Österlund's varroa resistance breeding population. However, in the winter of 2016/2017, they died off. I was left with only one colony, which I was unable to assign to any "source material" from memory.

In 2017, when rebuilding my apiary after losses I bought several bee colonies, used the local survivors for propagating and for a few years the situation seemed to be improving with this stock. Only after moving to a less favourable location which also had a constantly growing bee population and worsening of the forage opportunities did I enter another crisis period.

I must admit that as a hobby beekeeper, I did not try to control the pedigree of the various bee colonies I had purchased and this has been repeatedly brought up against me as an accusation.

I consider it worth propagating those colonies that start the season in good health, although they are not treated, regardless of whether I know their pedigree and detailed history or not.

---

231 T. Seeley, Progress report on three years of treatment free beekeeping, including a test of three types of queens: wild colony, Webster Russian, and VSH Italian, American Bee Journal 2020, 8,

232 A beekeeper from the Świętokrzyskie region, from whom I bought them, when asked what kind of bees they were, replied: "These are pre-war bees, such as those that used to be found here in tree hollows." And this name stuck to them.

In my apiary I also have colonies that originated from the "Fort Knox" project. These are lines that have not been treated in our apiaries for at least several years.

I put the greatest emphasis to breeding from my local surviving stock. From the best colonies I try and produce at least 4-6 daughters per colony. Most often, in May, they reach the strength of 3-4 boxes in my "eighteen" frame hive (that is at least volume of around 70 – 100 litres or denser with bees).

Most often, I take an old queen from the colony with an "artificial swarm" and then I allow the workers to raise emergency queens. Around the 10th or 11th day after making the colonies queenless, I split the colony into several parts, using the queen cells. If the colony gets into a swarming mood and establishes queen cells, I use them most willingly. Sometimes I create very small splits, sometimes I leave colonies stronger – I have never noticed that the strength of the colony after splitting is the key for their further survival.

If a colony has survived the winter on just a few frames and has not grown as well as the better colonies in the apiary but is not holding back in development and is not showing signs of disease, I try to split it at least in half, in the manner described above. The weaker and smaller colonies and those with problems are not split but I may induce a brood break.

I do not usually replace (kill) queen bees[233]. I reject this method to rejuvenate queens in colonies or genetic exchange. In the apiary I allow for a supersedure. It happens that some of the old ones perish together with the dying nuclei/colony.

New colonies do join my apiary, most often these are captured swarms. Such colonies undergo a kind of 'genetic quarantine' in my case. I treat them like all the others, but I wait at least two seasons without treating before I propagate them. However, I use the resources of these colonies. Just like with other colonies, I provide them with a brood-break by making artificial swarms with the old queen. However, I give orphaned nucs queen cells from the apiary's top performing colonies. If the bees manage to survive two consecutive winters, then I treat them in the same way as the others in

---

233   I do not consider killing, re-queening of a colony as a proper procedure (especially as a standard procedure in apiary management). I only do it when the queen is not (and will not be) able to lay eggs, e.g. she has hatched with a dislocated wing or without it, which means that it will not be able to make a mating flight. The lack of legs or part of a leg does not prevent young virgin queens from making a mating flight or laying fertilized eggs. There were situations when the bees did not want to replace such queens (which was also in line with my intention). Sometimes I kill a queen who stops laying eggs altogether, but such a decision is preceded by at least several weeks of observation. I must be sure that leaving the queen would lead to the colony's death.

the apiary. If they survive in good health, I propagate them in the same way as those whose "chemical-free" pedigree goes back to earlier years.

I hope that one day the resources of my apiary will allow me to operate with much greater freedom and I will split colonies into fewer but bigger nucs; maybe I will also start allowing some of the swarms to escape. However, I would prefer that instead of going to apiaries, they would end up in wild nest sites. Beekeepers most often subject the swarm to treatment and replace the queen in it. I do not want to waste my own selected resources on such activities.

I have not been trying to get any new genetic lines of bees for my apiary. The necessary genetic diversity is provided by free mating with local drones.

The experience of many beekeepers from Poland shows that new bees (brought or caught swarms) do not show a satisfactory level of varroa resistance. Even with my poor survival rate I see the probability of successful over-wintering is lower than in bees originating from colonies that have lived for some time in my apiaries. Of course, even those that have been in our apiaries for many years die due to mites or other diseases, as the previous seasons in my apiary have shown. However, the probability of their survival is higher than in the case of "random" colonies.

Expansion model beekeeping

In the winter of 2013/14, while searching for a suitable method of running an apiary and selection for mite-resistance, I came across information about how a break in brood rearing can limit the population growth of varroa[234]. By spitting a bee colony, we also divide the varroa population. At the same time, having more nucs, we gain more chances in this unequal fight. So, I decided that splitting colonies would allow me to select for varroa resistance without the need for any procedures, even biotechnical ones[235]. This strengthened my conviction that selection should be entrusted to the laws of nature. Even then, I believed that nature could handle selection much better than I or any other beekeeper (and I have not changed my mind).

---

234   See: https://www.resistantbees.com/

235   Splitting of the colony is also a specific biotechnical procedure. However, I draw attention to procedures that would involve the complete removal and killing of some part of the mite population (e.g. removal of the brood).

A few months later, on Solomon Parker's website, I found a detailed description of the method, which the author calls: Expansion Model Beekeeping[236].

Although it seems to me that the author's intentions are better translated as: expansive model of beekeeping. The term illustrates very well what this type of selection involves. In the first period, it is assumed that the number of bees will grow faster than the mites which is often referred to as "outbreeding the mites". Dividing the colonies into splits also allows for the fragmentation of the mite population. Using this method, we are theoretically able to split the colonies in such a way that they are biologically capable of surviving the winter. At the same time, the scale of the infestation is small enough not to pose a mortal threat. A similar mechanism, albeit in nature, has been observed in populations of wild bees that occupy small nest sites and swarm in large numbers. According to the assumptions, confirmed by the experience of beekeepers, the colonies that survive begin to show adaptation characteristics after a few years[237]. For the first few years, at least in theory, the survival of some colonies can be explained by "outbreeding" the mite population in terms of development. Varroa will sooner or later "catch up" with the colonies that do not have resistant characteristics. It could therefore be assumed that the entire "expansion model" is a specific biotechnical procedure. It differs from other methods primarily in that we do not use any of the methods of killing (controlling) the mite, all mites remain in the apiary, although they are sent to new colonies[238]. This model, by its nature, is not conducive to honey production. I try to obtain small amounts of honey from colonies that are gaining strength from the early summer nectar flows, before I perform splits[239]. The key to success here is the multiplication of colonies. This should be done reasonably as much as possible, depending on the number of queen cells and biological condition of the split colonies (allowing them to survive the winter).

---

236 The original description of the method can be found by entering its name in an internet search engine; see also: http://pantruten.blogspot.com/2015/08/moj-plan-czyli-pszczelarski-model.html; http://pantruten.blogspot.com/2020/11/model-ekspansji-rewizja-po-latach.html; accessed: 18 November 2022.

237 What is important is whether these adaptations will become apparent enough to give us reason to assume that colonies will survive, especially over the next few years.

238 Sceptics believe that professional beekeepers who do not treat varroa get rid of mites from their own apiary along with the sold nucs. In my opinion, this is only a small part of the truth (if that were the case, they would have to get rid of 80-90% of mites each year).

239 The situation is not satisfactory: intensive late summer harvests would undoubtedly help in obtaining larger quantities of honey.

In the peak season, a lot of nucs and hives ready for splitting appear in the home apiary, photo by author.

The intensity of splitting colonies is different every year. However, I am not afraid to make very small splits, because I know that if they are healthy, they will have time to grow by the end of summer and survive the winter. Healthy bees can survive winter even when the colonies are not large. Those that are developing very well, I can divide them into 6-7 splits using this method, although the size of these splits cannot even be compared to those created in intensive farming. Some can build up the strength needed to survive the winter on their own, while others require intervention or attention. I make fewer splits from a colony that is in worse condition, but I divide even quite weak colonies, if only to create a break in brood production. I believe this has a positive effect on the health of bees.

The classic Solomon Parker's method does not use interruptions in brood production, only the splitting of the colony with a new queen at the start of a new race with the mite.

Every year I manage to enlarge the apiary (the numbers of wintered colonies) by an average two to four times (depending on the season and health of the colonies). There were years that I gave some of the colonies to other beekeepers. Most often, by the end of August, my colonies reach the strength of 6-8 frames (which is a size of about 30 litres), which is more than enough for them to survive the winter (at least the size is not the issue here). However, I have overwintered both weaker and stronger colonies. The Expansion Model Beekeeping and my implementation of it have often been the subject of criticism. Beekeepers claim primarily that colonies created in this way are too weak to be able to function properly. This argument is not without foundation (and wintering success in my apiary may also be the anecdotal evidence, that the criticism may be valid). Many natural beekeepers believe that for colonies to perform hygienic duties, they must be at a strength capable of overwintering, have a large supply of food, and be able to rebuild the colony. Only then can they take care of, among other things, detecting and removing varroa. My experience suggests however that the relationship is not so obvious (again: this may be the truth, but other factors may be more important). I have had quite large colonies left after making splits (totally self-sufficient all the season), which, however, did not manage to survive until the next spring. I have also noticed that when assessing a colony, beekeepers very often consider the strength needed to produce honey. However, the method described is intended to serve, on the one hand, selection, and on the other hand, propagation of bees, which go through cycles of natural selection, not honey production. It is based on a specific balance on the border idea of the biological self-sufficiency of colonies, and not on keeping them at a strength that is optimal for honey production.

I treat the expansion model beekeeping as a permanent method. Its essence consists, among other things, in creating a reserve of colonies (so it's in the form of extensive farming). The number of multiplications may be smaller or larger and depends, for example, on the flows in the season or the losses to be replenished. I intervene in the development of colonies only when I consider it necessary. There is no rule here, because colonies grow differently, and it is not always related to their strength.

If a colony cannot cope with gathering food, I give it sugar syrup (this is necessary in most colonies almost every season during the summer nectar dearth). If it is unable to gain the appropriate strength by itself before wintering, I add a frame with brood from one of the stronger colonies. I do not combine colonies in which the queens are laying. I do this only if one of them loses its queen. Then I try to add such a colony to one of the weakest in the apiary. I believe that I am not able to harm them more (assuming the risk of transferring parasites and pathogens)[240]. The direction of the "flow" of bee resources (frames of food or brood) is always the same: from a colony whose condition I assess as very good to a weaker colony, never the other way around. In this way, I minimize the risk of transferring the problems of a weaker colony to others. I do not consider any of the consequences of expansion, i.e. artificial splitting, feeding, and reinforcement by other colonies, to be a good solution for the health of the bees (they also increase the labour intensity of the method). The expansion model, in a way, forces us to compromise. However, I decided that this type of compromise best corresponds to both my beliefs and the selection goals. I also believe that it helps to develop a comprehensive and long-term ability for the bees to survive. An alternative would be some chemical treatment or biotechnical procedure that removes varroa, and I want to avoid this. My experience shows that after giving up chemical treatments, it is difficult to determine which colonies have a good chance of survival. In spring, the situation is very often surprising. It is easiest to point out colonies that have the least chance of survival, but even here there is a risk of error. It often turned out that average colonies, in relatively good health, which I did not give much chance of survival, were gaining strength, and those that I assessed as the best, died early in the winter or even late autumn.

Sometimes colonies that I do not split excessively are not able to survive (perhaps along with their splits I don't free them from excessive mite populations), while small nucs perform their spring flights with vigour. Sometimes I also lose colonies that developed without problems throughout the season, which did not require any intervention or feeding and gathered a good food supply for overwintering. Sometimes those survive that required much help throughout the summer. Nature has its own key to survival.

---

240   One is weak (probably too weak to survive the winter in good condition) and requires help, the other does not have a laying queen, which without external intervention would soon lead to the death of the colony.

The expansion model beekeeping practised in the so-called transition period also has its drawbacks. These are:

▶ High mortality of bees subjected to natural selection (especially in the first years).

▶ Low profitability of running an apiary (in a small amateur apiary it is difficult to find a resource of bees that would ensure both development, replenishment of losses, and many strong colonies for honey production).

▶ "Blind selection" - it is impossible to direct the selection of bees without interfering with natural selection.

▶ Laborious, you need to look after and feed small colonies at the end of the season.

▶ No guarantees regarding the condition of the colonies, which may emerge weak after winter and develop slowly at the beginning of the season, such colonies are difficult to prepare for spring and sometimes even early summer honey harvests.

▶ Poor health of some of the bees subjected to natural selection, such colonies are not only unsuitable for honey production, but also for further propagation.

▶ The need to have a lots of beekeeping equipment (small mating or nucleus hives, standard hives, frames etc.).

The main advantages of the expansion model beekeeping selection include:

▶ Relatively quick selection (in my opinion, this method promotes the quickest selection for mite-resistance).

▶ Providing the most favourable traits needed by bees to survive through natural selection.

▶ Maintaining high biodiversity (on the one hand, natural selection rapidly reduces the genetic diversity of the apiary, on the other hand, by reproducing bees at such a rate, we can preserve a large part of it).
Acquiring the ability to quickly propagate bee colonies.
Rapid local adaptation of bees.

▶ Ensuring comprehensive selection of the entire ecological system (host - parasite - pathogens – micro-organism environment).

▶ Ecological benefits (we do not pollute the hive environment with toxic substances).

I am afraid that convincing most beekeepers to give up their income from apiaries for a few years and passively watch their bee colonies weaken, and die is practically impossible. However, when taking aside the honey production, I do not know of any other way to effectively and quickly, and at the same time comprehensively, select both mite-resistant honey bees and hive ecosystems.

The described model is just the way of keeping bees in an extensive way. I know that my personal example is not convincing, since it did not prove to be successful in my apiary. I strongly believe however that it's not because of the flaws of the model, but a very difficult external situation that I mentioned in previous parts of this chapter. In this region even treated bees die in unreasonable numbers, and this extensive model proved to be successful in many apiaries – including those of my friends and acquaintances in Poland. Taking that into consideration I strongly believe that this method of selection is sound and reasonable, because it eliminates the least adapted genetic lines and allows for the propagation of the most resilient colonies. It needs however enhancement by cooperation of beekeepers to solve the varroa problem in a broader sense. My example shows that you cannot solve the problem by being a small "island" when conditions are unfavourable. Whichever bees survive the natural selection, they need to be propagated in the surrounding apiaries in which they must pass further local adaptation process.

Artificial tree hives

In my opinion, we should promote the propagation and creation of wild and feral bee populations. Some consider such actions to be harmful due to the potential epidemiological threat. I completely disagree with this idea. Wild colonies can have a very positive impacts on the local population through their genes via their drones. Of course, this impact will be negligible if there are few such drones, and the surroundings are dominated by bees from managed apiaries.

That is why I treat attempts to recreate the population of wild living or feral bees as a supplement to our own selection. I also keep my fingers crossed for each new tree-beekeeping initiative, especially where mite treatment is abandoned, and the bees are allowed to be selected in accordance with the principles of natural selection.

Wild bees have very modest living conditions today. Forests, for the most part, are young[241]. For example, cutting down old trees, often of monumental size, is incomprehensible. This happens not only in forests, but also along roads and in urban areas. There are too few areas free from human interference. National parks in Poland occupy about 1% of the country's area (and even there, limited forest management is carried out). This is not much. In Germany, it is about 3% of the area, and in the Czech Republic about 2%. In commercial forests, logging is very often carried out without paying attention to the natural values of the trees. In social media, there is increasing discussions about tree rescue operations, including those with circumferences exceeding 2.5 m, marked for cutting. These attempts rarely end in success, most often only the tree stump remains. As a result, we will not find many old trees with suitable cavities in them, which deepens the nest cavity deficit, not only for honey bees, but also for other species of fauna. According to statistics presented by the State Forests department in Poland, forests are getting older, their biodiversity is increasing, as is the amount of tree biomass (the annual biomass of the felled trees is less than the growth of the remaining, unfelled trees). More and more nectar- and pollen-giving trees are being planted, which serve not only honey bees but other insects. Additionally, one may get the impression that private forests, of which there are quite a few in Poland, are gaining increasing natural significance.

Some of the private forests are being used less now for economic purposes[242]. Often no plantings are carried out in them, in accordance with the adopted forest policy, and there is no management in them whatsoever, so the vegetation regenerates naturally, which in the long term may support the creation of locally sustainable ecosystems[243]. Perhaps they will become real natural competition for state forests. According to statistics, everything is heading in the right direction and yet the current situation still seems unsatisfactory, at least from the perspective of benefits for honey bees.

---

241   I do not want to deal with forest policy or criticize forest management as such, because it also has positive sides: without wood we cannot function.

242   The situation in them differs much, because with private forests not everything depends on the state policy, but also on individual decisions of owners. Some of them do not carry out logging at all. On the other hand, it is sometimes said that in private forests that are used, the management is often much less sustainable than in the State Forests. It is difficult to say what the situation of private forests looks like in reality, because most data are based on estimates.

243   Sometimes they may be less diverse (at least in terms of tree species dominating the forest). Sometimes, nature, left to its own devices, locally reduces diversity by eliminating species from ecosystems that cannot withstand the competitive pressure in each environment.

The advantage of my boxes is that they can be made quickly and cheaply. Their low weight makes them easy to hang on a tree, photo by author.

Forests are changing, but many of them are still monocultures, which are unable to sustain the ecosystems needed by bees (e.g. pine forests or dark beech forests with poor undergrowth)[244].

Climate change is also a threat. Warming may cause the reconstruction of forests. This means that in the coming years, forests may be rejuvenated again because trees that cannot survive the changing climate and extreme weather events will not withstand the pressure of droughts or parasites. However, that is a completely different story.

In 2016, I decided to start a completely amateurish and very modest project consisting of hanging bee boxes in the local forest. The aim was to increase the number of potential nest sites, which are undoubtedly lacking in the local forest. I called these boxes "artificial tree hives". They are certainly not ideal nest sites for bees: neither their shape nor properties have anything in common with bee logs or, even less so, natural tree hollows. However, they provide an alternative for honey bees in the area. I paid for the materials for the boxes from private funds, and I hung the boxes in my free time. For several years, I annually added a few new boxes. I am helped by my friends with whom I started cooperating to help in the natural selection process of bees. In 2020, four years after the start of the campaign, among other things due to the COVID-19 pandemic (there was a ban on entering forests in the spring), we did not hang the boxes for the first time. Unfortunately, it was similar in the following years. In 2022 I managed to place only one box on a tree. However, I hope that we will be able to return to the tradition, especially since several ready-made boxes are waiting to be hung. This is the right step towards rebuilding the wild/feral bee population,

---

244   A commercial forest grows for about a hundred years, while a balanced forest ecosystem takes
at least three or four times longer to form. An ageing forest gains diversity (not only in terms of
species richness, but also landscape), some grow old, some die, each time their species composition
is reconstructed. These phenomena, apart from a few exceptions, in very limited areas, cannot be
observed in Poland, because most forests are young and are cut just as they become old enough so
new process of forest reconstruction could start. It is unnatural that the trees are relatively uniform
in age, which causes the crowns to tightly block the sun (the so-called dark phase of the forest).
In an ageing forest, when old trees die, natural clearings, marshes sometimes form (due to fallen
trees, watercourses, beaver activity, etc.) and such a state can persist for several decades.
In such places, the fauna and flora are completely different than in the dark phase forest. I have the
impression that a large part of our forests is treated as a tree plantation, and not as an area of
a balanced forest ecosystem. Phenomena such as pest outbreak are normal for the creation of
balance and reconstruction of natural ecosystems from the natural point of view. It is not surprising
that we deal with them in monoculture tree plantations, artificially planted in places where a different
tree species naturally occurred. They can also occur to some extent naturally and spontaneously,
e.g. in connection with climate change. Meanwhile, in our country, there are basically no areas where
such phenomena are not counteracted by people (see, for example, the recent situation related to the
occurrence of a bark beetle outbreaks in the buffer zone of the Białowieża Forest).

although it is a proverbial drop in the ocean.

The "Artificial Tree hive" project is far from the actions included in the formal framework and financed from external funds (state, EU or any other), such as "Bartnicy Sudetów" (https://bartnicy-sudetow.pl) or "Pszczoły wracają do lasu" ["Bees return to the forest" - this is the project of State Forests institution of Poland]. Many beekeepers operate on a much larger scale; the nest sites they create are also more valuable than my boxes. From this perspective, these actions seem extremely modest. However, I believe that they are in the right direction: as it turns out, each of us can do something good for bees, for example, by spending a few days a year in the forest and devoting a small amount of our funds.

I make the boxes with a capacity of about 35-50 litres from unprocessed wood. The significantly larger ones (80 litres) built in the first year of the project turned out to be a complete failure. They were very heavy and unwieldy, so firstly it was difficult to transport them and place them on trees, and secondly, more importantly, they turned out to be unattractive to bees (probably due to their large size) and none of them have been occupied by bees to this day. Each stage is carried out in consultation with the local forester. I consult with him on both the number of boxes and their approximate location, and sometimes even the specific tree. We place the boxes in areas that will not be cut down soon, so that the lumberjacks do not disturb the bees, and they in turn do not hinder their work.

The artificial tree hive is to serve local wild or feral honey bees. I hope that the boxes will become bee nest sites, and then have an impact on the local bee population since natural selection will take place without any intervention[245]. However, I have no illusions that just a few artificial hives will change the situation[246]. The effect should therefore add up to the selection carried out in our apiaries. Time is also important. After all, the selection process takes many years. So far, I have managed to hang around 20 boxes. In 2017, I placed artificial swarms in two of them. They survived until the spring of 2021, completely without help[247]! I believe that

---

245   Other forest inhabitants also use the boxes - birds have settled in several of them, and we also found boxes occupied by hornets.

246   For the effect to be noticeable for the local population, the bees would probably have to inhabit at least several dozen boxes each year, because so far it has not happened that all of them have been inhabited; the number of boxes should be around 200-300.

247   I must admit, however, that in the spring of 2018 I had no opportunity to check whether the bees were alive; the boxes were still populated in the summer of 2018, so if the artificial swarms that had been placed did not survive the first winter, the boxes were repopulated. I am also unable to say what the condition of the bees was during that period, or whether they produced swarms.

this is a very good result, confirming the ability of bees to survive in wild nest sites. I tried to carry out inspections of the boxes at least twice a year, before the swarming season (i.e. until mid-April) and after it (July-August). In the summer, I wanted to check in each location whether the boxes have been inhabited, while in the spring, I only visit those locations where hives were occupied. This is a way to gain high probability about the continuity of the colony. In 2019, I hung one of the boxes containing bees again. These bees, left to their own devices, survived at least two winters. When I visited them in the spring of 2021, I could confirm that they were alive[248]. The swarms colonised several boxes on their own. In 2019 and 2020, I recorded three such cases. Most often, however, the swarms die within the first winter.

One of them survived the first winter and died during the second. Before the winter of 2020, there were up to seven colonies (the largest number) in those boxes. Unfortunately, only one survived until spring 2021. Two boxes, including bees, were taken from the trees by somebody. One of them was stolen together with the swarm that survived its first winter. The other was lying under the tree where I hung it, and inside I found a few small slices of white wax, I therefore assume that it was taken from the tree shortly after the swarm entered. I can find no explanation for such conduct[249].

Finally, it is worth adding that the "Artificial Tree hive" project can fit in the activities that the Swiss association of 'Free the Bees' defines as "diversified beekeeping". It is intended primarily to implement sustainable apiary management, as well as to support the adaptation of bees. Apiary activity has been divided into several areas. The first assumes no interference by the beekeeper, including treatment against varroa; it is mainly based on creating nest sites. The second area is beekeeping this is close to natural, in which small hives based on natural comb are used, and all treatments are limited to the occasional use of essential oils. The third field of activity is extensive beekeeping, and the fourth - intensive beekeeping. Beekeepers can draw income from the last two areas of their activity and practicing the first two serves to support the creation of local and adapted bee populations.

---

248 The last few years have been exceptionally busy for me, in 2022 and 2023 I managed to check only some of the boxes, so I do not know their population status. What is worse, the lack of regularity in inspections results in a lack of information on the length of the period these colonies survive.

249 The only explanation I can think of is that the colony was taken by the rightful owner of the bees, who was chasing his own swarm. Article 182 of the Civil Code states: "A swarm of bees becomes nobody's property if the owner has not found it within three days from the day of swarming. The owner may enter someone else's land in pursuit of the swarm, but he must repair any damage resulting from it." In this case, however, the damage was not repaired, because the box was not returned to the tree.

Implementation of the concept of diversified beekeeping by most of beekeepers could serve to balance the natural needs of bee populations and the economic goals of people.

Some boxes were filled with bees using so-called artificial swarms, photo by P. Słotwiński.

When hanging the boxes, I was usually helped by Łukasz Łapka, Marcin Zarek and Mariusz Uchman, photo by author.

Ethics of not treating for varroa

Refraining from treating bees for varroa remains a controversial topic, also on an ethical level. So, I thought it was worth mentioning. However, I would not like to enter the area of moral evaluations of specific actions, but rather to show how relative judgments can be.

In the past, when I was active on various internet forums, I often had to deal with hate. It happened that I or my colleagues would be addressed with vulgar words. Beekeepers rarely apologized for the outbursts of emotion (maintaining the damning assessment of our actions), sometimes the entries were deleted by the portal administrators. It is a fact that discussions about abandoning treatment are a hot topic. The basic argument of opponents in the discussion is that since we take a creature under our care, we should take care of it and treat it if it is sick. In a way, I agree with this position. However, the case of the honey bee is specific, one could even assume that every action is to some extent morally questionable, so the least ethically questionable would be to stop breeding and even keeping honey bees at all, at least until the populations deal with their health problems. This is the only way we could, so to speak, wash our hands of it and avoid engaging in ethically questionable activities.

As I have already mentioned, the honey bee cannot be isolated from the outside world. Meanwhile, the practice of apiary management, with an emphasis on selection for exceptional productivity and gentleness (with complete disregard for mite-resistance and the ability to cope with other bee pests) caused complete dependency of bees on the so-called care of beekeepers and their inability of coping with the outside world. However, we must reckon with the fact that swarms will escape from hives and settle in unattended/unmanaged/wild nest sites, and this is tantamount to sentencing them to death. Is it therefore moral to send animals to death if we are unable to completely control them? I do not understand why in the era of bee health problems; beekeepers even consider whether it is ethical to import queen bees of a foreign subspecies (race). After all, this action serves to perpetuate the problems of bees and deepens their problems in adapting to the local environmental conditions. Not only because of genes that weaken local adaptations, but also because of the transmission of pathogens. Isn't it the case that many beekeepers are guided only by productivity indicators, while the moral aspect remains largely in the background?

Another problem is that we regularly place toxic substances in hives, even though we have no doubts about their negative impact. They shorten the life of bees (including queens), impair their immune system, damage their natural protective barriers, and at the same time promote the growth of pathogen virulence, which not only attack other honey bees, but also other bee species. Of course, bee selection could be carried out in a different way than by subjecting bees to natural selection. There are apiaries where work on selection for immunity is successfully carried out. However, most often these activities are associated with the replacement (shifting) of queens; often, to move away from chemicals, bee brood (mainly drones) is removed (physically eliminated), bees are drowned in alcohol solutions (to determine the degree of infestation, etc.), or queens are "imprisoned" in isolators. All in the name of the best intentions and the so-called lesser evil.

In my opinion, it is much more ethically questionable than natural selection to make an organism dependent on the intervention of a beekeeper. Besides, as experience shows, these actions are not a long-term solution to the problem[250]. I believe, therefore, weighing the moral relativity of this situation, that an action that is less ethically questionable is to subject bees to natural selection to develop the necessary adaptations for independent survival.

In David Heaf's apiary, as well as among his neighbouring beekeepers, bee losses are negligible. The average percentage over the last ten years is 8% (like pre-varroa losses). It is possible that the reason is that most beekeepers in the neighbourhood are hobbyists who interfere with bee colonies to a minimal degree[251]. According to many estimates, the mortality rate of bees there is lower than in the apiaries of beekeepers who treat bees. Despite this, Heaf was sometimes attacked, for example by organic beekeepers from Germany who accused him of not taking proper care of the bees under his care. This is a good example to show how difficult it is to reach an agreement between people with different ways of looking at the same problem. If we look at the issue in isolation from the view of honey production, it may seem quite strange that it is ethical to start a fight with one living organism (pest) to save another. We evaluate the role of one organism as important for the ecosystem (pollination, food production), while it is difficult to find justification for the existence of another organism (pest)[252]. However, we know that over time bees can develop a balanced relationship with varroa (if we allow them to). It therefore seems obvious that we can use the services of bees in the ecosystem without having to fight the mite. Implementing the process of shaping the adaptations of bees is less profitable in the short term than fighting varroa. That means we would either must show a greater selection effort or tighten our belts and reduce honey production for several years. Thus, it turns out that it is moral to eliminate individuals of one species to, in the short term, profit from the work of the other. In this way, however,

---

250   This is perhaps the key argument in this discussion: if an action causes harm (on various levels) and does not solve the problem, then continuing it seems questionable both ethically and pragmatically.

251   It is difficult to say whether bee losses would be higher if intensive beekeeping methods were used. The example of such an apiary, owned by Clive and Shân Hudson from Gwynedd County in Wales, shows that it does not have to be this way. I talked about the Welsh case with David Heaf, cf. https://youtu.be/1fdfg1mpWN4).

252   We also create different moral standards for different (although often very similar) organisms. There are people who have no qualms about using, for example, sticky traps intended for wild rodents (pests), which die glued to them from hunger and thirst, unable to move. However, we would consider deliberate starvation of a mouse kept in a terrarium to be barbaric.

we would reduce the ethical problem to economics. Sometimes I have the impression that this is exactly the case with bees.

Professional beekeepers do not have any ethical objections to taking the bees to late autumn nectar sources (e.g. heather), although they are perfectly aware that the bees will return to the apiary in a state that does not guarantee survival through the winter[253].

There is also the ethical issue of infecting other beekeepers' bees (this in a way also brings the problem back to economic aspects). Treated bees can potentially become infected by untreated bees, which will expose the treating beekeepers to losses or reduced income, as well as emotional damage. Professor Seeley believes that in the environment with other apiaries, keeping bees without treatment requires a "different responsibility"[254].

---

253 A large proportion of beekeepers have no problem making choices that do not serve the health of their bee population or individual bee colonies, if it translates into higher honey production (e.g. taking bees away for late harvests or migratory beekeeping in USA). In principle, the entire beekeeping community accepts this behaviour, in fact scientists involved in beekeeping say directly in their lectures that these colonies often simply must be written off. While I have heard many times from different sides that my practice "is the action to the detriment of bees", I have probably never (!) heard that taking bees to the heather (resulting in the necessity of writing them off as losses) was called an action to the detriment of bees. I assume that the number of bee colonies taken away for late harvests is far higher than the number of colonies deliberately left untreated by beekeepers who decided to refrain from performing treatments as part of leaving them to natural selection and working on the population's immunity. Meanwhile, few people notice any problem (including ethical ones) in the methods of commercial beekeeping, which is only based on profit – this is in fact called rational apiary management(!).

254 See: interview with T.D. Seeley, "Beekeeping Today Podcast", season 4, episode 48, https://www.beekeepingtodaypodcast.com/bee-hunting-with-dr-tom-seeley-archive-super- episode-s4-e48/, accessed: November 24, 2022). I believe that the professor is not referring to the ethical concerns related to the death of a colony due to varroa, but to the problem of spreading pathogens. After all, he promotes "Darwinian beekeeping", one of the assumptions of which is to refrain from treating bees (if the level of mite infestation exceeds the accepted norm, he even advises to pre-emptively kill the colony). I believe that the problem should be viewed in yet another light. The question is: why should only non-treating beekeepers demonstrate this special responsibility? I believe that everyone who undertakes any actions that may affect the health of bee populations should demonstrate it. I include here primarily: bringing queens from distant geofigureal regions; lack of work on the local adaptation of bees (not only among breeders, but all beekeepers); conducting migratory beekeeping (especially the transport of late flows, where problems from the entire season accumulate, with a large mite infestations); overuse of disinfectants (which facilitate the development of pathogen resistance and virulence). The current epizootic situation of bees is, in my opinion is very bad. It could probably be compared to the peak of the SARS-CoV-2 pandemic in the fall of 2021. For this reason, bringing in bees (queens) from outside or even more so transporting bees for nectar could be compared to organizing tourism or mass events at the peak of the pandemic. Meanwhile, few people see a problem with this. In my opinion, to be in line with the principles of special/different responsibility mentioned by Seeley, until bees develop immunity, we should treat the area of our country (or more broadly: regions where honey bees have problems with varroa, i.e. for sure all of Europe and North America) as an infected area, with all the consequences and regulations of quarantine, like the case of American foulbrood.

However, many beekeepers cope with the problem of mite re-invasion or mite bombs.

Especially professionals include this problem in their current practice, and this is regardless of whether they have non-treating beekeepers next to them or not[255]. If only the apiary procedures in nearby apiaries are not synchronized, problems due to the so-called mite bombs and reinvasions will be inevitable.

When it comes to ethics, it is difficult to justify one's own actions by the actions of others. Despite this, I believe that the causes of the current bee health problem (and the consequences resulting from the lack of adaptation to current environmental conditions) should be sought in the fact that over the last four decades, the issue of solving the problem has not been given due attention (the negative effects are only masked by regular killing of the mite). It is not the lack of treatment, but rather non-resistant bees that are the greatest threat to neighbouring apiaries. The problem of varroa would probably disappear quite quickly if the bees were locally adapted and if it was possible to limit the horizontal transmission of pathogens over vast areas (migratory beekeeping, importing queens of foreign races, trade in bees between distant locations, etc.). The spread of pathogens and parasites is therefore primarily facilitated by modern apiary management. In my opinion, after the stage of collapse and recovery, the problem would become completely marginal, similarly as with the general decision to breed bees for resistance which would lead to a type of mass immunization.

Therefore, I believe that it is unfair to blame the current situation on the beekeepers who have attempted to develop deeper adaptation of bees to local conditions and the presence of varroa mite. The problem lies elsewhere. Non-treating beekeepers simply see the conflict of values in these issues differently. The most ethical action would be to limit the import of bees and for beekeepers and to undertake a joint effort to breed mite-resistant and locally adapted bees. It depends only on us whether we will live in a state of ethical relativism, or whether we will manage to achieve such a state that will allow us to conduct apiary management without having to wonder whether we are harming the bees in the process of our beekeeping or not.

---

255 Hence, for example, treatment in late autumn or early winter, which was probably rare or an exception a decade ago, is now standard for many.

BEES AND PEOPLE

CHAPTER 5

BEES AND PEOPLE

In the 1990s, I was trying to find a way to select, and I decided that the best thing would be just to stop treating (...). When we were using amitraz (...) every time I had bad headaches. I thought it was only psychological until I went to bee meeting, and one of the beekeepers said "I just finished treating my hives with amitraz and I have a bad headache". The second beekeeper also said he had a bad headache. Then I thought it maybe wasn't psychological, but physical or chemical damage. I made a very simple calculation, calculated how much it costs new human brain for me versus the bees going through the process of natural selection. And since I couldn't find anyone selling human brains at that time I decided to go with natural selection.

JOHN KEFUSS[256]

Bee populations living around the world without varroa treatment

Wild-living populations - is this a myth?

The history of the *Apis mellifera* species is exceptionally long. The adaptations of bees allowed them to survive for hundreds of thousands, even millions of years in a slightly changed form and to naturally inhabit vast areas on three continents. With the development of civilization, humans moved honey bees to other continents, where escapees from apiaries began to create wild-living (feral) populations.

---

256   Mite black holes with Dr John Kefuss, https://open.spotify.com/episode/3VZ641VGB10IXcvGGI4m58; accessed: January 31, 2025.

The western honey bee *Apis mellifera* has settled in almost all corners of the world, and colonies are able to function independently wherever they find flora capable of supporting their existence and a suitable nest cavity. In a world dominated by the varroa, people often deny that bees are capable of surviving on their own. Meanwhile, free-living populations of honey bees, despite adversities, still live in many regions of the world. They also have their own ways of defending themselves against many threats, and often do well alongside managed apiary populations that are unable to survive without the help of a beekeeper.

I will mention only a few of the known and described by scientists varroa-resistant bee populations from all around the world; reports of the discovery of new resistant populations appear every now and then. My choice is subjective, but I was guided primarily by the fact that in the case of those mentioned, we can use scientific research that unequivocally confirms that bees are independent, create stable populations, and have various traits and properties (described in the studies) that allow them to survive despite being infested with varroa[257]. They are therefore excellent evidence that the treatment of bee colonies is not about saving the species from extinction, but a purely economic decision made by the beekeeper. Scientific research, as well as testimonies of practitioners, lead me to draw several basic conclusions.

Firstly, bees have the greatest health problems where intensive industrial scale beekeeping is carried out (Europe, North America - mainly the USA). Secondly, wherever bees had the conditions to undergo the process of natural selection after the invasion of varroa, the problem of varroosis does not exist or is marginal.

---

257   List of studies on which the knowledge presented in this chapter is based: T.D. Seeley, The Lives of Bees. The Untold Story of Honey Bee in the Wild, Princeton University Press, 2019; Y. Le Conte, Geofigureal distribution and selection of Honey bees resistant to *Varroa destructor*, Insects 2020, 12; B. Locke, Natural Varroa-mite surviving *Apis mellifera* honey bee populations, Apidologie 2015, 12; T.D. Seeley, Life-history traits of wild honey bee colonies living in forests around Ithaca, NY, USA, Apidologie 2017, 6; T. Seeley, Honey bees of the Arnot Forest: a population of feral colonies persisting with *Varroa destructor* in the northeastern United States, Apidologie 2007, 1-2; I. Fries, Survival of mite infested (*Varroa destructor*) honey bee (*Apis mellifera*) colonies in a Nordic climate, Apidologie 2006, 6; T.E. Rinderer, Resistance to the parasitic mite *Varroa destructor* in honey bees from far-eastern Russia, Apidologie 2001, 4; L.E. Brettel, S.J. Martin, Oldest Varroa tolerant honey bee population provides insight into the origins of the global decline of honey bees, Scientific Reports 2017, 4.

Arnot Forest, New York, USA

The wild-living population of honey bees in Arnot Forest, New York, has probably been the best studied of any in the world. For several years it has been continuously monitored by Professor Thomas D. Seeley of Cornell University, who describes it in many publications[258]. In the world of natural beekeeping, the author is an exceptional figure. As one of the first scientists, he began to study not only bees functioning within the apiaries, but also wild-living bees and their natural instincts, which won him many enthusiasts in the wild/feral bee and the natural beekeeping communities around the world.

The aim of one of the studies conducted in 1978 was to find wild-living bee colonies in the Arnot forest, an area located near the scientist's university. It turned out that nine wild colonies were found in half of the forest, hence the (estimated) conclusion that there were 18 colonies in the entire forest[259].

In 1992, varroa first appeared in the region, which caused similar consequences as the mite invasion in most regions of the United States or Europe that is bees deprived of treatment began to die *en masse*. In the 1990s, the belief that all wild-living populations had died off with the invasion of the mite became increasingly popular in the United States. Seeley also thought so, but around 2000 people began to say that some honey bee colonies could have survived, so he decided to investigate these ideas. In 2002, when he covered a similar part of the forest that had studied previously, he confirmed the presence of eight wild-living bee colonies (again, he assumed that he had not managed to find all the colonies and that there might be twice as many of them in the whole forest). It turned out that the population was doing just as well as in 1978[260]. What helped it survive?

---

258 For more, see: T.D. Seeley, The Lives of Bees. The Untold Story of Honey bee in the Wild, Princeton University Press, 2019. I consider this book to be the most important book of our time on bees, as it shows their other face, unknown to most modern beekeepers. Seeley is also the author of several other fascinating books on honey bees, e.g. Honey bee Democracy, where he describes the ways in which bees (primarily swarms) make decisions.

259 The population of the Arnot forest is not large, according to estimates, more than a dozen colonies living at a density of one colony per square kilometre (on average). However, the area under study is also small The forest is located within the Appalachian Mountains, where there are probably other bee colonies living in the neighbouring forests, which cope in a similar way to those from Arnot and together they form a larger population. The one studied is certainly self-sufficient and genetically distinct in some way. However, it is not fully isolated. Apiaries are sometimes set up in the Arnot forest area. The professor fears that, with the potential increase in the density of bred and managed by humas bees, their impact on the wild population could be devastating.

260 The professor concluded that the difference in population size compared to data from twenty years ago is statistically insignificant.

Scientists from Cornell University catch swarms from the Arnot Forest to test for resistance to varroa. In the photo, PhD student David T. Peck, under a so-called swarm trap protected from bears, photo by T. Seeley.

At first, he thought that the colonies were able to avoid the varroa thanks to the fact that the bees lived in isolation. However, his research showed that all the colonies were infested by varroa. It also turned out that due to varroa the honey bees went through the so-called evolutionary bottleneck. The bees collected before the arrival of varroa (samples taken in 1978) clearly showed that the bees originated from 11 maternal lines, while the research in 2002 proved that there are only now three pedigree lines. The research on the DNA of the bees, both mitochondrial and nuclear, allowed us to draw the conclusion that the colonies inhabiting the forest today are descendants of the colonies studied in the 1970s from which the samples were taken. Interestingly, the analysis of nuclear DNA did not show any long-term limitation (reduction) in their genetic diversity. Over the years, the population has changed genetically, only to a small extent.

A tree hollow inhabited by a wild bee colony in the Arnot forest, photo by T. Seeley.

It is still dominated by genetic material from ecotypes brought from Europe, namely: *Apis mellifera mellifera* (Central European), *ligustica* (Italian), *carnica* (Carniolan), and *caucasica* (Caucasian). However, studies have revealed an admixture (less than 1%) of genes from two other ecotypes: *Apis mellifera scutellata* (Southwest Africa) and *Apis mellifera yemenetica*

(Sub-Saharan Africa and the Arabian Peninsula). Seeley links this fact to the influence of Africanized bees on populations bred in the southern United States. From there, packages are imported *en masse* to the northern states. However, since the admixture of African genes is small, the scientist rules out the possibility that the bees owe their immunity to this combination.

In his opinion, the ability of the Arnot bee population to survive is due to both environmental factors and genetic traits. The local bee colonies have very good living conditions, and it would be in vain to look for such in apiaries. They live in isolation, at a distance of at least several hundred meters from each other, which helps minimize the risk of threats from epidemics. They occupy natural nest sites: tree hollows with a unique micro-climate and good thermal insulation properties. Their small volume (between 20 and 60 litres) favours bee swarming, and thus the division of the mite population, which in the long term promotes those genetic combinations that can maintain the level of mite infestation at a level that is safe for them.

Probability of colony swarming (Sw)	Events post-swarming	Probability of the Event (E)	Probability of survival (S)	Overall probability (Sw x E x S)
0,87	Mother-queen occupies new nest	1	0,23	0,2
	Daughter-queen inherits old nest	1	0,81	0,7
	Daughter-queen leaves in afterswarm #1	0,7	0,12	0,07
	Daughter-queen leaves in afterswarm #2	0,6	0,12	0,06
Probability of not swarming (nSw)				Overall probability (nSw x E x S)
0,13		1	0,81	0,11

The probability of a colony swarming and the probabilities of various events thereafter, as survival to the following summer for various kinds of offsrping colonies.

The probability of bee swarming and subsequent events as determined by Prof. Thomas Seeley in the bee population of the Arnot forest, table by M. Uchman based on The Lives of Bees by T. Seeley.

Studies have shown that these Arnot bees have traits and characteristics that help to limit the mite population. The results of tests on the level of traits such as grooming (active cleaning of the bees from the mite), recapping (a phenomenon related to uncapping and capping the brood) or hygiene (cleaning freeze-killed brood) show that they perform significantly better than the ordinary population used in professional apiaries and only slightly worse than populations selected for these features in the breeding process. However, when they were subjected to the regime of modern apiary management, it turned out that they did not cope with the mite sufficiently to satisfy beekeepers.

Seeley calculated that bee colonies occupy a nest for an average of 1.7 years, which results from the poor ability of young colonies to survive the first winter. The probability of survival of a primary swarm (with an old queen) is only 0.23, while for the afterswarms it is only 0.12[261]. Death of these colonies (swarms) occurs mainly due to starvation, as they are unable to build the nest (draw combs) and gather adequate supplies in time for winter. Therefore, they have significantly lower chances of survival.

Contrary to popular belief, colonies in an old nest (which, according to beekeepers, are full of mites and infected by pathogens) can live quite a long time, on average as much as 5.2 years. Those that manage to survive the first winter are also able to survive subsequent winters, with a very high probability of 0.81. It turns out therefore (despite the concerns of beekeepers) that an old nest, even if full of mites, is not the biggest problem for bee colonies. A much more dangerous killer is starvation. However, this applies primarily to young colonies, i.e. swarms in their first year of life.

Since 2010, Seeley has been checking the continuity of the nesting holes three times a year to eliminate the risk of an error in estimating the lifespan of a colony (i.e. to rule out the possibility that one of the colonies dies and is replaced by a new swarm). Each time, he considers the presence of bees carrying pollen, to reduce the probability that the flying bees are robbers or scouts looking for a new home. It turned out that one of the nest sites observed by the professor was constantly occupied for seven years. Monitoring shows that the population in Arnot Forest is therefore more or less stable.

---

261   Probability is defined in the range from 0 (impossible event) to 1 (certain event). The probability of survival of bee colonies at the level of 0.23 means that statistically, out of 100 colonies, 23 will survive (23%).

Pennsylvania, USA

Wild bee colonies in Pennsylvania were studied, among others, by Margarita Lopez-Uribe's team. However, the work did not concern the description of the population (the distribution of bees, their nest sites, etc.), but primarily genetic features, or rather the immune response of bees to pathogens. It is therefore difficult to refer to the population in detail. However, the research by Lopez-Uribe's team shows that there are wild-living bee colonies in the state that differ in their characteristics from the managed bees, which suggests that this population is self-sufficient and distinct from apiary population (although it is not physically fully isolated from it). Pennsylvania borders New York State (Arnot Forest is several dozen kilometres from the state border). Both states lie within the Appalachian Mountains, a mountainous and forested region. It is therefore possible that the studied bee population was created in a similar way as that in the Arnot Forest and perhaps they even inter-breed.

Lopez-Uribe stated that the mortality rate of this population is comparable to the annual mortality rate of domesticated bees in the United States, which is about 40%. A large part of the colonies dies each year, but many of them can survive at least three winters. The population ensures the continuity of generations. Comparative studies of these bees with non-resistant bees from apiaries have shown that wild colonies from Pennsylvania have a much better immune response, both to pathogenic bacteria and DWV. It is thanks to this feature that they can cope much better with the pressure of pathogens and survive, despite not being treated for varroa.

Arizona, USA

In the 1990s, there was a great fear of invasion of the so-called Africanized honey bees, in the southern United States.

Africanized honey bees are a cross-breed between European subspecies and other subspecies from Africa (mostly *Apis mellifera scutellata*). This mix was created in South America in the 1950s, when a few people knew about the varroa mite. The swarms escaped from the experimental apiaries, and the cross between the bees turned out to be exceptionally aggressive[262].

In connection with this, in Arizona, in the years 1992-2000, monitoring of about 250 nest sites (mainly in rock crevices) occupied by wild bees began. These observations coincided with the invasion of varroa. It turned out that the appearance of the mite caused the local population of European bees to drop from 160 colonies to about 10 in six years but since then began to recover.

The influx of Africanized bees occurred at a time when most of the local colonies had been lost because of varroa but due to the Africanised bees' exceptional resistance to the mite the wild-living population of bees was recovered in Arizona. It seems that the march of these bees north was easier precisely because the field was cleared for them by varroa, which almost eliminated the competition (local bees), reducing the number of drones from these colonies almost to zero[263]. It is difficult to say what happened to the remnants of the European subspecies of bees. Perhaps they

---

262 In South America, the crossbreeds of subspecies brought from Europe used there were not performing well enough according to local beekeepers (which didn't check in their economic assumptions) because these bees were accustomed to the changing seasons from cold to warm, not hot and dry to humid and wet; they were not also accustomed to the (sub)tropical climate of some regions of Brazil and neighbouring countries. Therefore, in the mid-20th century, as part of work on increasing the productivity of bees, it was considered rational to import honey bees from Africa (from regions of similar latitude). In this way, African bees, including the subspecies *Apis mellifera scutellata* (dominant among those imported), was introduced into South America. Crossbreeds of European and African subspecies were called Africanized bees (as opposed to African bees, i.e. those that did not have an admixture of European subspecies genes). According to one account, the African breeds imported to Brazil (carefully selected) had to be disposed of (killed) due to legal/procedural problems. The next batch was supposed to be imported in a hurry, hence the crossbreed's produced bees that should rather be called aggressive than defensive, see: https://www.beekeepingtodaypodcast.com/dr-warwick-kerrs-passing-and-his-africanized-honey-bee-legacy-with-jim-tew-011/, accessed: November 18, 2022. I have not found sources that clearly describe how the bees got out of the researchers' control, some hypotheses assume the escape of swarms, others talk about sharing genetic material with local beekeepers, which turned out to be more productive in that climate than the European subspecies bees used in apiaries. Undoubtedly, however, Africanized bees achieved exceptional evolutionary success in that region, in some areas completely replacing European hybrids. Some claim that sexual selection in Africanized bees leads in the long term to the removal of European genes from the population genotype (thus Africanized bees become African bees again). This is not ruled out since similar selection characteristics have been confirmed in some ecotypes of Central European bees (*Apis mellifera mellifera*). I do not consider myself a specialist in Africanized bees, so I will leave any verification of these anecdotes and hypotheses to the readers. It could probably be an interesting excursion through the history of beekeeping on both American continents.

263 Drones from European subspecies colonies remained in the area, after all they were kept in local apiaries. Beekeepers in the southern states of the USA claim, however, that despite their efforts, they have difficulty preventing the genetic traits of the more aggressive Africanized bees from penetrating their apiaries. It is claimed that these drones may be more attractive to queen bees than of European races; it is also said that they are smaller, faster and more agile in flight, which may increase their competitiveness with males of the European subspecies.

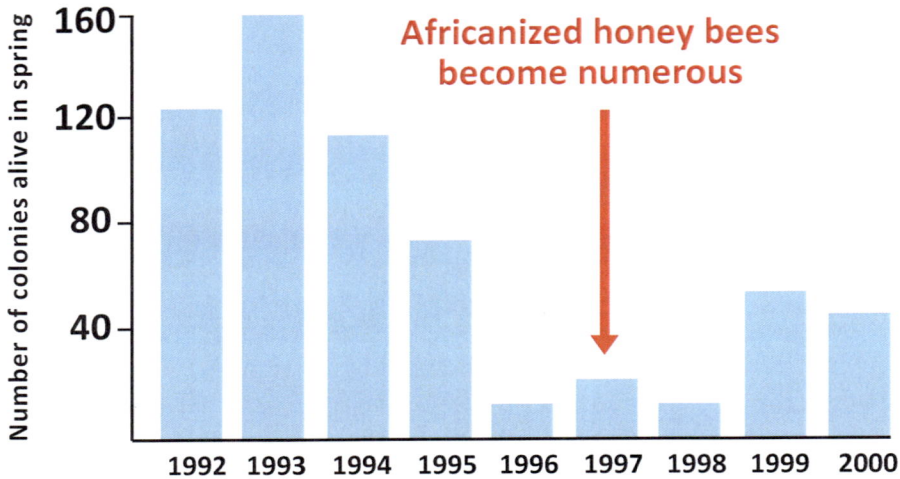

Africanized honey bees become numerous

Number of colonies alive in spring (y-axis: 40, 80, 120, 160)

1992 1993 1994 1995 1996 1997 1998 1999 2000

Monitoring results of known bee nest sites in the mountains of Arizona (USA) documented the collapse of the wild-living bee population with the invasion of varroa mites, figure by E. Luty based on The Lives of Bees by T. Seeley.

died out over time, perhaps they mixed in with the Africanised population. Although monitoring of this population was stopped after a few years (so we do not have documented sources regarding its current size), numerous opinions indicate that the population is doing well and is resistant to the mite[264]. Local authorities and beekeepers decided to remove these bees from unmanaged nest sites as soon as they learn about them, especially when the swarm settles in a place where it could pose a threat to people. Sometimes colonies are killed, or beekeepers try to replace the queen with one from their gentle bred lines. Some try to minimize the impact of drones from wild-living colonies in their own ways (using various methods of queen insemination). However, there are also those who accept the reality and try to run apiaries based on Africanized bees, even though this method is more difficult and usually involves working in thick overalls, which is a challenge in the hot climate of Arizona. They also claim that they use additional apiary equipment such adhesive tape, which they use to carefully seal any leaks in their suits[265].

---

264 As far as I know, the topic is not the subject of scientific research.

265 It is believed that over time, the Africanized bees have become gentler. However, there are still cases of stings and even deaths (most often due to the lack of due care by people allergic to bee venom).

Cuba

A team of scientists led by Anais R. Luis confirmed that Cuba has the largest population of bees resistant to varroa, consisting exclusively of colonies originating from subspecies introduced from Europe (*A.m. mellifera, A.m. ligustica, A.m. carnica, A.m. caucasica*)[266]. This fact contradicts the opinion of many beekeepers who believe that the European subspecies of the western honey bee are less resistant to diseases (especially varroosis) than the subspecies originating from Africa and are unable to develop full resistance to the disease. In Europe and the United States, only relatively small populations of bees resistant to varroa have been observed, while in Cuba such population is not only large (estimated at 220,000 colonies!), but also occurs throughout the island (area of over 100,000 square kilometres)[267]. Scientists confirm that no acaricides have been used in apiaries in Cuba for at least two decades (in the first period after the varroa invasion, the so-called drone brood removal was used). Despite the presence of the deformed wing virus (DWV) and the Korean haplotype of *Varroa destructor* in Cuba the population is resistant, having a well-developed recapping trait and the ability of bees to remove a large percentage of mites from the brood (it has been estimated that on average, workers are able to remove 81% of infected cells). A low ability of mites to reproduce was also observed in female mites. Scientists point out that Cuba is an excellent example of how bees can naturally adapt to the presence of varroa with minimal human interference.

Subspecies *Apis mellifera scutellata, Apis mellifera capensis*

The subspecies *Apis mellifera scutellata* and *Apis mellifera capensis* originate from southern and southwestern Africa (south from Kenya to South Africa). It is estimated that in Africa, wild populations significantly dominate over those kept in apiaries. Barbara Locke from the Swedish Agricultural University in Uppsala estimates, for example, that in the Republic of South Africa, where the *A.m. capensis* ecotype dominates, 99% of the bee population lives in the wild in its natural state and only one percent is kept in apiaries. Thanks to this, after the invasion of varroa, these populations underwent

---

266   A.R. Luis, Recapping and mite removal behaviour in Cuba: home to the world's largest population of Varroa-resistant European honey bees, Scientific Reports 2022, 9.

267   The second largest resistant population of European honey bee subspecies is that found in North Wales (estimated at several hundred bee colonies).

a process of natural selection in a way that was not possible in Europe, for example. It is reported that in many regions of Africa, beekeepers did not even notice that the bees had become infested with the mite. Thus, the mite did not affect the productivity of the bees in any way.

Stephen Martin also claims that in some bee colonies from the South African population there could be as many as 20-30 thousand mites[268] and yet these colonies did not show any symptoms associated with varroosis, which, according to the researcher, can be explained by the lack virulent forms of DWV at that time. Subsequently, the bees have developed certain behaviours that allow them to limit the mite population growth.

The African and South American populations of the subspecies *Apis mellifera scutellata* have practically the same characteristics: they have a very low mite population growth, and if necessary, they are more likely to leave an infested nest (which is called absconding; the Eastern honey bee *A. cerana* has a similar adaptation). *A.m. scutellata*, similarly to *A.m. capensis* has a shortened pupal development phase (sealed brood), which significantly reduces the reproductive success of the mite. This bee also builds smaller comb cells, which, according to scientists, may in some situations make it difficult to for the young female mites to be mated[269].

Africanized bees from Brazil began to displace European breeds because they were better adapted to local conditions and varroa.

---

268 L.E. Brettel, S.J. Martin, Oldest Varroa tolerant honey bee population provides insight into the origins of the global decline of honey bees, Scientific Reports 2017, 4.

269 Within the subspecies *A.m. capensis*, a very interesting, unique and parasitic survival strategy has evolved, consisting in taking over the nests of bee colonies of another subspecies (*A.m. scutellata*). The so-called pseudo-clones, derived from a common ancestor from the capensis subspecies, can lay diploid eggs (i.e. containing a double set of chromosomes, which leads to the development of female, worker bees; in contrast to haploid eggs, i.e. with a single set of chromosomes, from which only males (drones) hatch. All pseudo-clones are genetically identical. The strategy consists of the pseudo-clones (sometimes even in small numbers) invading a bee colony of *A.m. scutellata*, then laying eggs in it, from which new 'capensis' workers hatch, having the same ability to lay more diploid eggs, and over time their number grows, and they take over the nest along with all the resources gathered in it. Quite quickly, they eliminate the (*scutellata*) queen from the colony, which eventually leads to the death of the colony (by depopulation of the workers). Scientists have observed that when larger workers (by an average of 8% than workers of *scutellata* bees) of pseudoclones of *A.m. capensis* lay eggs in very small cells built by *scutellata* workers, the male varroa are "trapped" in the upper parts of the worker cells by the much larger larva and cannot reach the females to fertilize them. In this way, due to the parasitic behaviours of this honey bees, the reproductive success of varroa is greatly reduced. Although this stops the development of the varroa population, it does not affect the survival of the colonies, because they die due to the invasion of the parasitic bees: S. Martin, P. Kryger, Reproduction of *Varroa destructor* in South African honey bees: does cell space influence Varroa male survival? Apidologie 2002, 1-2. Scientists still argue whether the small cell effect is also beneficial in other populations. Regardless of the issue of resistance to varroa, the example of pseudoclones of capensis bees shows how diverse adaptive adaptations can be even within the same species, and that evolutionary mechanisms affect not only entire bee colonies, but also individuals.

The population of wild bees in South America is now dominated by hybrids of the subspecies *A.m. scutellata*. It was observed that in Africanized bees in South America, in the first phase of the mite invasion, only 50% of the parasite's females were fertile. Later, this value increased to over 80%, which was associated with a change in the haplotype of the mite from Japanese (which originally infected this population) to Korean. However, this did not significantly increase the mortality of bees.

Tunisian Honey bees

Apis mellifera intermissa is a subspecies found in northern Africa, including Tunisia. These bees, like the South African ecotypes, do not require treatment with chemicals. They were characterized by a high level of grooming and hygienic behaviour. The subspecies was used for breeding crosses, for example by John Kefuss. It also retained its resistance in crosses with other breeds.

Fernando de Noronha, Brazil

The population of *Apis mellifera ligustica* (Italian honey bee) on the Brazilian island of Fernando de Noronha consists of several dozen bee colonies living in several apiaries, and an indeterminate number of wild-living colonies (also several dozen colonies). It has been infested with varroa since 1984. The infestation is periodically very high, reaching even several thousand individuals (initially it was estimated at a level of even 25%), and despite this, it does not lead to an increase in colony mortality (i.e. the bees have no health problems).

Professor Martin believes that the survival of bees in conditions of high infestation of the mite was possible due to the absence of virulent strains of DWV. The studies detected very small amounts of the virus's genetic material, and no detectable levels of the virulent versions of DWV. Recapping or mite removal was not studied in this population.

The bees living on Fernando de Noronha island are resistant to varroa locally. However, the colonies originating from queen-mothers from this population were tested in comparison with the stock subject to the mites in Germany. In case of both groups, the results of the experiment were similar: the bees turned out not to be resistant, and the lack of treatment resulted in their death.

Professor Martin believes that the population of Fernando de Noronha is doomed to inevitable extinction. It is only a matter of time (it could happen very quickly or in a few decades) before the small population is (indirectly) destroyed by the mite. The bees will die if there is a mutation (or introduction) of a virulent strain of Deformed Winged Virus on the island, which would then be able to multiply uncontrollably[270].

Primorsky Krai, Russian Honey bees

The Maritime Krai (Russian: Primorsky Krai) is a part of the eastern coast of Russia. It lies in the northern part of the natural range of the eastern honey bee (*A. cerana*). Western honey bees (*A. mellifera*) were transported there, probably from Eastern Europe, in the mid-19th century. It is therefore believed that the first contact of the honey bee with the *V. jacobsoni* may have taken place in the Primorsky Krai region. It cannot be ruled out that it was then that the first host change of the honey bee parasite, varroa, took place. The Primorsky bees (also called Russian bees in the English-speaking world) probably owe their resistance to the process of natural selection that shaped this line.

The breed was brought to the United States by the United States Department of Agriculture. After research confirmed its greater resistance to varroa, it was then made available to commercial breeders[271]. Russian bees are quite often kept in apiaries in the USA, where the bee mite is not controlled. They are bred, for example, by Kirk Webster[272]. They are characterized by a high level of hygiene and grooming instinct. It was also found that their brood is less attractive to the mite and almost every second female varroa is infertile[273].

---

270 Professor Martin suggests that the population is probably too small to survive infection with a virulent form of the virus and develop immunity.

271 To this day, Russian bees are the subject of research and breeding work in the facilities of the US Department of Agriculture; this approach to the issue should be, in my opinion, a model for many European facilities (including certainly Polish ones).

272 According to beekeepers, these bees are similar in their behaviour to Central European dark bees, from which they in a sense originate. Beekeepers, especially professional ones, do not like some of their characteristics. This may be the reason for their relatively low popularity in apiaries. Beekeepers prefer to use chemicals than to keep and breed bees whose characteristics they do not like.

273 Characteristics, history and attempts to select both immune and economic traits of Primorsky bees, see: T. Rinderer, S.E. Coy, Russian Honey Bees, Salmon Bayou Press, 2020 (unfortunately, I have not managed to obtain this book so far). K. Webster also mentions this breed many times, see: Feral bees; https://kirkwebster.com/feral-bees/, accessed: February 2, 2025.

Gotland, Sweden

In 1999, a very interesting experiment was started on Gotland, the aim of which was to check whether bees can develop a balanced relationship with varroa under conditions of partial isolation and complete abandonment of treating procedures with minimal or no management. In eight locations on the southern tip of the Swedish island, 150 colonies were placed in hives with a volume of about 80 litres. Each of them was infected with several dozen mites[274].

The place where the research was conducted is surrounded on three sides by the waters of the Baltic Sea, and on the northern side the headland is connected to the main part of the island by a narrow isthmus.

At the beginning it was claimed that the experimental population was isolated from other bees. However, it turned out that at least since 2009 (10 years after the beginning of the research) other apiaries, where varroa treatments were being carried out, had been located about 10 km north of the headland. I was unable to find information whether the bee colonies were placed there earlier or only in 2009. I also do not know whether they were part of the assumptions of the main experiment, but it is certain that samples of varroa were taken from them later for research (as a comparison group).

The experiment used bees from standard apiaries, which were kept in hives but were left to their own devices, limited only to feeding colonies that had not managed to gather sufficient supplies for the winter. At first, the colonies were not very infested with varroa. In the second season, the colonies were still strong and swarmed, but the mite infestation began to increase rapidly, leading to the collapse of about 30% of the population. In the third season, the health of the colonies deteriorated further, the bees were too weak to produce swarms. After the winter, about 80% of the remaining colonies collapsed. Only 21 colonies survived until the fourth season, but more than half of them did not survive the following winter. By 2003, only nine colonies remained. However, it turned out to be a breakthrough year, since from that period on, the health of the colonies began to stabilize. First, the level of varroa infestation decreased significantly. In addition, one of the colonies produced a swarm. In the following years, swarms also appeared in other colonies, and the population mortality was relatively low.

---

274   I. Fries, Survival of mite infested (*Varroa destructor*) honey bee (*Apis mellifera*) colonies in a Nordic climate, Apidologie 2006, 6.

The smallest number of colonies was recorded in the spring of 2004, only seven, of which only six were original bees and one swarm from 2003. In 2005, the experimental population consisted of 13 colonies: five colonies from the initial population (3.33%), a swarm from 2003, three swarms from 2004 and four swarms from 2005.

For the next 10 years after the end of the first phase of the experiment (2005-2015), the colonies were left to their own devices. During this time, the population remained at a stable level of 20-30 colonies and had become completely self-sufficient.

Bee mortality in the Gotland experiment in the first years

Yer	Parameter	Original colonies	Swarms 2000	Swarms 2001	Swarms 2003	Swarm 2004	Swarms 2005	Total number of colonies
1999	Wintered	150	–	–	–	–	–	150
2000	Survived	142	–	–	–	–	–	142
2000	Wintered	130	16	–	–	–	–	146
2001	Survived	92	12	–	–	–	–	104
2001	Wintered	91	12	17	–	–	–	120
2002	Survived	21	8	8	–	–	–	29
2002	Wintered	17	4	4	–	–	–	21
2003	Survived	9	0	0	–	–	–	9
2003	Wintered	8	0	0	1	–	–	8
2004	Survived	6	0	0	1	–	–	7
2004	Wintered	6	0	0	1	4	–	11
2005	Survived	5	0	0	1	3	–	9
2005	Wintered	5	0	0	1	3	4	13

Table by M. Uchman

In 2006, comparative studies were conducted on the resistant bees that survived the experiment and bees from other parts of Sweden. They also used varroa from different sources to make sure that the parasite on Gotland had not changed its virulence enough to affect the results. It turned out that in the "genetics" of the Gotland bees, the population growth of the mite was on average 82 percent lower compared to other colonies, regardless of the origin of the mite, which gave grounds for the conclusion that the ability to limit the rate of reproduction of the mite in bee colonies from Gotland was related to the adaptation and development of the bees' resistance.

Studies have shown that bees have developed mechanisms that limit the reproductive capacity of varroa. For example, it was found that only 50% of female mites produced fertile and viable offspring, while in a normal (i.e. normally maintained) apiaries were treatments

VARROA MITE SAMPLES OBTAINED FROM THE APIARY OF THE UNIVERSITY OF UPPSALA (CONTROL GROUP)

EXPERIMENTAL APIARY UNTREATED SINCE 1999 (RESISTANT BEES)

CONTROL GROUP (SUSCEPTIBLE BEES)

For comparative studies of mite-resistance, samples were taken from the experimental apiary, from the apiary across the isthmus, where mite populations are treated, and from the university apiary in Uppsala, figure by E. Luty.

are used this value reaches 80%. Probably the most important in terms of the ability to reduce the mite on the island was recapping. It was also found that the surviving bees from Gotland formed smaller colonies than before and were more prone to swarming.

The Gotland experiment also showed that bees can remain in a crisis caused by varroa for several years but then emerge from it and develop healthily. During this time, they look weak, they may even have problems with gathering food for the winter, they emerge from the winter weakened, which is why beekeepers often combine them and kill the queens. Meanwhile, such colonies may have the potential, to overcome the problem and develop resistance to the mite and immunity to the pathogens.

I think that every beekeeper who has tried his hand at beekeeping without chemicals could have observed[275] such colonies in his own apiary.

Equally interesting are the results of the study of the mites that lived on the Gotland bees[276], which used samples of the parasite secured in 2009, 2017 and 2018, both in the experimental population that had not been treated since 1999 and in those colonies undergoing treatment that lived about 10 km north of it, across the isthmus. Mites from the Uppsala University apiary were also included as a comparison group from outside Gotland. It was found that over almost a decade the genes of the mites in the experimental population had changed. In 2009, they did not differ much from the comparison group from across the isthmus. During the first decade of the experiment, the mite probably did not change significantly. However, it turned out that in the second decade the parasite in resistant colonies underwent greater genetic differentiation, while in the comparison population it remained relatively uniform. There was also strong selection against the most common genotypes of the mite. The authors suggest that the parasite, despite its low genetic diversity, quickly adapted to coexist with the honey bee, which in effect increased the mite's chances of survival in the new conditions. The change in the ecological situation therefore influenced the process of co-adaptation of the bee and the parasite.

After about a decade of stability, the bee population on Gotland began to decline around 2017. Why? It's hard to say.

Scientists have undoubtedly confirmed the presence of mite resistance traits in the Gotland bees. For some reason, this turned out to be insufficient: the bees were no longer able to cope with the mites to a degree that would allow them to maintain a good condition. The population began to shrink, and the colonies began to have health problems. It was therefore decided to carry out treatments to combat the mites. It was decided that for the good

---

275　More than one colony in my apiary managed to get out of the crisis. One year, a micro-colony survived the winter in just one frame of bees, and in the next season it reached the strength of four boxes of my hive (around 100 litres in volume).

276　A.L. Beaurepaire et al., Population genetics of ectoparasitic mites suggest arms race with honey bee hosts, Scientific Reports 2019, 8; see also studies where differences were found in the genome of the virus revealed in a bee colony from southern France, where varroa was not treated for 17 years. The researchers suggest that characteristic parts of the virus genome could have been shaped in the process of co-adaptation of bees and the pathogen, A. Dalmon et al., Evidence for positive selection and recombination hotspots in Deformed wing virus (DWV), Scientific Reports 2017, 1.

of science it would be better to keep the bees alive to continue research on the resistance mechanisms they had acquired. In my opinion, this disrupted the continuity of the experiment. I believe that this population was valuable precisely because it was able to function without treatment for such a long period[277]. Moreover, I find it incomprehensible that throughout all these years the Gotland bees were not widespread in breeding and production apiaries. After all, it was possible to work on their productive features while maintaining their resistance characteristics[278]. This unfortunately shows that, on the one hand, the interest in "resistant genetics" among beekeepers is negligible (otherwise they would force its dissemination), and on the other, that the achievements of scientists do not always translate into practice. Unfortunately, this fact fits into the picture of the passivity of the beekeeping community in the field of keeping and breeding resistant bees.

It is difficult to say why the bees from Gotland stopped coping with varroa. At most, we can venture some hypotheses. It seems that the reason could be that the partially isolated population was severely limited by the "bottleneck effect", which over the years led to negative effects of inbreeding. It may also be the result of the growing popularity of beekeeping on Gotland. The beekeepers there treat the bees (they do not believe in the possibility of keeping them alive without treatment, even though for almost two decades there were bee colonies in the neighbourhood in which the mites were not controlled), so the non-resistant genes could have become dominant over time.

---

277 I find it difficult to understand the reasons for the decision to treat the experimental population from Gotland. If we were to consider that further studies were to be carried out on its abilities and genetic traits responsible for fighting the mite, it would have been sufficient to raise the queens from this population and place them in other colonies on which the studies could be continued. Of course, it would have been important to create as many such colonies as possible to eliminate the threat of inbreeding and to provide a maximum number of potential genetic combinations to illustrate the complexity of the experimental population. Then it would have been possible to observe the fate of the bees on Gotland, in accordance with the original assumption, which was to see whether the bees would cope on their own. In my opinion, from a scientific point of view, the collapse of the entire population would be as valuable an outcome as any other. However, if non-genetic factors influencing immunity (e.g. the micro-biome of untreated bees) were to be studied further (i.e. after the treatment), then such treatments would have completely distorted the potential results. Any results of such studies suggesting some kind of resistance based on non-genetic factors could then be questioned at the methodological level.

278 In a similar way to how the US Department of Agriculture with cooperation with breeders distributed the imported Primorsky bees, still researching them and working on their characteristics as low swarming and higher productivity.

Avignon and Le Mans, France

In the 1990s, a decision was made in France to collect bee colonies that had survived the first phase of the varroa invasion despite not being treated. They were collected in two centres: Avignon, in the south of France, and Le Mans, in the north-west. They were mainly wild colonies or those from abandoned apiaries. Bees that had owners were also included in the project, provided they had not been treated for at least two years. In this way, 82 colonies were collected: 52 in Avignon and 30 in Le Mans. Both populations were managed with minimal intervention, limiting the management to collecting honey. It turned out that between 1999 and 2005, mortality in both groups were the same as in the colonies in their vicinity which were susceptible to varroa and treated. In the population of resistant bees, the mite infestation was recorded three times lower than in the treated colonies, but honey production was on average twice as low. Studies of the population in Avignon conducted in terms of the gene expression for adaptation to environmental factors showed that the local bees respond more easily to local conditions. In addition, they are effective in detecting and cleaning mite infested cells. It was assumed that the resistance of bees depends primarily on hygienic behaviour, and especially the VSH trait (Varroa Sensitive Hygiene)[279]. In the Avignon bees, the reproductive success of the mite was 30% lower than in the group of non-resistant insects. Also, the percentage of infertile females of varroa mite, was higher than found in the Gotland population. French populations not subjected to treatment, maintained resistance, even though they were not isolated from other bees. The disadvantages of the selected genetic material were lower honey production and greater defensiveness.

The British Isles

Beekeeping in the British Isles is specific in many ways, which of course translates into the uniqueness of the local population. It is not possible to discuss all its aspects, one can only attempt to show the potential, although this will necessarily be only a sketch, because the British population is not a monolith. There are many beekeepers in the Isles who do not use any

---

279   This is a type of hygienic behaviour (cleaning out brood); however, the specificity of VSH is that the behaviour is aimed directly at the mite infested brood (only pupae infested with reproducing mites are removed), and not, for example, bacterial or fungal diseases, as in the case of standard hygienic behaviour.

treatments against varroa[280]. These are largely people fully devoted to natural beekeeping, hobbyists who take special care to provide bees with good living conditions and development. More and more voices are coming from the beekeeping community in Great Britain testifying to the very good condition of

Przemysław Kasztelewicz has been involved in the breeding of local bees in Cambridgeshire, UK, for many years, photo: P. Kasztelewicz.

the bees without the need to apply any treatments[281]. And although these are primarily anecdotal data (most often beekeepers' have detailed accounts of how they run their apiary), in my opinion these voices should be considered completely credible, if only because they are really numerous. It is also worth adding that the official statistics of the British Beekeepers' Association show that during the 2021/22 wintering period, nearly one third (33.1%)

---

280   I have met at least a few beekeepers from the UK: e.g. Heidi Herrman, Jonathan Powell, Peter Brown, John Harveson; I have corresponded with David Heaf, Clive Hudson, Steve Riley and the Ibbertson brothers, among others; none of these people treat bees, most are hobby beekeepers. Other non-treating beekeepers include Ron Hoskins and Phil Chandler. All the people I have met or corresponded with claim that there are other non-treating beekeepers in their neighbourhoods.

281   There are many publications, and they are easy to find (on the Internet www.varroaresistant.uk or in the trade press, e.g. Natural Bee Husbandry).

of beekeepers did not perform any treatments against the pest[282]. Among those who do not apply treatment, there are professional beekeepers, but there are also those who do not collect honey at all (only from dead colonies).

Others only collect the excess honey after accounting for what the bees need for overwintering. Many of these beekeepers observe free-living colonies in their area, claiming that there are a lot of them (some observe even a dozen or so in the immediate vicinity). These colonies often function completely without human interference for five, 10 years, or even longer. Moreover, due to the way some of those people keep their colonies in their apiaries, we could classify at least some of the bees as free-living – the management is minimal or even no-existent.

One of the beekeepers in the British Isles who does not use treatment against varroa is Przemysław Piotr Kasztelewicz[283]. He has been around bees since childhood, under the watchful eye of his father, at the Sądecki Bartnik Apiary Farm in southern Poland, so setting up an apiary in England, where he lives, was somewhat natural. He began his beekeeping adventure in the British Isles in 2011 and immediately decided that he would not treat the bees. Like most beekeepers who make such decisions, he encountered critical voices from all sides.

At that time, he decided to cease treatments a significant part of beekeepers in the British Isles (and this is still the case in continental European countries) believed that bees would not cope without human help, and that those living outside the hives were only one-year swarms. Przemysław Kasztelewicz, like many beekeepers who care about propagating the resistance of bees, built his apiary on the potential of captured swarms from wild-living colonies. He also acquired bees from unmanaged nest sites, but he only transplanted them into hives when the situation required it (e.g. the owner asked for intervention because he did not want to have bees under the roof of his house, the wind blew down a tree with a bee nest in it). Over time, Kasztelewicz got to know both new people and the area. It turned out that bees were everywhere: in chimneys, roofs, walls, old tree hollows. By 2022, he had located more than 50 free-living bee colonies in the Cambridgeshire area and began monitoring them.

---

282 BBKA data [The British Beekeeping Association], https://www.bbka.org.uk/news/84-of-beekeepers-who-replied-to-survey-had-no-winter-losses, accessed 2 February 2023.

283 Owner of Cambridge Honey bees, a company engaged in the keeping and breeding of local bees and the production of nucs; www.cambridgehoney bees.co.uk.

Losses in the apiary, even though its owner did not use treatment against varroa, usually did not exceed 10% per year, and the bees remained in good condition. For this reason, the number of colonies in the apiary in recent years grew so quickly that there was often a shortage of equipment. It soon turned out that the apiary already had 200 hives. At that time, Przemysław Kasztelewicz began to wonder where other beekeepers' problems with keeping their bees in good health come from. In his opinion, the bees owe their good condition to natural selection and a favourable environment. Cambridge's home gardens, where no chemicals are used, provide a rich and diverse food supply for the bees, and the large number of available nest sites in which the bees can live without the supervision of beekeepers allow unhampered natural selection. Przemysław Kasztelewicz also emphasises the importance of introducing bee-friendly breeding methods, such as stimulating propolis production or using natural combs or foundation with a small cell size[284].

It turns out that in such conditions varroa may not be such a big problem for bees.

It is hard to say what is the secret of bee survival in the British Isles, many people from continental Europe use similar methods to those of British beekeepers, but they do not have such good results. Beekeepers in the UK believe that the good condition of bees is largely due to the friendly management, such as little interference in the nest, acceptance of the natural life cycle, acceptance of escaping of swarms, etc. However, these are only hypotheses. I think that the density of the bee population, is also important. According to official data, on average there is no more than two bee colonies per square kilometre in the British Isles, which is several times (4-5 times!) less than in Poland[285].

---

284 P. Kasztelewicz most often uses frames with a strip of foundation or, at most half a sheet of foundation. This allows for maintaining straight combs and gives the bees a lot of freedom.

285 This is a very important factor from the point of view of parasite pressure. A study conducted in Germany showed that the pressure of varroa from outside (invasion from other colonies) is much lower in areas with lower colony densities (Black Forest) than higher colony densities (in the southern part of Baden the population density is more than 4 colonies per sq. km); E. Frey, P. Rosenkranz, Autumn Invasion Rates of *Varroa destructor* (Mesostigmata: Varroidae) Into Honey Bee (Hymenoptera: Apidae) Colonies and the Resulting Increase in Mite Populations, Journal of Economic Entomology 2014, 4. During the 3.5 months of the experiment, in the colonies located in regions with low bee density, the level of invasion was found at the level of 72-248 mites, while in regions with high bee density it was 266-1171 mites. In the untreated colonies from both groups, in November, the average autumn infestation was found at the level of 340 in the first group and over 2000 in the second group.

The lower bee-density significantly reduces horizontal transmission of parasites and pathogens. It is possible that the good condition of the bees is due to the insular nature of Great Britain, because it seems to me that external trade in bees (importing queens) may be difficult[286]. For this reason, colonies are better adapted to local environmental conditions and local strains of pathogens. Many representatives of beekeeping in Great Britain believe that the condition of their bees is due to the high tolerance of bees to viruses, especially DWV.

The image of beekeeping in Great Britain is also based on the accounts of scientists and is reflected in their research. One example is the apiary of the late Ron Hoskins from the Swindon Honey bee Conservation Group[287]. The study of bees from this apiary confirmed, among other things, the principle of the displacement of more virulent forms of the DWV virus by less virulent ones[288].

A major contribution to scientific work on varroa and the pathogens it transmits is also made by the British scientist Prof. Stephen Martin, who distinguishes the development of natural resistance to mites (Natural Resistance to varroa) from the resistance selected by beekeepers.

According to the English researcher, such mite resistant populations are growing in Europe, with the largest number in Great Britain. Stephen Martin believes that the bees owe their good condition, among other things, or perhaps even primarily, to the low population density (beekeeping is mainly practiced by hobbyists, who usually keep no more than a few colonies, and bees are not transported long distances, as is the case in the USA). What else determines the good condition of bees in Great Britain? There are several reasons, explains the researcher:

"There's a strong possibility we have an increasing number of people now that just don't treat. We've got one area in North-West Wales with about 100 beekeepers keeping 500+ colonies that have not treated for 16 years now. The problem in the UK is that its government policy remains that you must treat. So, there are a lot of people who just don't mention it. And every

---

286   Queen bees are still being imported to the British Isles, much to the opposition of supporters of natural beekeeping and the local black bee. Natural beekeepers had hoped that thanks to the so-called Brexit, importing bees might become more difficult and marginal - however, it seems that the UK's exit from the EU has not had a noticeable effect on the scale of importing queens from outside. Still, I think this is much lesser an issue than what is happening in continental Europe, especially Poland.

287   See: www.swindonhoney beeconservation.org.uk.

288   G.J. Mordecai, Superinfection exclusion and the long-term survival of honey bees in Varroa-infested colonies, The ISME Journal 2015, 10.

time I go around to give a talk to beekeeper groups, I get people coming up afterwards saying, "Oh, I didn't like to say but I haven't treated for 10 years." There's even a group called the Naturalist Beekeepers who refuse to treat. And I always thought, "Well, they'll just die out, varroa will kill them." But they're thriving and their numbers are growing. We've done an instructional leaflet and have a website (www.varroaresistant.uk) that lays out all the science. It is important to note if we just let Darwinian beekeeping go, you may just end up having too many mites, and basically, even if you had highly resistant bees, they would still probably die because you may have too many mites in the environment. But in the UK, I think in 10 years' time, most people will not treat for varroa. We know of the halo effect, we've seen it, so when people who are not treating, beekeepers nearby can benefit with their colonies becoming resistant as well. And all the selection happened in the wild colonies. Most of these beekeepers are just originally collecting them from the wild and then when they didn't die, they then split the colonies building up their stock from that"[289].

Some time ago, The Guardian newspaper reported on a discovery: in the oak woodland complex at Blenheim Palace in southern England, a unique population of wild-living bees was found, consisting of about 50 colonies[290]. This was considered exceptional for several reasons.

Firstly, the bees are in good condition, even though, naturally, no one is fighting the mite here. Secondly, it was discovered that the population has specific features of an evolutionarily formed ecotype characteristic of the British Isles within the Central European subspecies (*A.m. mellifera*)[291].

289   Natural Varroa-Resistant Honey Bees and Small Hive Beetles, „Two Bees in a Podcast", https://entnemdept.ufl.edu/media/entnemdeptifasuflecu/honeybee/pdfs/two-bees-in-a-podcast-transcription/Episode-35-Mixdown-PROOFED_otter_ai.pdf, Accessed February 02, 2025.

290   D. Ferguson, 'No one knew they existed': wild colonies of lost British honey bee found at Blenheim, https://www.theguardian.com/environment/2021/nov/07/no-one-knew-they-existed-wild-heirs-of-lost-british-honeybee-found-at-blenheim, Accessed Nov. 19, 2022.

291   Although almost the entire beekeeping community of the British Isles appreciated the significance of the discovery, for most natural beekeepers it was not really a surprise. Jonathan Powell suggests that the situation may not be as exceptional as it appears (https://www.freelivingbees.com/post/bees-dont-make-honey-teams-do, accessed: 19 November 2022). However, since it was widely believed that the tracheal mite (*Acarapis woodi*) invasion in the British Isles during the 1920's eradicated the local British bee, the discovery of this population was received with surprise. Today we have no doubt that representatives of local ecotypes survived the tracheal mite invasion from about 100 years ago and for the last about 30 years (not only in Blenheim) are successfully resisting the varroa mite. Powell emphasizes the importance of organisms coexisting with bees (mainly the microbiome) for their survival and health. He also warns that the use of resistant bee lines in other conditions (isolated from the environment in which the bees survived and coevolved for some period of time) may not bring the expected results, which is why it is important that the co-adaptation of all locally occurring organisms takes place without interference and disruption.

It turned out that after the varroa invasion, the bees not only managed to stay healthy and build a self-sufficient population but also maintained their genetic distinctiveness[292]. This was undoubtedly facilitated by both the low level of beekeeping in England, as well as the lack of apiaries in the immediate vicinity of the forest complex and the surroundings in the form of marshy valleys, which create specific conditions of isolation.

In 2021, the website "Free-Living Bees" (www.freelivingbees.com), founded by Jonathan Powell from the British organization 'Natural Beekeeping Trust', was created, where you can (and should) register wild bees that you know about.

"Bartnicy Sudetów", Poland [Sudety Tree Beekeepers Program]

The project of the Międzylesie Forest District of the State Forests (Kłodzko Valley) called "Bartnicy Sudetów"[293] was created by Konrad Zaremba, a beekeeper and forester. In the first stage, the initiative developed very quickly. The aim was to create numerous nest sites for honey bees in the surrounding forests. So far, over 1,000 log-hives have been placed in the forest. The aim of the project was to help honey bees find nest sites, which indirectly should lead to build up of the forest ecosystems diversity. Logs are also used by other insects and animals, so they can be treated as a substitute for natural tree hollows. Together with hanging the logs, it was planned to start observing the bee colonies settling in them. I would like to believe that the activity of "Bartnicy Sudetów" project will allow for the commencement of scientific research on the wild-living population of bees in Poland, of which there are still too few in our country.

It is still too early to draw any conclusions regarding the chances of creating a stable population of free-living bees in Polish forests, but we can already speak of the first successes: bee colonies have settled in the log-hives. As experience from around the world teaches us, it takes about a decade to develop stable populations of bees, if there are enough of bees' nest sites which are in areas sufficiently rich in nectar and pollen. I deeply believe that the example of the Kłodzko Valley will become proof that it is enough to create favourable conditions for bees to develop a balanced relationship with the mite under our Polish conditions.

---

292 In some ecotypes of Central European bees, a preference for sexual selection of individuals of their own subspecies was confirmed, and it cannot be ruled out that this was also the case here.

293 See: www.bartnicy-sudetow.pl, accessed: February 02, 2025.

The "Bartnicy Sudetów" project could also serve typical beekeeping purposes. Namely, there is a chance that over time a resilient wild-living local population will develop, which will be able to influence the genetic characteristics of local bees. I believe that in the future such a situation will allow local beekeepers to decide to abandon their mite treatments.

"Fort Knox" project, Poland

When we started working together as part of natural beekeeping groups a few years ago, we were aware that our amateur selection efforts against the mite might prove insufficient, and that small apiaries would not withstand the pressure of varroa. So, we were looking for a formula that would allow us to gather the largest possible resources to go through the most difficult selection period. This is how the idea of one of my colleagues, Marcin Zarek started the "Fort Knox" project[294]. The main goal is to encourage beekeepers to conduct their own selection. "Fort Knox" colonies are devoted to natural selection, they are not treated with any chemicals (including so-called natural chemicals) or biotechnical procedures aimed at reducing the population of pathogens and mites. We leave the fight for health to the bees.

Participants declare that they will aid other members of the group, guaranteeing the reconstruction of part of the apiary in case someone loses their colonies. If a colony covered by the project dies, the participant could count on receiving a free nuc from the project pool. The group is open to beekeepers who have at least minimal experience in keeping bees, are willing to cooperate and accept the rules of operation, declaring that they will not perform procedures to combat varroa in their colonies covered by the project and that they will create nucs from these colonies for other participants. Selected bees circulate between members of the group, which allows us to assume that from year to year, within our dispersed apiary, the genetic material we have is getting better and more resistant to the mites. It can be used, for example, to raise queens for your own apiary (even if treated for varroa). If, however, participants do not have colonies outside the project pool, they can count on their bees becoming increasingly resistant.

I had the pleasure of presenting "Fort Knox" at two international beekeeping conferences: first one in Neusiedl am See in Austria and the second one in Doorn in the Netherlands. The project aroused great interest among natural beekeepers. It later turned out that the initiative inspired

---

294   See: https://bees-fortknox.pl

beekeepers in Europe (even in Great Britain) to co-operate in the selection process, even though it is based on different principles than in Poland.

Unfortunately, the project is not developing as quickly as – in my opinion - it deserves. Many beekeepers, even those interested in the principles of natural beekeeping, are discouraged by the length of the selection process, so they prefer to reach for chemicals. Others are discouraged by the logistics associated with co-operation, and still others are not convinced that creating nuclei and transporting bees to other apiaries is the best solution. Well, it is impossible to disagree with many of these arguments. The project could work perfectly if many beekeepers form a local cooperation on such conditions. If however, there are only few cooperants scattered around large area (as a county) it may never outgrow from its beginning phase. However, local selection and changing genetic resources in apiaries to resistant ones is at least a start. After all, as Laozi, the Chinese philosopher, used to say, even the longest journey begins with a single step.

Beekeepers, enthusiasts, and bee lovers

Subjective ranking

The world is full of people who do not want to perform treatments on bee colonies. Often, they do not see such a need, because their bees cope very well without treatments. They also believe that using treatments is in the long run harmful to bees, the environment and even to themselves, because it carries the risk of contact with harmful substances. There are places in the world where refraining from using toxic substances almost always ends in the death of a large part of the bee population, in other places such a decision does not have such tragic consequences. Unfortunately, continental Europe belongs to the first group, and Poland is no exception.

Probably for this reason, it is not easy to find beekeepers who consciously give up treatment here. However, I would like to mention a few names. This subjective ranking is made up of people whose substantive work I particularly value, or those who are close to me for other reasons. Some are authorities on natural beekeeping, I follow their achievements, familiarize myself with the results of the research they conduct, read their blogs. I had the opportunity to meet some of them in person or during meetings on popular conference platforms on the Internet, and I maintain e-mail correspondence with others.

Michael Bush, Nebraska, USA

Michael Bush is sometimes called the most famous beekeeper in the world. It seems that this name is familiar to anyone who has ever had a brush with natural beekeeping, where no toxic or biocidal agents are used. Bush has been involved in beekeeping for several decades.

Michael Bush is a well-known figure in the world of natural beekeeping, photo by M. Bush.

When the invasion of varroa reached his apiary in the 1990s, he, like others, reached for chemicals. At that time, he was an unknown hobbyist with a few hives with bee colonies in his garden. Despite declaring war on varroa in his apiary by using many methods of control, the bees were regularly dying. This prompted him to look for an answer to the question, how to keep bees alive without treatment? The turning point was information about the so-called small cell and the suggestion to breed bees naturally adapted to local conditions. When he decided to base his own beekeeping on natural comb and swarms from a feral population, bee mortality dropped significantly.

Bush willingly shares his knowledge and experience on his website[295] and on beekeeping forums. He is also the author of the book 'The Practical Beekeeper; Beekeeping Naturally', in which the reader will find not only information on how to abandon the use of chemicals, but also other practical

---

295   See: https://bushfarms.com/bees.htm. Bush's texts have been translated into many languages (his website or part of it is available in several languages).

advice. For many natural beekeepers, this monofigure is like a "beekeeping bible". Bush is also a publisher. He has published many important titles for beekeepers (including the work of the Swiss beekeeper and researcher Francis Huber, and a collection of texts by Dee Lusby). The American beekeeper also gives lectures at numerous beekeeping conferences in the United States. He also organizes so-called beekeeping camps, where participants gain knowledge and experience by taking part in current apiary work.

Bush is a proponent of so-called lazy beekeeping. According to this concept, bees should take care of themselves, and the beekeepers' task is to organize the work in such a way as to limit their interference in the life cycle of the honey bee colony to an absolute minimum. This approach is justified not only for economic reasons (minimizing work input) but also seems to improve the health of the bees.

A beekeeper from Nebraska practices beekeeping in modified Langstroth hives. He uses one type of frame throughout his apiary, the so-called medium. To reduce the need for lifting, he narrowed the boxes to the size of eight standard frames (because he uses narrowed comb axes, he places nine frames in a box). Bush uses only upper entrances in the hives, located in the roofs. This solution has many advantages, but it is difficult to say whether it is ultimately beneficial for the bees. It undoubtedly leads to faster cooling of the nest during wintering, but Bush claims that it works in Nebraska. In this US state, the climate is like continental Europe, with hot summers and frosty winters. Average annual losses of bees usually do not exceed 20%, but during severe winters they can be greater.

Dee Lusby, Arizona, USA

Dee Lusby and her husband, Ed[319], are considered pioneers of beekeeping that do not control varroa mite populations. They are believed to have been the only professional beekeepers in industrialized countries who did not take up the fight against varroa when the invasion occurred (so they never treated). In Arizona, they ran a professional apiary consisting of almost a thousand colonies. In the first phase of the invasion, the mite caused huge losses with 80-85% of the colonies died. After a few years, everything began to function almost as before the mite invasion. Since her husband's death, Dee Lusby has been running the apiary alone. For several years, due to age and health problems, she has begun to reduce the size of the apiary.

She is one of the main proponents of the so-called small cell. In her own apiary she uses only 4.9 mm cell foundation. It was her example that the proponents of the so-called small bee followed. Thanks to Dee Lusby, this movement gained popularity.

Kirk Webster, Vermont, USA

Kirk Webster is a professional beekeeper from the northeastern part of the United States who manages an apiary with several hundred bee colonies. One part is used for honey production, while the other is used to produce nuclei and raising queen bees for sale, because he is a supporter of diversifying the sources of income from his apiaries. Webster sells not only honey, but also nuclei and queen bees. This management model has allowed

Kirk Webster (left) during the visit in Sweden in an apiary with Marcus Nilsson, photo by M. Kranshammar.

him to survive more than one bad year, including the mite invasion, which he has not treated for two decades now. The decision to abandon treatment was preceded by an initial selection of a bee on which his beekeeping could be based. He associated his beekeeping with the Russian bees, which were imported to the United States from the Primorsky Krai. Webster claims that, alongside bees bred for resistance, swarms captured from feral colonies, this beeline offers the greatest guarantee of success in a country where varroa remains a major problem[296].

---

296   K. Webster is the author of many inspiring texts and articles, see: www.kirkwebster.com. Webster promotes a retreat from the intensive principles of bee-farming conducted on an industrial model and a return to sustainable agriculture based on the bond between man and nature.

Sam Comfort, Florida and New York, USA

Sam Comfort's apiary[297] has about a thousand hives, but when asked how many colonies he would like to have in the future, the beekeeper invariably answers: "ten". His colonies live in many apiaries, in a semi-wild way. It seems that it is difficult to make a clear division into production colonies and nuclei, because the bees live quite freely, according to their natural cycle. The beekeeper does not use young colonies in his production, and often does not even inspect them, leaving them to their own fate. There are apiaries that he visits no more than two or three times a year, which results from his approach to leave nature to sort things out. In some apiaries, where he raises several thousand queens for sale each year, Comfort conducts a slightly different type of beekeeping with more intensive breeding methods. His apiary grew systematically, he started with a small number of hives, which he maintained while working for other professional beekeepers.

He relied on breeding his own queens, as well as those coming from selected sources (using natural selection or breeding), hence his apiary is very diverse, and the bees are constantly subjected to natural selection.

The beekeeper bases his business on hives of very simple construction. During the first period, the so-called Kenyan top-bar hives dominated, then he became a big supporter of the Warré hive, modified according to his own ideas. But in Comfort's apiary we can also find Langstroth hives, which are standard in the United States. The beekeeper builds bee boxes from the cheapest materials, including raw, untreated wood, often of lower quality, which is not suitable for other purposes. In most hives, he also does not use frames but either ordinary slats or bamboo sticks, to which the bees built (of course without foundation) their combs from.

Salomon Parker, Oregon, USA

Salomon Parker has now given up beekeeping[298]. He published his decision on his blog and announced it online in January 2022, after this part of this book about his work had already been written. However, I have decided to include Parker because of his great influence on the community of non-treatment beekeepers. The Facebook group Treatment-Free Beekeepers,

---

297   See: www.anarchyapiaries.org

298   He said however, that he might remain a "bee-haver", meaning he would leave one or more hives in his garden.

which he founded, exceeded 90,000 members from all over the world[299]. Parker hosted a beekeeping podcast and a YouTube channel, which regularly gained him many new subscribers world-wide. He invited guests with different approaches to beekeeping to talk[300].

David Heaf, Great Britain (Wales)

David Heaf observes the entrance to the Warré hive, photo by D. Heaf.

David Heaf is a hobby beekeeper, with a PhD in biochemistry from the University of North Wales in Bangor (1976). He resides in Gwynedd, North Wales. He has only managed a dozen or so hives (and now decided to reduce even more). In his apiary we could find mainly Warré hives, although he also had so-called national hives, i.e. structures with dimensions like the Langstroth hive, and finally hives like our traditional Polish "Warszawski" hive (a modified hive designed by the Russian beekeeper Fedor Lazutin, which has a vertical Dadant frame). Dr. David Heaf is the author of many publications on beekeeping[301].

---

299   The group was created so that beekeepers could have an opportunity to discuss without the pressure of other people to impose treatments, any posts encouraging treatment are deleted by moderators.

300   Parker decided to leave his own materials online; see: https://www. Youtube.com/c/ TreatmentFreeBeekeeping. The podcast site is no longer available, but still You can find Solomon Parker's talks on many platforms e.g. Spotify.

301   David Heaf (with his wife, Patricia) is also involved in translations, including beekeeping texts and books; he has written several books on bee-friendly beekeeping: Natural Beekeeping with the Warre Hive. A manual; The Bee-friendly Beekeeper. A sustainable approach; Treatment-free beekeeping, where he also mentions my apiary; see also: conversation on the YouTube channel: https://www. youtube.com/watch?v=5mxSNTBzK9I, accessed: February 03, 2025; Heaf is the author of many interesting summaries and overviews, which are a compilation of knowledge gained on the basis of research (http://www.bee-friendly.co.uk).

Bee losses in David Heaf's apiary compared to losses according to the BBKA* (British Beekeepers Association).

Winter	Number of wintered colonies	Number of surviving colonies	Loses in percentage	Average losses by the BBKA*
2007/2008	6	6	0	30,5
2008/2009	11	6	45,5	18,7
2009/2010	12	10	16,7	17,3
2010/2011	12	4	66,7	13,6
2011/2012	15	13	13,3	16,2
2012/2013	12	12	0	33,8
2013/2014	15	14	6,7	9,6
2014/2015	14	13	7,1	14,5
2015/2016	12	9	25	16,7
2016/2017	13	13	0	13,2
2017/2018	13	11	15,4	25
2018/2019	10	10	0	8,5
2019/2020	9	8	11,1	17,3
2020/2021	9	9	0	18,6
2021/2022	8	6	25	16
Totals	171	144	15,5 (altogether)	18,1 (average)

* BBKA British Beekeepers Association https://www.bbka.org.uk/news/84-of-beekeepers-who-replied-to-survey-had-no-winter-losses
Table by M. Uchman based on Treatment-free beekeeping by D. Heaf

The type of beekeeping which he runs is a perfect example of what a bee-friendly apiary should look like. He limits interference in the lives of bee colonies to an absolute minimum. He also allows the swarms to leave, accepting that some of them will not be caught. In return, he acquires others, which make up for the (small) losses in his apiary. Since he abandoned treatment, overwintering losses have been around 16% per year. Since 2011, which ended the most difficult period of the initial selection, the mortality rate in his apiary was less than 9% (100% of colonies survived the winter of 2020/2021).

Data from Shan and Clive Hudson's (Wales) survey

Winter	Number of respondents	Number of reported colonies	Treated colonies		Untreated colonies	
			Colony numbers	Loses in percentage	Number of colonies	Loses in percentage
2010/2011	14	71	44	27	27	11
2011/2012	40	355	180	8	175	7
2012/2013	53	251	75	41	176	32
2013/2014	65	396	81	9	315	6
2014/2015	77	500	97	8	403	8
			477 (altogether)	19 (average)	1096 (altogether)	13 (average)

Table by M. Uchman based on Treatment-free beekeeping by D. Heaf

Heaf is not the only non-treating beekeeper in his county, in fact most beekeepers in Gwynedd do not treat for varroa. Between 2010 and 2015, Clive Hudson and his wife, owners of 20-25 colonies, decided to conduct a study to verify the annual mortality of local bees. Several dozen beekeepers signed up and filled out a survey on the mortality of bees treated against the mite and those that were not. A total of 477 treated bee colonies and 1,096 colonies that were not treated were included in the study.

The results were surprising. Firstly, it turned out that in their area most beekeepers do not use any treatments against varroa. Secondly, the average mortality rate of bee colonies not subjected to treatment was lower than that of colonies in which varroa was controlled, 13 and 19%, respectively. Data collection was stopped in 2015. Most bee colonies in Gwynedd are still not treated[302]. Many local beekeepers know where at least a few or a dozen long-lived bee colonies live without being managed. Each year, they cast new swarms, which supplement the losses caused by the death of colonies in apiaries. Interestingly, tests of bee samples taken, among others, from the apiary of David Heaf and Clive Hudson, as well as from wild-living colonies, confirmed a very high level of DWV. Despite this, the bees cope perfectly well without treatment, demonstrating a high resistance to viruses.

---

302 This is the county that Stephen Martin had in mind when he said that only one beekeeper currently uses treatments against the mite

Jonathan Powell, Great Britian

From left: Heather Swan, Thomas D. Seeley, Johannes Wirz, Jonathan Powell during the conference in the Netherlands in 2018, photo by author.

Jonathan Powell is a member of the Natural Beekeeping Trust. This British group of honey bee enthusiasts regularly submits valuable initiatives promoting natural beekeeping and organises international conferences. Powell is involved in projects aimed at developing populations of wild honey bees (so-called rewilding projects). He also runs workshops for those interested in tree beekeeping. Near Powell's house, we can find log-beehives, but in the apiary there are only a few hives, in which live on their own (usually for 3-5 years). He only collects honey when the bee colony dies. Powell does not like to call himself a "beekeeper" or his own activity "beekeeping" as he most often refers to himself as an apiologist, so a person who studies bees. Together with Michael Thiele from the USA, he organizes online meetings called Arboreal Apiculture Salon[303], during which they both promote primarily tree beekeeping (broadly understood), as well as projects aimed at returning honey bees to nature. They also promote bee-friendly ways of keeping bees. Their project brings together a considerable group of beekeepers and bee enthusiasts from all over the world.

Norbert Dorn, Austria

Norbert Dorn is a hobby beekeeper from Austria who has not been carrying out any treatments against the bee pest in his apiary for several years. He was the organizer of the first beekeeping conference in Europe, entirely

---

303   https://www.freelivingbees.com/the-salon, accessed: November 26, 2022.

devoted to beekeeping that does not control for varroa. It was attended by, among others: John Kefuss, Erik Österlund and Juhani Lunden, but also Heidi Herrman from the Natural Beekeeping Trust (UK) and Torben Schiffer.

Erik Österlund, Sweden

At one time, the Swedish beekeeper was said to be the best prepared in the world for the *Varroa destructor* invasion. In the 1970s, the mite reached Europe, and it was only a matter of time before it appeared in Sweden. In March 1989, Erik together with two beekeepers from Sweden and one from the Netherlands, went to Kenya, to the region Elgon Mountain in Kenya, where he obtained genetic material from the *Apis mellifera monticola*[304] subspecies. From the Dutch beekeeper, Michael van der Zee, he got genetic material from the *Apis mellifera sahariensis*, obtained during an expedition to Morocco. Then Österlund mixed the African genes with the genes of the Buckfast bee[305] and in this way a crossbreed was created in Sweden, later called the Elgon-bee (after the place where the genes were obtained)[306]. Erik had the right to believe then, that the Elgon-bee would be more resistant to varroa than the local bees and even may be mite resistant.

In preparation, following Dee Lusby's advice, Österlund started using a small cell (4.9 mm) throughout the apiary. Until the mite appeared in his region (around 2007), the breeder did not waste time: he sent his own queens to those parts of Sweden where the mite had already appeared (the south of the country and Gotland). He also carried out selection based on feedback from beekeepers who had tested his queens in mite infested apiaries. He assumed that this would ensure that he would be able to run

---

304 Africanized bees in South America proved to be resistant to mites; at the time, it was suspected that the resistance might be a genetic trait, and Österlund and other beekeepers tried to obtain the genetic material from Africa.

305 The proportions on some levels of breeding were: 70% - Buckfast, 25% - *A.m. monticola*, 5% - *A.m. sahariensis*. Österlund wrote to me that when the genes of *A.m. monticola* bee were introduced, several (5-6) colonies of this bee (reasonably pure-bred) were observed for their mite population growth rates, and compared to ordinary, locally used crossbreeds. They were found to have a lower mite infestation (probably due to a shorter period of pupae development); the colonies were smaller, probably due to the later start of the spring development period. In the first-generation crossbreeds (so-called F1), the queen-mothers often laid eggs throughout the winter (although the bees did not allow the larvae to develop later), the bees were unable to form a winter cluster properly.

306 It should be noted that although Erik Österlund's breeding priority is bee resistance to varroa, as a professional breeder he also considers other characteristics needed in apiary management: above all productivity (honey production) and gentleness of bees. However, he only looks for the latter in bees that demonstrate the ability to limit the growth of the mite populations in his colonies. See: www.elgon.se.

his apiary without having to control the mite (when it appeared). However, it turned out that the scale of the varroa invasion surprised him. He also had to admit with humility that he had been wrong, as his bees, by no means, were resistant to varroa. They began to weaken, almost half of the colonies died. Then he decided to treat them with thymol. However, he did not stop the selection, during which he treated the bees when they needed it, but he completely abandoned preventive treatments[307]. During the selection, he paid attention not only to the resistance but also to the productivity of the bees.

The members of the expedition to Kenya in 1989, from the left: Dr. Bert Thybom, Erik Bjorklund, Erik Österlund, the Nyongensa family (the hosts of the expedition in Kenya), Michael van der Zee, photo by E. Österlund.

---

307   The breeder did not carry out the treatment like most of the beekeeping community, guided primarily by the season, he did not treat the bees, for example, before wintering or during the nectar gaps. He performed the treatments only when the bees had a high level of mite infestation or DWV symptoms, regardless of the season or year.

I visited Erik Österlund's (first from the left) apiary in 2019, on the right Sibylle Kempf, photo by the author.

In the first phase of selection, when he saw the symptoms of DWV infection (he searched for deformed bees thrown out by the colony on a mat placed in front of the hive entrances) he treated those colonies within the week and then eliminated queens from them from breeding program. He also assessed the condition and development of the colony. For some time, he also observed the bees in terms of the occurrence of the VSH trait (his bees had this trait developed at a level of 50%), but with time he abandoned it, considering that the method was too laborious, and does not translate into significant progress in selection[308]. He also noticed that in colonies that had high values of the VSH trait, symptoms of DWV infection occurred to a greater extent. Incidentally, I will admit that at first Österlund's observation seemed downright strange to me. I wondered whether such a correlation was possible. I found a probable explanation in a scientific paper from 2021[309]. It turned out that the cause of the phenomenon may be cannibalism

---

308   This does not mean that this must be the case in every population. In some US centres, selection for the VSH trait yields excellent results.

309   F. Posada-Flores, Pupal cannibalism by worker honey bees contributes to the spread of deformed wing virus, Scientific Reports 2021, 4.

of bees, consisting in the removal of pupae infected with the virus (VSH). Bees that remove (cannibalises) an infected pupa become infected with the DWV virus from it, and then via trophallaxis transmit the infection to other adult workers, as well as larvae, during feeding[310].

After several years of work, Erik Österlund changed the selection criterion and began treating those colonies where the mite infestation rate was higher than 3%.

To estimate this, he used to place 300 bees in an alcohol solution (technique known as flotation or alcohol wash). However, the beekeeper did not stick to this criterion rigidly. In exceptional cases, if he did not notice any symptoms of disease in the colonies, he did not treat them, even though the infestation exceeded the assumed tolerance threshold (then the colony was under close observation). There were also opposite situations: the breeder claimed that the health condition of some colonies, despite a low degree of infestation, may be poor (many bees with symptoms of DWV infection).

Erik Österlund's selection accelerated rapidly about a decade after the invasion of varroa, which is about how long the breeder needed to acquire enough of the bee resistance mechanisms in the population. In 2018, Österlund treated less than 50% of his population, in 2020 only 20% (he performed the treatments with very small doses of thymol), and since 2021 he has not used any treatments in any of his apiaries. In the winter of 2021/22, only three out of about 100 colonies died (several others came out of the winter without queens). For many years, the mortality rate in his apiary has been very low remaining within the range of 5-10% annually.

Also interesting is the initiative of joint breeding effort of bees, which Erik Österlund presented to beekeepers in his neighbourhood[311]. Thanks to making the breeding material available to others, an area was created

---

310   This does not mean, however, that selection for the VSH trait is counterproductive in the long term, that it threatens the process of selecting resistant bees. It seems (provided that the bees are sufficiently resistant to the virus) that to prevent the colony from collapsing, the virus infection should decrease along with the removal of many infected larvae and mites from the hive, at least to a level that does not threaten the life of the bee colony. Studies show that without many mites in the colony (the presence of intensive VSH behaviours favours this), the virus should not be a real threat to the colony.

311   Beekeepers from the local association are involved: the eight most active (including Österlund) have an average of one hundred colonies; there are also those who have almost 200 colonies. These beekeepers record similar losses of bees, most often 5-10%. Österlund claims that beekeepers who also have other lines, e.g. from the *A.m. mellifera* subspecies, have much higher losses. Beekeepers themselves assess which colonies require treatment, in recent years 5-20% of the colonies are treated (with very small doses of thymol), but the percentage is systematically decreasing. Four beekeepers have completely given up the treatment. See the talk I recorded with the breeder in summer 2024: https://youtu.be/3_YWBJaO1pE; accessed: February 3, 2025.

in which several hundred selected colonies live. All beekeepers involved in the project note with satisfaction that the percentage of colonies requiring treatment drops significantly each year. This is the best proof that cooperation with others and consistency in action bring results. Österlund encourages maintaining the greatest possible diversity of bees in apiaries and making splits from many of the best colonies, which should raise their own queens. Erik Österlund has been retired for several years (he is over 70 years old). For this reason, he reduces the size of his apiary every year and now keeps about 60-70 bee colonies (at the peak he had around 200).

Juhani Lunden, Finland

Juhani started beekeeping in the 1970s as a teenager. Varroa came to his bee yards in 1996, and a few years later (in 2001) he started his program to breed bees for mite resistance, which would allow him to stop all chemical treatments[312]. He took a slightly different approach than most of the beekeepers mentioned above. At first, he decided to conduct selection by consolidating the desired traits in the population using controlled isolation apiary matings. This era stopped in 2014 because of severe bear damage. After that, for two years (2015-2016) he used open mating, but during this time he almost lost all his colonies. Therefore, he changed to use instrumental (artificial) insemination.

When starting his selection in 2001, he divided his apiary into two groups. In the 70-nucleus group, he stopped treating from the beginning. In the other group consisting of 150 normal sized hives, he immediately stopped summer treatments (August), and only treated with oxalic acid dribbling in late October. He also systematically reduced the use of oxalic acid and sugar solution. He counted all the dead varroa on the bottom board after the treatment and made his breeding decisions mainly according to low mite numbers, and from colonies he considered the best.

The experiment with the first group, in which all nucs have been left without treatments, lasted for 5 years, then the last one died. Their genes however were introduced to the main population (about 150 colonies), which the breeder stopped treating in 2008. From that time on, he began to choose for instrumental insemination or isolation mating queen-mothers and drone lines from the same group of bees. At the same time, he decided

---

312   See: https://naturebees.wordpress.com/. I have visited Juhani Lunden's apiary during the cycle tour around Baltic Sea – then I've recorded a short talk in his apiary: https://youtu.be/c9FwZUGYq28.

that regardless of the results (including losses) he would make no more than one split from each bee colony that survived another winter. He did not want to weaken them excessively, and besides, he wanted to be able to observe their production values. For the first years, the Lunden's population systematically shrunk. Periodically, the mite infestation in the population increased and after that decreased. Losses went similarly up and then the following years down. Increasing mite infestation caused bees to become edgy (and so more defensive) for a couple of weeks. After that they calmed down in the periods of good flows and forage. Mite infestation levels in the beginning years varied from 3 to 10%, but everything above 5% proved to be lethal. After about 10 years, only a dozen or so bee colonies remained in Lunden's apiary. It turned out that despite careful observation of traits and their fixation (using instrumental insemination), this type of selection method is not easy. According to the Finnish breeder this may be because some of the genes associated to varroa resistance might be recessive. Since then, however, the population has started to grow systematically (currently it has about 50 colonies plus 40 nucs). Today, the average mite infestation level in his stock is about 1,5%. Lunden also stated that his bees overcome problems which could have been due to inbreeding or brood diseases.

It is worth mentioning that bees in Juhani Lunden's apiary were originally Buckfast bees from Sweden and Luxembourg. During the years his population was crossed with imported stock: in 2002 with Primorski bees, in 2006 with bees from Columbia, in 2015 with new untreated material from a Finnish beekeeper Heimo Kangasaho, and in 2016 with one queen from Josef Koller. Juhani Lunden believes that when using instrumental insemination or isolated apiary matings the stock may suffer from inbreeding, especially if there is a large evolutionary bottleneck – as it happened in his apiary. Introducing new queens from time to time allows to broaden genetic diversity, and fix different traits of mite resistance better into the local population.

Juhani Lunden also claimed that customers appreciated his bees, also for their productivity. His stock was also available from different bee-breeders from Europe (e.g. from Paul Jungels' apiary), where it has often been appreciated as bees that do not need treatments!

Torben Schiffer, Germany

Schiffer owes his popularity in the natural beekeeping community to his research on the physics of bee rest sites and the book scorpion[313]. He conducted his work under the supervision of Professor Jürgen Tautz, an undisputed authority in the world of beekeeping. As far as I know, Schiffer gave up on purely scientific activities and focused on popularizing his earlier discoveries and promoting the need to restore the honey bees' ability to live naturally.

Torben Schiffer during a lecture at the "Learning from the Bees" conference, Netherlands, 2018, photo by J. Ruther.

Despite this, for some beekeepers, including some of the so-called natural ones, Schiffer has become a controversial figure. It happens that his ideas (or maybe the way he expresses them?) are treated as an attack on the entire beekeeping community, as well as people who work with honey bees in a different way. In Schiffer's statements, they read calls to give up beekeeping to save bees. Many people also perceive his words as an accusation against beekeepers, as if they were unscrupulous, greedy people who, through their own actions, destroy the adaptation of the *Apis mellifera* species. Well, it is impossible to disagree with many of Schiffer's ideas. It is true that the way beekeeping is conducted it is very often not sustainable, which is the cause of many health problems for bees. Professional breeders and many hobby beekeepers do not work to ensure the species' resistance and independence, and their activities mean a systematic weakening of

---

313   See: www.beenature-project.com.

the bees' environmental adaptations (which is what this book is about). However, I believe that beekeeping can return to the principles that will allow us to keep bees in a way that is beneficial to their health and their adaptations.

John Kefuss, France

An American by birth (from Ohio State), a doctor of entomology[314], he is a colourful character with an interesting story[315]. He keeps bees in southern France, where he has lived and worked for many years. After the mite invasion, he worked on developing recommendations and methods for using pesticides to combat the mite, but he was one of the first in the world to undertake - successfully - the challenge of breeding bees resistant to varroa[316]. In the 1990s, Kefuss also used the African subspecies from Tunisia (*Apis mellifera intermissa*) in his own breeding crosses, but he also noticed a certain degree of resistance in Carniolan bees. However, he found that it was not only specific subspecies or ecotypes of bees that were characterised by resistance to the varroa mite (although the response of some may be stronger). According to Kefuss, individuals that can cope with the mite can be found in every population – on average valuable characteristics can be found in every tenth colony.

The impetus to begin breeding resistant bees was the episode of brood bacterial diseases that had taken hold of many colonies in the beekeeper's apiary. At this stage, selection for hygienic behaviour became the most important trait. The procedure brought the expected results as the bees got rid of their health problems quite quickly. Then the breeder decided to subject the bees to natural selection, which resulted in the creation of very resistant crossbreeds, later used in many commercial projects around the world. Interestingly, Kefuss claims that the results of hygiene tests of bees that coped perfectly with mites are often worse now than years ago.

Problems with the mite are common and affect all beekeepers. While most of them believe that there are too many mites, Kefuss complains that he has often too few of them. Since his bees generally cope very well with

---

314   His doctoral thesis concerned the problem of photoperiodism in honey bees (the influence of the length of day/night on the development of the bee colony and, above all, the initiation of egg laying by queen bees).

315   I met John Kefuss during a conference in Austria in 2018.

316   Since 1998, Kefuss has completely stopped carrying out treatments in his apiary.

the mite, the American is unable to maintain a high parasite pressure in the apiary to be able to constantly verify the effects of breeding work. There were times, as he claims, when he bought mites (more precisely bees or heavily infested frames with brood). He even considered keeping non-resistant bees just to have easier access to large quantities of mites.

In the beekeeping community, the offer of Kefuss is also well known. He once declared that he would pay a cent for each mite found in his resistant population, and then, he said with amusement, he had never had such cheap labour before. Well-known beekeepers and scientists also got involved in the search for mites, and thanks to their work, he was able to make breeding decisions.

John Kefuss is credited with the term "Bond test" (a paraphrase of the title of one of the 007 films: Live and Let Die) and "varroa black holes". The latter is used to describe mite-resistant colonies since mites disappear in them as in black holes of time and space.

According to Kefuss, every beekeeper can conduct breeding for resistance, but on one condition: one must be able to count from one to ten. Reproduce those colonies in which there are the fewest mites, and if in each sample of worker bees, he sees an infestation of more than 10 mites, he will eliminate the colony from the breeding selection process. Kefuss used to say: "If I could do it, that means anyone can do it!"

Other amateur beekeepers

I have met many amateur beekeepers from all over the world. Some of my acquaintances started on internet forums. Often these are people who do not have great achievements behind them. But it is precisely such people who make up the natural beekeeping community, they are the ones who initiate discussions that often inspire others. One thing unites us, we try, although not always successfully, to find our own and unique way of running an apiary in harmony with nature. We all share the desire to restore free living bees to nature.

In April 2018, during a conference in Austria, I met several natural beekeepers, including those from Germany. One of them is Jeurg Ruther, who uses Warré hives in his small apiary. There was also Sibylle Kempf, who at that time ran an apiary on the German side of Lake Constance, where, like us in Poland, she tried to persuade beekeepers to cooperate in a so-called good cause. A few years ago, together with her husband, Wolfgang, she decided to move to Sweden, where she bought a small house, not far from Erik Österlund's home. She gave her colonies to a beekeeper from Germany, and in Sweden she obtained bees from Österlund. Sibylle's apiary is small, currently it has a dozen or so bee colonies originated from Elgon stock. In 2019, my wife and I visited her and Wolfgang during a cycling holiday in Scandinavia [and then again in 2024 during our three-month long cycling tour around the Baltic Sea]. We also paid a visit to Erik Österlund and a Czech named Radim[317], who, together with his wife, Hana, moved to Sweden, where he had been building an apiary for several years, to take up beekeeping professionally.

Behind the scenes talks during the conference in Neusiedl, from the left: Juhani Lunden, Jürgen Küppers, Jeurg Ruther and Piotr Piłasiewicz, photo by S. Kempf.

The natural beekeeping community in Sweden is very large. One of them is Hannes Bonhoff, who loves creating nest sites for bees that are like natural ones (bee logs). I met another amateur beekeeper, Marcus Nilsson, through social media. He previously ran his own apiary of around 60 colonies, and worked in a beekeeping company, with several hundred colonies. However, over time he began to have doubts as to whether this was a sustainable type of business. He decided that there had to be another way of keeping and breeding bees, one that was closer to their needs. So, he gave up intensive beekeeping methods and now takes care of around 20 colonies (in accordance with Darwinian beekeeping principles) and runs a permaculture garden. He recently bought a building that used to belong to a local school, which he is transforming into the first Swedish centre for teaching the principles of beekeeping close to nature. Marcus Nilsson also tries to promote bee-friendly breeding methods by writing articles and translating publications on this subject into Swedish. He also encourages

317 Radim Gavlovsky actively participates in the selection of bees initiated by Österlund. When I met him, he had about 200 bee colonies, of which he treated on a very small percentage, when required.

local beekeepers to cooperate in the selection process, organizing a group of beekeepers whose aim is to return beekeeping to nature.

Thomas Gfeller during a bicycle trip through Greece, photo by T. Gfeller.

During the beekeeping conference in Doorn I also met Thomas Gfeller, a beekeeper from Switzerland. When he lost all his bees one year, he felt the need to deepen his knowledge of natural beekeeping. He acquired it in a rather unusual but interesting and close to my heart way: by travelling around Europe by bicycle, stopping at apiaries where he could learn something about the principles of beekeeping where no biocidal chemicals were used to treat varroa[318].

---

318   Gfeller met a large group of beekeepers who did not use treatment not only in Switzerland, but also on the Iberian Peninsula, France, Germany, the British Isles and the Balkans. In his film "Has Varroa lost its sting?" Shan and Clive Hudson from Gwynedd, Wales, present a short history of their apiary. In 2024 Thomas published his diploma thesis of the Swiss Federal Beekeeper Certificate Course, entitled "Treatment-free beekeeping – A systematic analysis" based on the survey made by over 60 chemical-free beekeepers from all over Europe.

His compatriot, Andre Wermelinger, whom I met during a conference in Austria, is the co-founder of the dynamic Free the Bees association[319]. Andre is engaged in hanging beehives and observing wild bees. He admits that beekeeping without treating varroa is not easy, but he believes that if it were possible to change the methods of beekeeping and restore the free-living population, returning to healthy beekeeping without chemicals would not be a distant prospect.

Poland

I do not know of any breeder or professional beekeeper in Poland who would manage their apiaries without using toxic or biocidal substances or without resorting to biotechnical procedures to control varroa. I have also not heard of a breeder who would seriously try to take up such a challenge[320]. Yes, there are beekeepers who deal with breeding queen bees and declare that they have queens from resistant lines (imported from various regions of the world), but most often they do not talk in detail about their own methods of maintaining and developing their resistance traits[321] and rather encourage others to effectively treat bees.

There are also those who demand the approval of new active substances (synthetic pesticides), the use of which is currently prohibited in Poland. Apparently, still only few people believe that it is possible to do beekeeping without using chemicals, and that bees can survive despite being infested with varroa. This lack of faith translates into the reality of apiaries. I would say that we function within the framework of a self-fulfilling prophecy, if no one seriously works on obtaining bee resistance, then they are not resistant.

---

319 The association promotes bee-friendly farming methods, issues publications, runs workshops for beekeepers, and cooperates with a large group of scientists; see: https://freethebees.ch.

320 I am waiting for breeding programs and scientific projects focusing on bee immunity and resistance to appear in Poland. Or maybe the first signs can be seen today? In the end of 2024 some bee breeders – finally! - publicly declared taking up selection of resistant bees. Polish beekeeping community seems to be at least a decade behind some of the western communities. See also: Jaroński J, Hodowca pszczół świadczy usługi dla pszczelarzy. Musi odpowiadać na ich potrzeby oraz nowe wyzwania, rozmowa z Przemysławem Szeligą [A beekeeper provides services to beekeepers. He must respond to their needs and new challenges, interview with Przemysław Szeliga], Pszczelarstwo 2022, 4-5. I hope that the publishing profile will also change, today dominated by textbooks on apiary management and guides on how to set up an apiary based on honey production.

321 Rather, having them is a marketing strategy, they treat them just as queen-breeders to reproduce queens and sell their daughters to other beekeepers, treating them like any other bees, and without any serious work on local adaptation (sometimes they are not even tested locally, just mass-reproduced even the same year they are imported into foreign environment).

As a result, bee colonies deprived of treatment mostly die. While in western Europe and the USA varroa resistance traits and bee immunity has been the subject of research for many years, in Poland until recently the problem did not attract the interest of scientists. A certain breakthrough occurred recently, when the topic of bees resistant to varroa is starting to appear at some beekeeping conferences (unfortunately, it is most often presented as an anecdote or an exotic curiosity). Most often, however, these are reports from research conducted outside our country[322]. I hope that this is the first step towards long-term research and selection and scientific programs that will contribute to progress in this field.

During the hanging of the "artificial tree hives"; on the left Łukasz Łapka, on the right Marcin Zarek, author's photo.

Still, the questions remain unanswered, why is the scientific community not involved in popularizing methods that are conducive to the health of bee populations? Why is it not obvious to abandon the import of foreign bee ecotypes? Why are breeders not working on obtaining increasingly resistant bees and genetic material adapted locally? Everything indicates that the current situation has been accepted as the norm. Since activity is most often limited to promoting increasingly better methods of treating bees, amateurs are trying to change the situation, although the community is small, fragmented and does not have sufficient resources to have a big impact. Only a few operate in groups, others work on their own.

---

322   Possibly initiated in the EU, in which Polish scientists participate.

One of the beekeepers with whom I have established cooperation is Łukasz Łapka from the Świętokrzyskie province[323]. For many years, he has been beekeeping without using chemicals, and he does not carry out biocidal treatments. He shares the results of his work and experiences on his own blog (www.llapka.blogspot.com), which is a source of knowledge on breeding and raising resistant bees. Łukasz has been using Dadant hives, without foundation for several years. In the first phase of selection, he imported queen bees from several European beekeepers working on bee resistance. For years, however, his apiary has been based on bees that survive without treatment. Unlike myself, Łukasz has never experienced total die off of his untreated apiary. He had some years with major losses, but never had to buy bees, always being self-sufficient in his project.

For several years now I have also been working with Marcin Zarek on organizing the community of non-treating beekeepers (we co-create the "Bee Brotherhood" group) and the "Fort Knox" project. Marcin runs an amateur apiary in the northern part of the Małopolska province. He usually prepares about 10-20 bee colonies for overwintering.

The group of beekeepers who do not use chemical treatments also includes Mariusz Uchman from the Podkarpackie province. He cooperates with us, helping to create a place where beekeepers who do not treat for varroa will be able to exchange experiences[324].

I also exchange experiences with Kamil Bućko from the Lublin province, who runs an apiary (similar in size to mine) and has not treated bees for about the same period as I have. Kamil can, however, boast a higher survival rate of bees; he also has better honey production. Similarly to Łukasz, he never experienced losses closing to 100% and never had to buy bees for his apiary.

The "Bee Brotherhood" group and "Fort Knox" bring together many amateur beekeepers. Among them are those who are more cautious, i.e. they leave only some of their colonies without treatment, and observe others but subject them to treatments, most often using so-called soft or ecological chemicals. One of such people is Patryk Słotwiński, a photofigureer by

---

323 We started discussing beekeeping without chemicals in 2014, later we participated in establishing the Natural Beekeeping Association "Wolne Pszczoły", the "Fort Knox" project, and then the group "Bee Brotherhood". Łukasz Łapka is the author of several publications in the monthly "Pszczelarstwo" on the natural beekeeping and the so-called small cell.

324 He is also a figure designer, helping the author of several of the infofigures for this publication.

profession, who is also involved in social activities, e.g. he co-creates educational apiaries, primarily for children and young people[325].

There are also beekeepers who operate outside the structures mentioned. It is impossible to mention all of them, I hope that those which I have not mentioned here, will not hold it against me. Interesting activities are undertaken by, among others, Piotr Piłasiewicz from the Augustów Forest region, one of the founders of the "Bractwo Bartne" Foundation ("Treebeekeeping Brotherhood - Fratrum Melicidarium"), which deals with restoring beekeeping traditions. It is also the platform to cooperate with tree-beekeepers from Lithuania and Belarus. He also organizes tree beekeeping workshops, where one can learn the secrets of making a log tree hive and traditional forms of bee care[326]. I hope that the natural beekeeping community in Poland will start to gain strength at a much faster pace.

Patryk Słotwiński mows the grass in the educational apiary, photo by P. Słotwiński.

---

325 Recently, he also became a councillor of one of the communes in the Silesian Voivodeship and a vocalist of a music band called "Saw-dust".

326 For some time now, Piotr Piłasiewicz has also been involved in the production of mead, co-founding Augustowska Miodosytnia [Augustów Meadery].

Mariusz Uchman's drawings, full of humour, often express thoughts much more accurately than words and shorten the path to acceptance, figure by M. Uchman.

AT THE CROSSROADS, OR THE PATH TO BEEKEEPING WITHOUT TOXIC SUBSTANCES

CHAPTER 6

AT THE CROSSROADS, OR THE PATH TO BEEKEEPING WITHOUT TOXIC SUBSTANCES

> *Owing to this struggle, variations, however slight and from whatever cause proceeding, if they be in any degree profitable to the individuals of a species, in their infinitely complex relations to other organic beings and to their physical conditions of life, will tend to the preservation of such individuals and will generally be inherited by the offspring. The offspring, also, will thus have a better chance of surviving, for, of the many individuals of any species which are periodically born, but a smaller number can survive.*
>
> **CHARLES DARWIN**[327]

---

327   Ch. Darwin, The Origin. op. cit., page 77

Tools, characteristics, properties of bees, resistant traits

Hygienic behaviour

The hygienic behaviour of bees is directly related to the care of brood health. It involves the removal of pathogen-infected, impaired or dead larvae and pupae, thus helping to minimize the risk of an epidemic within the hive (i.e., they allow bees to better cope with potential bacterial or fungal infections). Bees coming from hygienic lines are therefore better able to take care of their colony health. Hygienic behaviours are quite often selected in various breeding projects. For this purpose, basic tests[328] have been developed, which involve killing pupae in the comb (sealed brood), which is most often done in two ways. The first one involves freezing a specific area of the sealed brood, e.g. using liquid nitrogen. The second, is the needle test that involves puncturing individual pupae through the cell cap. To consider the results as reliable, tests are typically performed on at least one hundred cells (in a compact area). After several hours, the test area is checked to see whether the brood has been removed from the cells. Bees that remove all pupae within 24 hours are considered hygienic[329].

There is a belief that pricking the brood is a more precise method than freezing the comb. Occasionally, the pupae are only injured, but even then they emit a specific odour that can be identified by the workers (it is possible that the same mechanism works when the pupa is punctured by a varroa mite), which also stimulates the bees to clean.

It is believed that bees with a high level of hygiene have a better developed sense of smell.

---

328 Some people combine hygienic behaviour with maintaining order in the hive (e.g. cleaning the bottom of the colony of wax or other organic particles); to test such behaviours, the basic, simplest and least invasive test used involves pouring various foreign bodies (torn sheets of paper, crumbs, etc.) into the hives. Those bees that remove foreign bodies the fastest will certainly take better care of the cleanliness of the nest cavity of their colony. However, it is difficult to say whether this type of behaviour translates into a real increase in disease resistance. Debris constantly lying on the bottom of the colony (e.g. a larva that died and was removed from its cell) may become the source of an epidemic. However, the essence of hygienic behaviour is not about cleaning the bottom.

329 As the selection progresses, the removal time will be shortened. Very hygienic bees remove dead brood even within six hours.

HYGIENIC BEHAVIOR

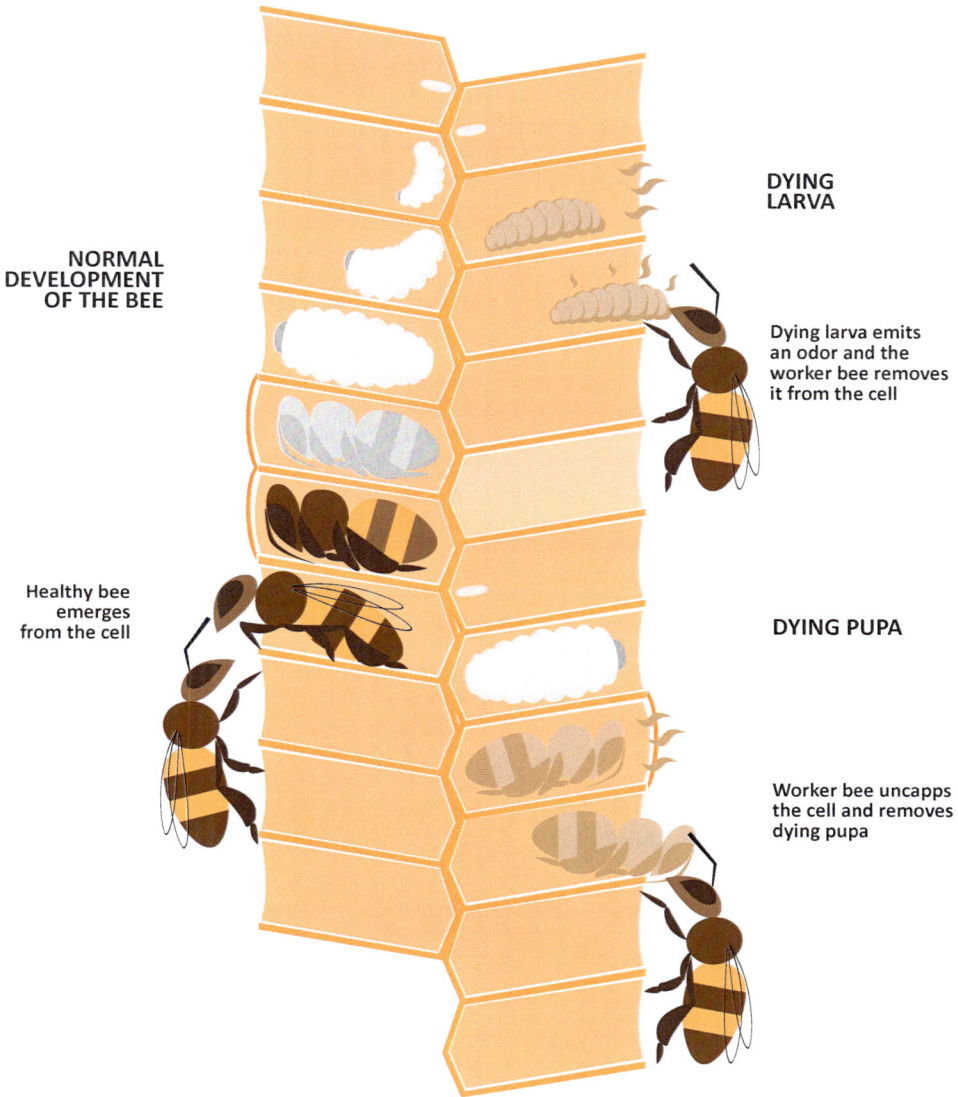

**NORMAL
DEVELOPMENT
OF THE BEE**

**DYING
LARVA**

Dying larva emits
an odor and the
worker bee removes
it from the cell

**Healthy bee
emerges
from the cell**

DYING PUPA

Worker bee uncaps
the cell and removes
dying pupa

Hygienic behaviour involves removing dead (sick) larvae and pupae figure by M. Uchman.

The pin test involves piercing one hundred cells of capped brood. After several hours in very hygienic colonies, the cells have been cleared of dead pupa and the new eggs have been laid, photo by Ł. Łapka.

Even though hygienic behaviours are not directly related to the bees' resistance to varroa (they are aimed at other diseases of the brood), the research found a relationship between the bees' ability to remove pin-killed brood and mite levels in the colony. That is the better the bees' results in the hygiene tests (the shorter the time to remove the killed pupae), the lower the mite infestation. It is believed that this correlation is related to the specific smell emitted by the infected (punctured) pupa. This fact also directly links the described trait with other bee behaviours, such as VSH and recapping, which, however, are considered separate properties or characteristics of bees.

Varroa Sensitive Hygiene (VSH), recapping

As I have already mentioned, one of the first characteristics of bees, which is strictly associated with the fight against the mite is VSH. It was only later that attention began to be paid to another feature called "recapping" or "uncapping-recapping". Most of the beekeeping community distinguishes the two mentioned behaviours, treating them as separate traits.

However, in some way, they are initiated by the same or similar odour-specific mechanism. Some people believe that it is just one bee trait (behaviour) that, depending on the situation, leads to different behaviours of the workers towards the damaged pupae. We will therefore discuss them together in one chapter, pointing out the differences between them, and the ways of observing and selecting them.

The colonies resistant to varroa often perform uncapping on the infested pupae which are sometimes removed (VSH), but sometimes they remain uncapped for some time, and then re-capped (recapping) photo by Ł. Łapka.

VSH was first described in the laboratory of the United States Department of Agriculture in Baton Rouge (Louisiana), where research was carried out on the phenomenon of limited mite fertility. Under specific conditions (see the section on SMR)[330] it was found that in some colonies workers can detect mite infested pupae through the cell cap[331], by uncapping them and

---

330   In the initial phase, mite population growth in short periods of 3 months, was observed in the group of studied colonies, then colonies with the smallest mite infestation growth began to be propagated. After several cycles, there was a significant difference in the rate of mite infestation growth in the selected group compared to the control group. Only then scientists began to monitor the bees for the emerging characteristics, trying to find out the reasons for the slower growth of the mite population. It was found that in the studied colonies the mites didn't reproduce as well as in the control group (they did not lay eggs or did so with a delay). This feature was called SMR (Suppressed Mite Reproduction). Only later it was found that the cause of phenomena in the studied population was VSH (removal) behaviour.

331   Chemical (odour) signals detected by workers are secreted by infested larvae/pupae that produce a specific hormone because of a stressful situation; K. Wagoner et al., Stock-specific chemical brood signals are induced by Varroa and Deformed Wing Virus, and induce hygienic response in the honey bee, Scientific Reports 2019, 6. How do bees sense mites inside the cell? According to J. Harris, a specialist in VSH behaviour from Mississippi State University, worker bees must touch the cap of each cell with their antennae to determine whether mites are breeding inside (see: "Two Bees in a Podcast", episode 33: Varroa Sensitive Hygiene & Various Beekeeping Climates; https://open. spotify.com/episode/odl1t43iTZmDi7B8ewP7j6, accessed: November 20, 2022). Another study suggests that bees uncap not only cells with mites, but also neighbouring cells, I. Grindrod, S. Martin, Spatial distribution of recapping behaviour indicates clustering around Varroa infested cells, Journal of Apicultural Research 2021, 5. This may be due to the fact that they are not able to precisely determine where a specific smell comes from (they are not able to precisely determine the cell in which the mites are located each time, only the general area). In the cell cap they create a small central hole and thus search for the infested pupa, and when they find it, they enlarge the hole in the cell cap.

then removing them from the cell (taking them outside the hive or eating them). Bees with the VSH trait can identify cells in which female mites are reproducing (which leads to intense feeding on the pupae), but they do not deal with cells containing mites that are not reproducing. In the first phase, the bees uncap these cells and remove any infested pupa. Behavioural signs are examined by uncapping the brood cells and checking the cell contents very carefully. There are therefore cells without mites, those containing only the foundress female mite, and those containing a foundress female, a male and female offspring. The value of the VSH feature is expressed as a percentage. Bees with a 100% VSH uncap the cells and remove all the larvae/pupae in which the mites were breeding (i.e. those containing mite offspring)[332]. Therefore, in the comb of such bees, single foundress female mites can be found (this does not reduce the value of the feature), but without offspring, so they are not contributing to the growth of the mite population.

Please remember that male and female offspring mites are transparent white when developing and can therefore be easily missed during examination. Therefore, the examination requires great concentration and good light (it is also worth having a magnifying glass).

Moreover, in bee colonies where there are few mites (because they cope with the parasite), it is very difficult to detect the VSH trait (it is difficult to properly assess the behaviour and calculate the true value of the trait). Of course, such bees are worth further breeding, regardless of whether they cope with the mite thanks to the VSH trait or due to some other behaviour or characteristic. However, to be able to study and consolidate the VSH trait in the population, breeders very often provide colonies with frames with brood from hives where the level of mite infestation was very high. This allows you to quickly determine whether a colony has a high level of VSH or not. As you can see, researching and selecting this feature is quite time-consuming and requires experience. However, people who have made this effort claim that it is worth doing. VSH bees maintain a low level of mite infestation and acquiring them to start our selection program can pay off.

---

332 Bees with 100% of the VSH trait can relatively quickly remove the entire population of mites from the hive (and then maintain this state). However, it remains difficult to maintain this feature in subsequent generations. Research on colonies with queens from a pure VSH line shows that those inseminated via open mating (with drones from unselected lines) can slow down the mite's growth in hives by at least one third compared to unselected colonies.

VSH — VARROA SENSITIVE HYGIENE

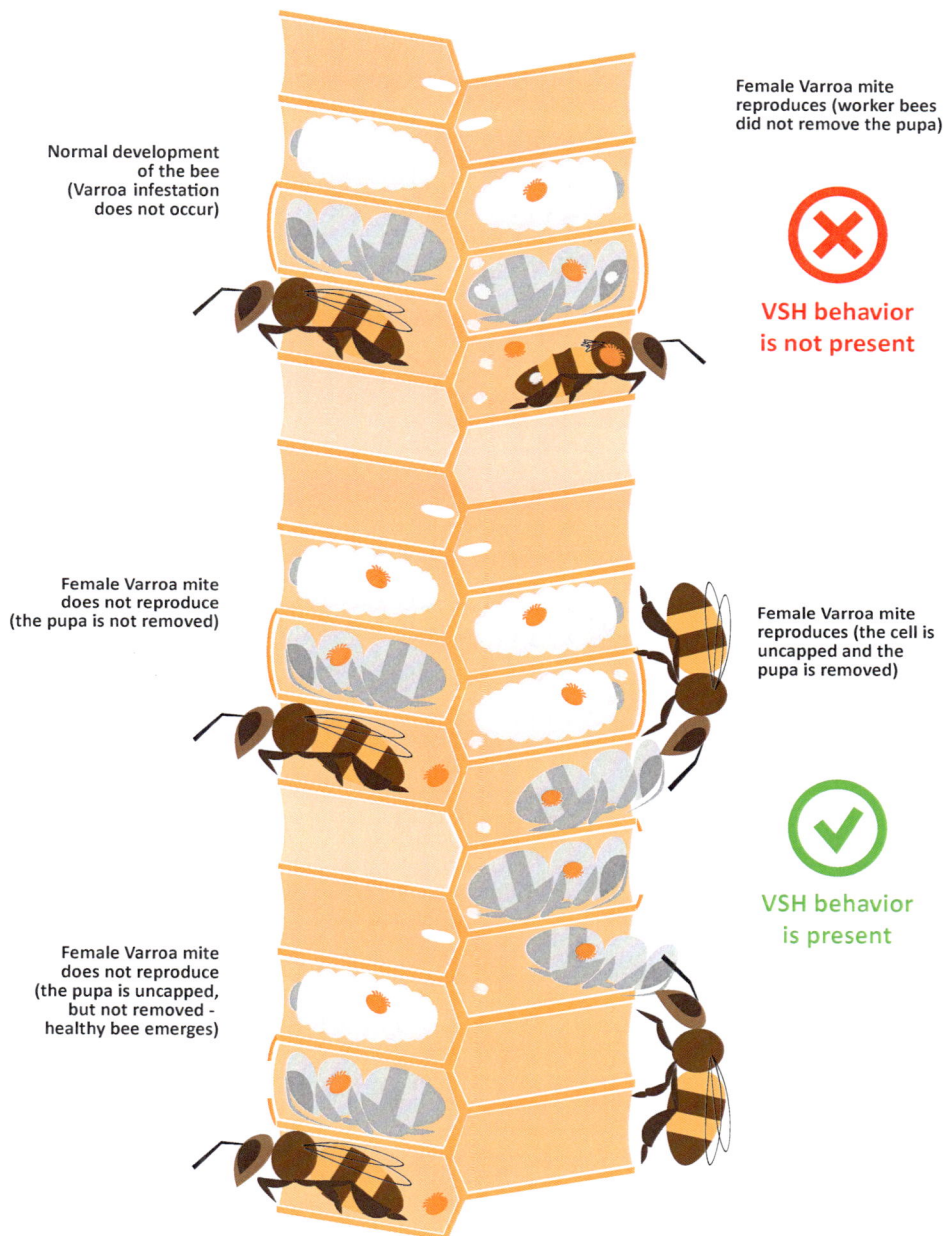

Female Varroa mite reproduces (worker bees did not remove the pupa)

VSH behavior is not present

Normal development of the bee (Varroa infestation does not occur)

Female Varroa mite does not reproduce (the pupa is not removed)

Female Varroa mite reproduces (the cell is uncapped and the pupa is removed)

VSH behavior is present

Female Varroa mite does not reproduce (the pupa is uncapped, but not removed - healthy bee emerges)

Varroa sensitive hygiene (VSH) involves removing the pupae from the cells where the mites breed. Workers most often ignore cells in which the mites do not produce offspring (such cells are at most uncapped), figure by M. Uchman.

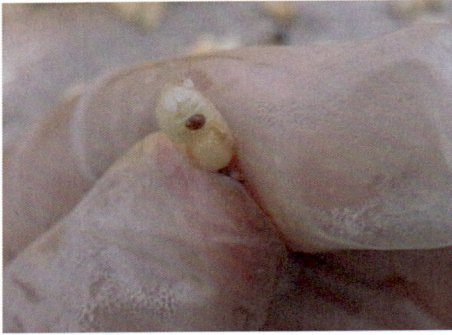

Brood analysis: the bee pupae show the mite family with a mature female and her offspring, photo by Ł. Łapka.

Bees equipped with the recapping trait detect mites in the cell in a similar way. However, they act differently with the cell in which they detect a mite. Worker bees remove the cap of the cell thus the pupa remains open (most often) for a few hours[333], and then it is sealed again. Sometimes the same cell may be uncapped and resealed multiple times during the brood development cycle. The workers do not kill or remove the developing pupa which has a chance to mature normally and eventually hatch. For the bee colony, this feature has a lower price than VSH, because the resources (food, energy, work) invested in brood rearing do not go to waste. It is therefore believed that in the process of natural selection this behaviour will be promoted more than VSH: the smaller the loss of resources, the greater the competitiveness of these colonies and therefore increasing their chance of survival[334]. The recapping feature has been confirmed in many wild and/or free-living bee populations. Many researchers also believe that this property may prove to be the most important in the fight against mites. There is also discussion in the beekeeping community about whether all populations of resistant bees in the world develop the same characteristics or whether they differ depending on the subspecies or region. The description of many populations shows that there are some differences, but they share the same features (which may have different intensities). Stephen Martin claims that

---

333 R. Büchler claims that the cells are sometimes sealed immediately, sometimes after a few hours, sometimes even after a few days. The larva/pupae can develop well even without a cell cap, since it is only necessary during metamorphosis, when it begins to form a cocoon.

334 A bee that was fed upon by a mite during its development phase may be weakened compared to a healthy pupa, and this affects its lifespan. However, if the virus infection is not significant, its impact on colony development may be only small. If the mite infestation is high, this may translate into many such virus infected insects. However, it is difficult to say how important this is in a multi-thousand-superorganism of the western honey bee colony.

the recapping trait has been diagnosed in all resistant bee populations studied[335]. It must therefore be important for the survival of untreated bee colonies in the presence of the mite. But it probably wouldn't be enough to overcome the mite if the bees had no other important traits.

So far, it has not been clearly explained what the mechanism is of reducing the dynamics of the mite's population growth resulting from recapping, in other words, why this behaviour serves to reduce the mites' population (does it affect the biochemistry of the mite or its behaviour?). The research continues[336]. However, it is known that the uncapping of cells is related to the reproduction of mites, and sometimes even stops it. Why? There are many hypotheses.

RECAPPING

Normal development of the pupa

No recapping trait - the worker does not detect the infested pupa

Cell only partially uncapped

Recapping trait present - the cell with infested pupa is uncapped

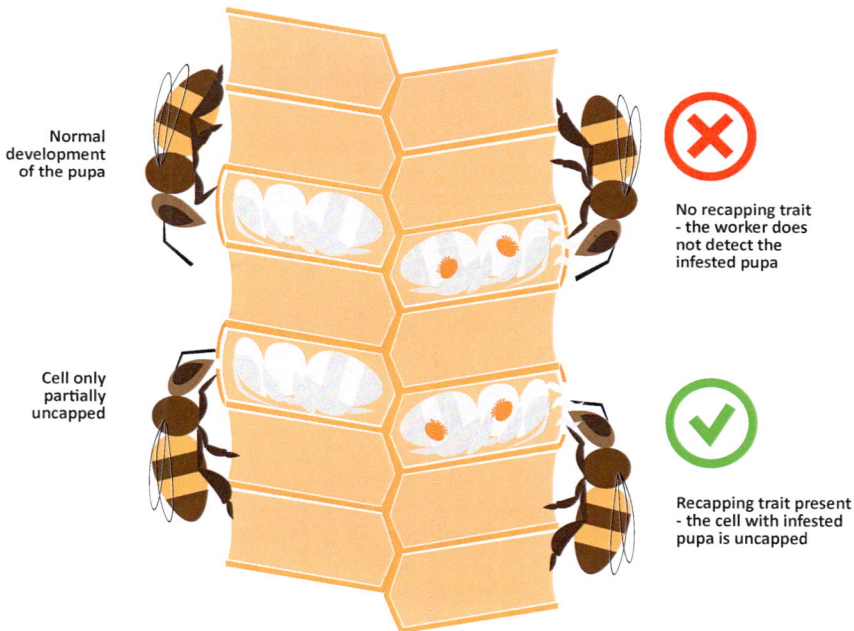

Recapping involves removing the cell cap of the developing pupae. Workers can make small holes in the cell cap of a healthy pupae (not infested with mites), and they enlarge the holes in the cell cap where mites are reproducing, figure by M. Uchman.

335   I. Grindrod, S. Martin, Parallel evolution of Varroa resistance in honey bees: a common mechanism across continents? Proceeding of the Royal Society B 2021, 8; S. Martin, *Varroa destructor* reproduction and cell re-capping in mite-resistant *Apis mellifera* populations, Apidologie, 2019, 12.

336   G.P. Hawkins, S. Martin, Elevated recapping behaviour and reduced *Varroa destructor* reproduction in natural Varroa resistant *Apis mellifera* honey bees from the UK. Apidologie, 2020, 3; M.A.Y. Oddie I, Reproductive success of the parasitic mite (*Varroa destructor*) is lower in honey bee colonies that target infested cells with recapping. Scientific Reports, 2021, 11.

Some people think that opening the cells in the later developmental phase of the brood may cause the males to leave the cell, and their loss will mean that they will not fertilize young females[337], Ralph Büchler adds that it happens that mature females enter the open cells, but due to this, their pupal development phase is shorter than the normal full cycle (from the initial sealing of the larva), this results in the young mites produced not having enough time to mature and so later they die. Therefore, we are dealing with desynchronization of the reproduction cycles of bees and their parasites[338]. However, it is believed that recapping has the greatest impact on the development process of mites (change in temperature, humidity, kairomone level[339] or other yet unknown reasons), which may lead to a reduction in their fertility.

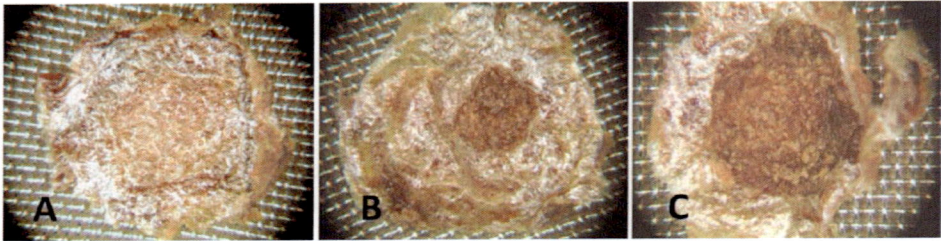

After sealing the cell, the bee larva inside spins a silk cocoon before transforming into a pupa. If the cell is then opened and closed again, there will be no characteristic silk cocoon attached to all of the underside of the cell cap, hence the visual difference between those cells that were uncapped and those that have not been disturbed. The photo shows the bottom side (underside) of the cell cap, from the left: uncapped, only small hole recapped; a larger hole recapped; photo by Agricultural Research Service (ARS), US Department of Agriculture (USDA), Baton Rouge.

---

337   It is believed that male mite is unable to live outside the bee cell. Perhaps the recapping effect could be enhanced by the so-called small cell; the less space there is in the cell for the male to reach the foundress female, the more likely it is that he will leave the cell and then die.

338   This argument does not concern the direct causes of the recapping behaviour (although it may be one of the reasons for the development of such an evolutionary mechanism), because the cells that are uncapped are the ones that the females have already penetrated. However, the fact that Büchler mentions, may have an impact on the success of mite reproduction, because if the parasite goes through a reproduction cycle in such a cell and the offspring does not have time to mature, then there will be one generation fewer mites in the hive. If this process happens more often, it will affect the dynamics of the mite growth in the bee colony.

339   Kairomone (from Greek kairos "benefits"), a semiochemical substance secreted by the body, which means a neutral or unfavourable signal for the sender, but a favourable signal for the recipient. Kairomone may be the scent of a predator sensed by its prey, which upon detection of such a substance may reveal defence mechanisms. Kairomones are also substances secreted by plants that attract herbivores. It happens that a substance is a kairomone for one species (attractant) and an allomone for another (repellent).

Recapping can be seen by carefully removing the cell cap (using a scalpel or strong adhesive tape) and looking at its underside[340]. If we find any irregularity (different colour, structure), it means that the bees are uncapping and re-sealing the brood cells. The more uncapped or recapped cells, the more this type of behaviour occurs.

It is believed that the recapping trait is easily inherited and therefore, selection may bring results[341]. Stephen Martin also believes that bees may learn recapping (as well as other behaviours) from each other, since workers are not genetically programmed automatons[342], but are also affected by epigenetic mechanisms.

Many beekeepers and breeders have been eliminating the VSH and recapping feature for many years. It was believed that an important feature in the selection of bees is the so-called compact (regular) brood pattern on the comb.

Therefore, queens were selected for breeding from colonies where the bees' natural abilities (including those to fight mites) were not respected. The direction of this type of selection was turned against their resistance! It is difficult to say to what extent the situation has changed in recent years because of new research.

We can also select the VSH and recapping features without the need to perform specialized tests, but we must be aware that the risk of errors increase. We can assume that mite infestation will be relatively high in late spring or early summer if we use standard bee stock without resistant traits. If during any inspection in such colonies we notice many uncapped brood cells, we can assume that the bees may have the described characteristics.

---

340   Changes in the cell cap from the outside can only be noticed in exceptional cases, because they are quite subtle. During a regular inspection, however, you can see patches with uncapped cells and visible pupae inside.

341   It should be noted that recapping does not always have to be related to the presence of mites, bees also display such behaviour in other cases (e.g. infection of the lesser wax moth, whose larvae attack the capped brood, drilling corridors under it). Some people believe that to select for the recapping trait, it would be necessary not only to examine the cell cap from underneath, but also to confirm that there was a female mite in the uncapped cell. This would mean that selecting for this trait would be extremely labour-intensive (as or even more labour intensive than selecting for VSH). It is probably important to do so during research, to be 100% sure that the feature is related to the presence of mites. In practice, however, during selection, it seems sufficient to confirm the presence of the feature only by examining the bottom part of the cap.

342   Genes are a specific code for the proper production of proteins in the body, they do not directly encode the behaviour or characteristics of organisms. However, our biology, through the genetic code, influences our behaviours and predispositions (genes are our building blocks, they are a kind of hardware; for the body to be able to perform specific tasks, it must have appropriate hardware, we do not expect woodworking from a laptop, or complex calculations from a drill). But we still need software that will activate specific mechanisms and allow us to use it.

However, if the brood is completely sealed, we can assume that the bees do not have the ability to detect mites under the cap[343].

Suppressed Mite Reproduction (SMR)

The SMR[344] phenomenon is responsible for the slowed rate of growth of the mite population in the hive, and concerns female mites which, although mature, remain infertile or their fertility is significantly reduced[345]. Observing symptoms of SMR in apiaries is very difficult[346]. However, in some resistant populations, scientific research managed to confirm its occurrence in colonies of specific ecotypes or bee lines that had a higher percentage of infertile female mites. The SMR phenomenon in Europe was studied by, among others, Ralph Büchler's team from the Kirchhein Institute

---

343 This is not a suggestion on how to conduct the selection! The point is rather that it is better to do something (and pay attention to certain behaviours) than nothing. If conducting specialised observations and research is beyond our capabilities, it is worth at least paying attention to bee behaviour during regular inspections, and after the season, after treatment, at least compare the number of mites in different colonies (you do not need to count dead mites but only estimate the size of their population).

344 In the literature on the subject you can also find the term Mite Non-Reproduction (MNR) which is sometime treated as the synonym of SMR. In other sources this mechanism is interpreted as a description of the cumulative effect of VSH, recapping and SMR.

345 F. Mondet et al., Evaluation of Suppressed Mite Reproduction (SMR) Reveals Potential for *Varroa Resistance* in European Honey Bees (*Apis mellifera* L.), "Insects" 2020, 9 (Polish scientists also participated in this study).

346 Some symptoms of SMR can be observed without specialized equipment (however, a microscope or a strong magnifying glass is recommended): we remove pupae from a patch of brood and calculate the proportion of cells (in percent) of those containing mite offspring and those containing non reproducing females (like the case of the VSH). Büchler suggests that to determine when reproduction is delayed, the cells must be opened at a specific time in the development phase (it is crucial to track the age of the larvae/pupae: only then, based on time left for the bee pupae to hatch, we can determine in what phase of development the mite offspring should be). Then examine the number of mite offspring and developmental phase of them, and then compare it with the expected phase. With such simple observations, however, it is impossible to determine whether the female mites that do not reproduce are mature enough to reproduce; the SMR trait applies directly to those female mites that are capable of breeding but do not do so. The scientist claims that for the results to be reliable, it is necessary to examine at least 35 cells containing mites – and so, depending on the degree of infection, it is necessary to open 300-500, and sometimes even a thousand cells with pupae. The study must be repeated at least 2-3 times during the season and carried out on at least several dozen colonies, which means a lot of effort. Therefore, it seems that the selection of this feature using the breeding selection method is very difficult to carry out (which is why this feature is rather studied only in scientific programs). Currently, there are genetic studies carried on, aiming at finding specific markers that would allow the scientists to determine whether bees have these abilities/characteristics (without the need for detailed brood examinations).

SUPPRESSED MITE REPRODUCTION - SMR

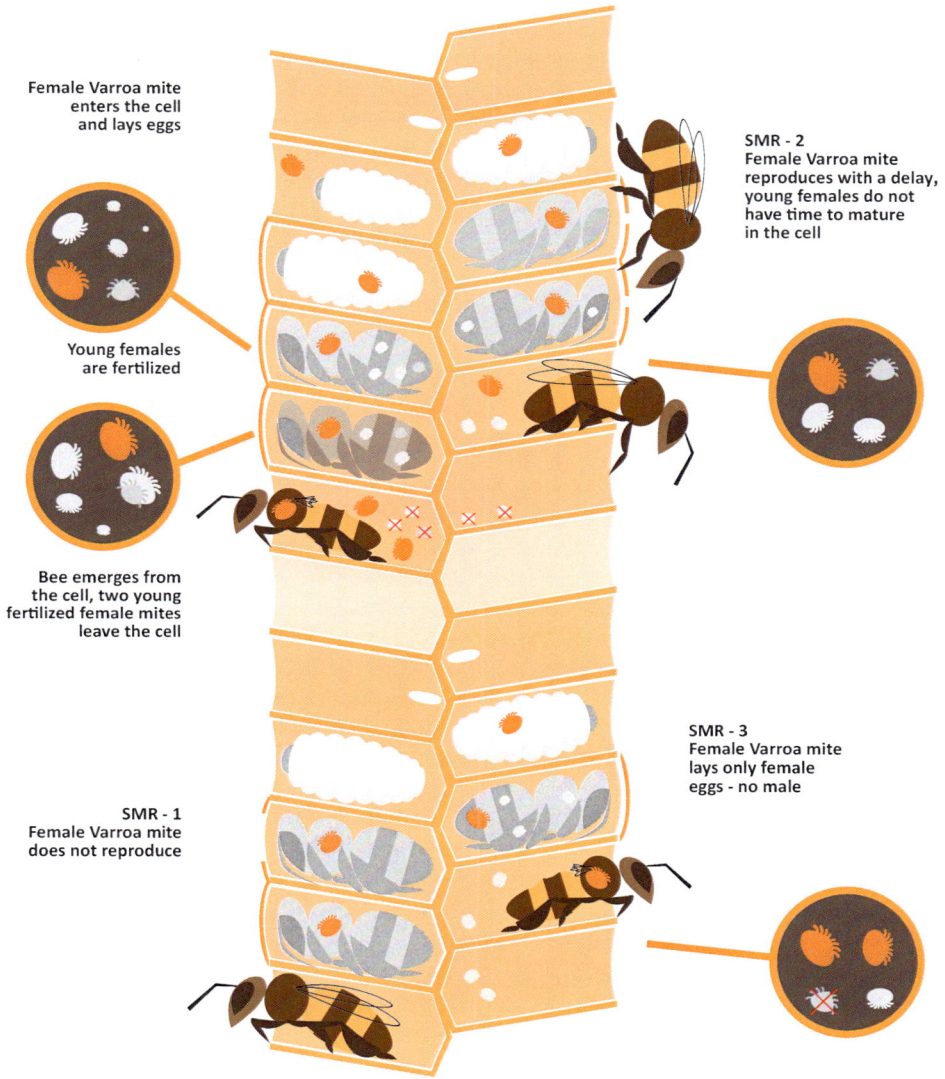

Female Varroa mite enters the cell and lays eggs

Young females are fertilized

Bee emerges from the cell, two young fertilized female mites leave the cell

SMR - 1
Female Varroa mite does not reproduce

SMR - 2
Female Varroa mite reproduces with a delay, young females do not have time to mature in the cell

SMR - 3
Female Varroa mite lays only female eggs - no male

The SMR characteristics can manifest itself in several ways, figure by M. Uchman.

325

in Germany[347]. The scientist suggests that in the case of bees having these characteristics, we have three possibilities:

1. The mites will not reproduce; 2. Their reproduction will be relatively normal, but will be delayed (mites lay eggs later, so the young offspring do not have enough time to mature before the bees emerge); 3. There may be no males among the mite offspring, which means that the young females will not be mated so unable to lay fertilized eggs, (as a result, they can achieve reproductive success only when a female mite enters a cell with another female and her son fertilizes the offspring of the other one). As a result, it was found that as 80-85% of mites in a population are fertile, than in bees with the SMR trait this percentage is often around or below 50%. The reasons for this phenomenon are not fully known. Ralph Büchler claims that one of the reasons for SMR may be longer brood-free periods in the bee colony. That is after the bees' development cycle is interrupted and also that of the mites. So, the decisive factor might be the mites physiological status in which the mites enter the brood but fail to reproduce which may be influenced by the time of the season).

Another hypothesis of Büchler assumes that the presence of SMR may be determined by bee behaviour, particularly VSH. Therefore, it suggests that in the process of shaping hygienic behaviour, bees may remove fertile females from the hive, which may result in an increase in the percentage of infertile females in the colony[348]. The researcher also does not rule out that the lower reproductive success (related to SMR) may be caused by a disruption of the mites' development cycle caused by recapping. As a result, the development cycles of bees and mites have become desynchronized. The researcher's last hypothesis assumes the influence of physical conditions (e.g. change in humidity, temperature) in the bee cell on the development of mite offspring

---

347   Scientific work at the Institute related to the selection of bees resistant to mites are carried out on 300-400 bee colonies at the Institute's disposal and in cooperation with bee breeders from Germany (primarily with the associations of Carniolan bee breeders and Buckfast): "Two Bees in a Podcast" episode 66, European Honey Bees & Sting Management, https://open.spotify.com/episode/4Kf2dcVpVCHaKl2m9BCwmd, accessed: February 20, 2025.

348   This would be consistent with what we know about VSH and the original assumptions and observations of researchers (John Harbo's team, where Jeffrey Harris worked, among others) who discovered a feature of VSH in Baton Rouge, Louisiana. That is bees remove mites from cells only when they breed and ignore other females. The scientist claims that the effect of VSH on the proportion that reproduce or are infertile is, however, temporary, i.e. it is not related to long-term selection, because only those individuals that have this ability reproduce (infertile individuals, obviously, are not able to transmit this feature). According to this hypothesis, there is no specific property, characteristic of bees, or phenomenon that we could describe as SMR; the relatively high number of mites that do not breed is a direct result of the removal of breeding mites as part of the VSH trait.

or the development of their reproductive abilities (also related directly or indirectly to recapping). It cannot also be ruled out that the SMR behaviour is related to some substance that would affect the fertility of female bees. For example, a hormone in bee brood (this phenomenon has been confirmed in the colonies of the Eastern honey bee)[349].

Grooming

This is how we name active removing of the mites by bees themselves. When worker bees remove mites from their own body, we talk about auto grooming, and when a bee is groomed by other workers it is called allogrooming.

Removal involves the bees biting the mite, which often mutilates (pulling out limbs) or kills the mite. The mite dies on the bottom of the hive as it is unable to stay on the bee. It is the damaged mites found on bottom inserts that indicate that bees are equipped with the grooming trait. The more damaged mites we find among undamaged individuals (those that died of natural causes – e.g. aging), the higher the value of this trait[350]. In outstanding colonies that have undergone selection for this type of grooming, the percentage of individuals killed by bees in relation to all dead mites may be over 90%, which means that only about 10% of the mites die of natural reasons. However, it has been found that honey bee grooming is not as effective as that of the Eastern honey bee. The results of research on the heritability of this trait are also not clear. According to some researchers, its heredity level is high, others say it is very low[385].

---

349   See: J. Woyke, Bee Breeding... op. cit.

350   Varroa mites – as all living organisms – are mortal. It is assumed that in the summer season the daily mortality of *Varroa destructor* mites is 1-2%. That is, if there are 100 individuals in the hive, statistically 1-2 mites per day die of natural cause. Finding dead mites on the bottom board is therefore completely normal. To check whether there is a grooming trait in the colony, we should collect fallen mites and, after detailed observation, determine the proportion of individuals injured by bees to those undamaged. If out of every 10 fallen mites, 8-9 are mutilated (e.g. missing a limb), it may mean that the grooming is highly developed. You should be also aware that mites can be damaged after death by other creatures feeding on dead individuals.

GROOMING

The worker bee itself can remove the mite from its body

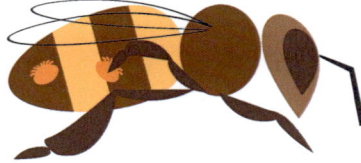

Other worker bees can remove the mite from the bee's body

Female Varroa mites are mutilated (usually their legs are bitten off), they fall down to the hive bottom unable to sustain life

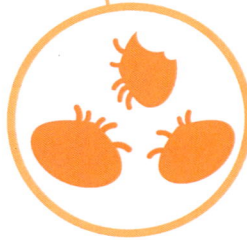

Grooming is the removal of mites from the bees, which are usually mutilated (e.g. Their legs are torn off), thus they are no longer able to feed on the bees and die on the bottom of the hive, figure by M. Uchman.

To analyze the natural fall of the pest, inserts with sectors are usually used to facilitate the counting of the parasites, photo by Ł. Łapka.

According to Torben Schiffer, the activation of behaviours related to grooming depends on the situation in the hive, on the conditions enabling specific biological priorities (in accordance with the so-called pyramid of needs), i.e. primarily supplying the colony with food[351]. Schiffer claims that until the bees have a sense of "storage security", so have accumulated a sufficiently large supply of honey, grooming behaviour will not be "activated" by them, even though this instinct is encoded in their genes. According to Schiffer, bees groom themselves most intensively when their nest cavity is small, then the colony has a sense of control over the space and can ensure that their biological needs are met. If the beekeeper increases the volume of the hive (by adding another empty body - supers), the workers will stop cleaning themselves and start collecting food.

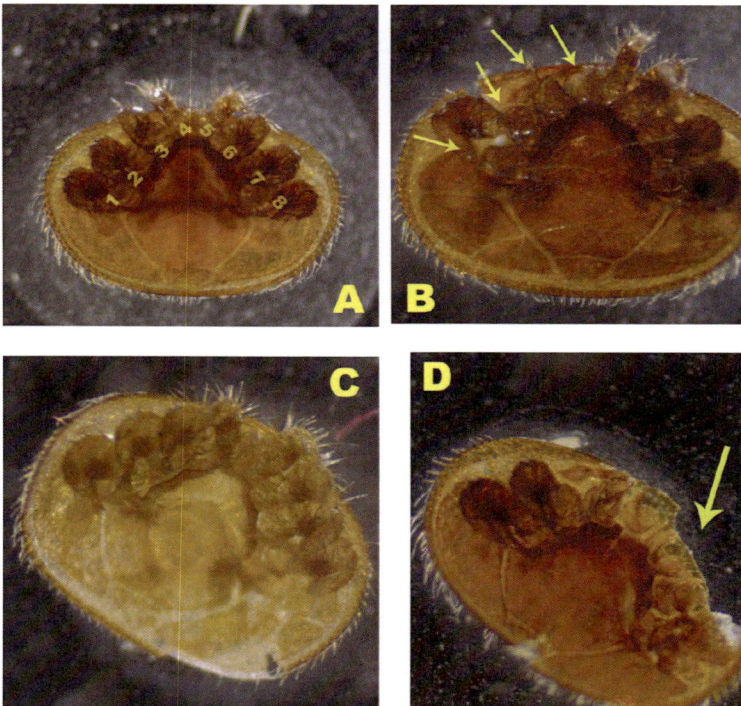

The female's mite body damaged because of grooming. Source: J. Smith et al., Morphological changes in the mandibles accompany the defensive behaviour of Indiana mite biting honey bees against Varroa destructor, photo by Frontiers in Ecology and Evolution, 2021, 4.

---

351   According to Torben Schiffer similar mechanism occurs in case of propolising the nest - bees will uptake it or intensify if they have "food security" ensured.

Attractiveness of the brood

At a certain stage of development (before sealing), the bee larva produces a hormone whose smell induces adult workers to seal the cell. However, this hormone is also an attractant for female mites, thanks to this, they know when they can hide in a cell with a larva. Research suggests that the brood of Africanized bees is less attractive to mites than some European subspecies. However, it seems that this feature is not important in selection for varroa resistance.

ATTRACTIVENESS OF THE BROOD - SMELL

Not all maturing bee larvae emit the same odour (kairomone). The odour of some larvae is more attractive to mites, which means they may be more likely to become infested with mites, figure by M. Uchman.

Shortening of the sealed brood phase

Female Varroa mite lays eggs

timeline · day 0

timeline · day 20 · day 20 i 12 h · day 21 · day 21 i 12 h · day 24

The period of development of a worker bee (especially after the cell is sealed) may affect the reproductive success of varroa. The longer the development of the pupa lasts, the better the chance that more of the female mite offspring have time to mature, figure by M. Uchman.

Some bee subspecies are characterised by a shorter development phase of the sealed (capped) brood period (this can also lead to shortening of the complete development of the bee brood from the egg to the hatching bee is shorter than the standard 21 days). The growth of the mite population in these colonies is lower, because the developmental phase of the pupa on which the mite feeds is insufficient to allow some varroa females offspring to mature (thus fewer female varroa can fully mature)[352]. In European subspecies, the variability of this characteristic is very small in most bees. That is the development phase of the capped brood lasts approximately the same time (longer than in other subspecies of bees, e.g. those of African origin). Probably, if variability was higher, varroa pressure and so natural selection would quickly lead to an increase in the reproductive success of bees with a shorter pupae development phase. Attempts to select this feature did not bring any positive results, since any change in mite population growth was not significant. This feature is also not easy to identify in the population (i.e. very difficult to select). It is believed that the use of a small cell may help to shorten the pupal phase in bees. This way, we could stimulate bees regardless of the feature resulting from the bees' genetics, shortening the pupal development phase.

---

352 See: J. Woyke, Bee Breeding, op. cit.

Influence of natural brood breaks on mite population growth

Varroa is only able to propagate in the worker and drone bee sealed brood. Therefore, if we extend the natural periods of breaks in bee brood rearing, we can cause the speed of mite population growth to significantly decrease. During the year, we can observe two natural breaks in the egg laying of the queens. The first is the winter break. It is difficult to say how long it lasts exactly as it depends on the subspecies, as well as the climate and weather. In the past (if we believe the traditions), queens from colonies locally adapted to our latitude in central-eastern Europe stopped egg laying at the end of September, they probably produced no eggs in October. They started laying eggs again in early spring, when the days became longer and warmer, most often in late February (maybe March), depending on the weather. This means that the brood-less period could last four or five months, sometimes longer. It is difficult to say how long the brood-free period currently lasts, but I think that in many places it does not last longer than two months (in our climate zone)[353]. In such a situation, varroa has excellent conditions for development, especially if the bees do not have any resistance traits. Within two months, the number of mites can increase two or three-fold. Early cessation of brood rearing in autumn can be selected by breeding. For example, such project is carried out in southern Poland with the "Dobra" line of Carniolan bees - genetics of colonies in which queens continue to lay eggs at the end of summer is eliminated from the breeding stock.

The second natural brood break is associated with swarming. Although it lasts much shorter than the winter break, it is also very important because it significantly shortens varroa's opportunity to reproduce. Seeley, during experiments conducted on bees resistant to mites, found that only those colonies that produced swarms in the second season lived for more than two years. The brood break associated with swarming usually lasts about three to four, less often five, weeks. Some (e.g. Ralph Büchler and Torben Schiffer) suggest that during this period mite population can be reduced even by 70%. This happens for several reasons. Firstly, it is believed that together with the hive swarm, up to several dozen mites will leave the hive.

---

353 This is what beekeepers who investigate their hives in winter say. Very often they observe the queen laying eggs in November (which usually means the need to replenish food supplies for the winter). Some people believe that bees need brood in winter, which allows them to regulate the humidity level in the nest (brood "absorbs" excess water), favouring good overwintering. Queen bees can also lay eggs, but the bees do not allow the larvae to develop (they eat them), see: https://www.youtube.com/watch? v=PokglAs-QRo, accessed: November 20, 2022.

Secondly, the natural mortality of the mite population is estimated to up to 2% per day (for 20 days it can be up to 40%, usually probably less). Thirdly, the brood break, unlike the winter break, falls in the period when the bees are active. In a situation, where there is fewer and fewer brood day by day and the number of young workers increases, the level of "unemployment" of the beehive increases, and the workers begin to sanitise the nest quicker[354]. It is also possible that a break in brood rearing may affect the physiological status of mites, which will result in a reduction in reproductive success[355], or at least a delay in the process. Torben Schiffer explains that mites that cannot reproduce due to the lack of brood, like mites in winter, fall into a specific physiological state until brood appears again[356], some mites must re-adapt to reproductive functions, which means that they are not able to reproduce when the brood first appears, but only in later appearing brood.

Seeley showed that the mite infestation of colonies producing numerous swarms (e.g. due to living in small hives), can remain at roughly the same level as in spring (or only slightly higher). The natural life cycle of bees is therefore a very important factor in the survival of free-living bees. It also seems that the beekeeper can artificially (and imperfectly) simulate that process by enforcing the brood-break (e.g. by making an artificial swarm with the old queen and not placing the young queen in the remaining hive). Although there is no doubt that the biological state of such an artificially split colony will differ from the one that cast a natural swarm, such a procedure may bring similar benefits to the bees. You can plan it so that the brood break falls before the main harvest. Thanks to the decreasing amount of brood in the colony (which consumes resources), higher harvests can be achieved.

---

354   Let's assume that the initial mite population in the hive is one thousand females. Up to 20-30% of the mites may leave the nest together with the swarm of (up to about 300 mites). Another several hundred (200-400) mites may die during the break in the brood rearing (natural mortality). Another 100-200 may be removed by bees because of intensified hygiene procedures (grooming, VSH). This means that in this hypothetical model, after the new queen begins egg laying, only about 300 of the initial 1,000 mites may remain. For this reason (if there are no other unfavourable circumstances, e.g. robbing, drifting), colonies that casted a swarm can survive without treatment significantly longer than those kept strong throughout the season.

355   I mentioned this in the chapter on SMR, suppressed mite reproduction.

356   This condition also applies to worker bees as in summer they live for a short period of time, but winter bees can survive at least 6-7 months (some say up to 10). Mites, like bees, do not age, but rather wear out when they are most active. On the one hand, inactivity prolongs their life, but it results in the modification of their physiological status.

The dynamics of population growth depending on the presence of resistance mechanisms in the bee colony.

Monthly growth factor of the parasite population:

▶ 2, 1.6, 1.2, assuming*: in January, there are 20 mother mites in each colony.
▶ the mite population decreases by 10% monthly during the brood-less period (natural mortality).
▶ after the swarm leaves, the pest population decreases by 50%.

1. Monthly multiplication of the parasite population by factor 2 in the absence of bee immunity mechanisms.

	Colony size (numbers of workers in thousands)	Number of mites	Mite infestation in percentage
JA	10	20	0,2
FE	9,5	18	0,18
MR	10,5	36	0,34
AP	25	72	0,28
MY	45	144	0,32
JN	50	288	0,57
JL	52	576	1,1
AU	45	1152	2,5
SE	33	2304	7
OC	24	4608	19,2
NV	15	9216	61
DE	12	8295	69

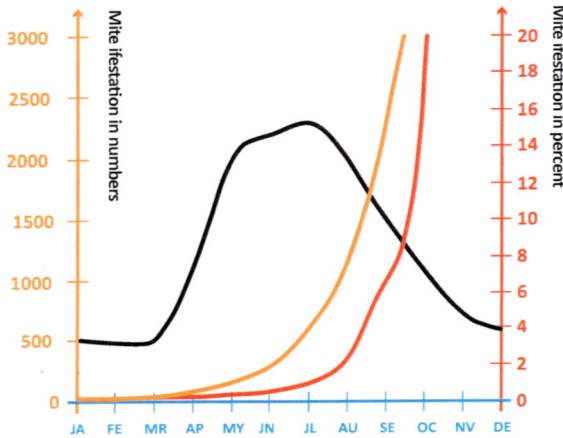

2. Monthly multiplication of the parasite population by factor 2, considering the swarm's exit.

	Colony size (numbers of workers in thousands)	Number of mites	Mite infestation in percentage
JA	10	20	0,2
FE	9,5	18	0,18
MR	10,5	36	0,34
AP	25	72	0,28
MY	45	144	0,32
JN	30	72	0,24
JL	43	144	0,33
AU	40	288	0,72
SE	31	576	1,8
OC	22	1152	5,2
NV	15	2304	15,3
DE	12	2074	17,2

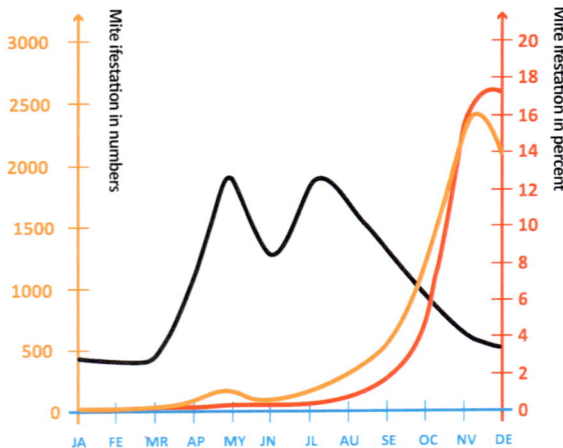

3. Monthly mite population growth by a factor of 2, considering the exit of the swarm and the delay in spring development.

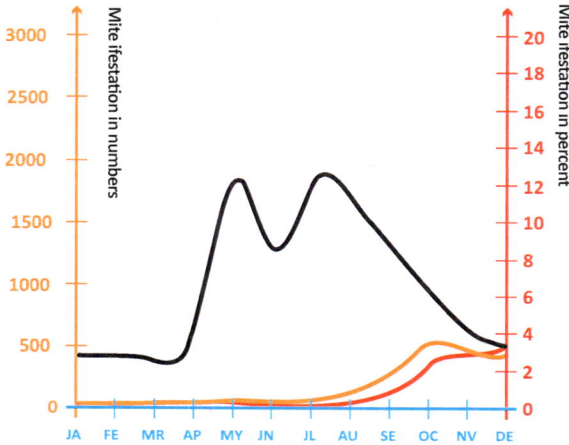

	Colony size (numbers of workers in thousands)	Number of mites	Mite infestation in percentage
JA	10	20	0,2
FE	9,5	18	0,18
MR	9	16	0,17
AP	15	32	0,21
MY	43	64	0,14
JN	30	32	0,1
JL	43	64	0,14
AU	40	128	0,32
SE	31	256	0,8
OC	22	512	2,2
NV	15	461	3
DE	12	415	3,4

4. Monthly mite population growth by a factor of 1.6 with moderately developed mite resistance mechanisms.

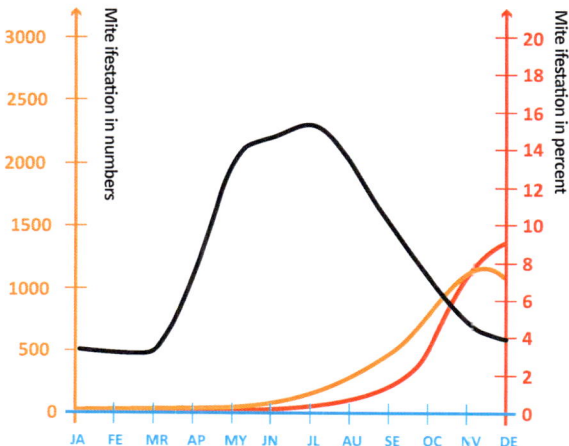

	Colony size (numbers of workers in thousands)	Number of mites	Mite infestation in percentage
JA	10	20	0,2
FE	9,5	18	0,18
MR	10,5	29	0,27
AP	25	46	0,18
MY	45	73	0,16
JN	50	117	0,23
JL	52	187	0,35
AU	45	299	0,66
SE	33	478	1,45
OC	24	765	3,2
NV	15	1124	7,4
DE	12	1002	9,1

Colony size (number of workers)
Varroa mite infestation in percentage
Varroa mite infestation in numbers

* The dynamics of the varroa population growth is rarely simple as in the hypothetical model (as in the examples above), especially in colonies with resistance traits, i.e. bees may allow the mites to propagate for some time with quicker rate, and then the activation of hygienic behaviour mechanisms may be caused by an internal factor (the condition of the bee colony) or an external factor (environmental). The bees then begin to reduce the mite population quickly to safe levels, figures by M. Uchman.

5. Monthly mite population growth by a factor of 1.6, considering the exit of the swarm.

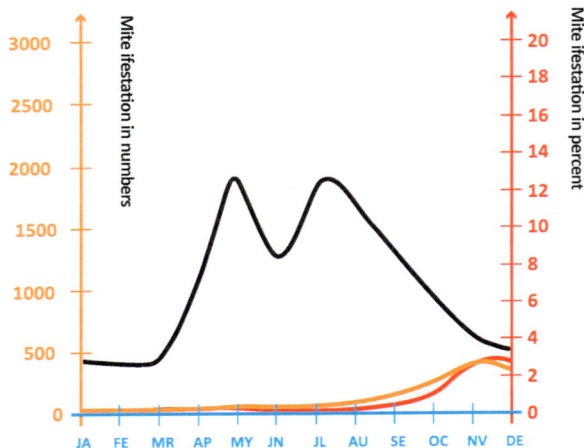

	Colony size (numbers of workers in thousands)	Number of mites	Mite infestation in percentage
JA	10	20	0,2
FE	9,5	18	0,18
MR	10,5	29	0,27
AP	20	46	0,23
MY	45	73	0,16
JN	30	37	0,12
JL	43	59	0,13
AU	40	94	0,23
SE	31	150	0,48
OC	22	256	1,1
NV	15	409	2,7
DE	12	359	2,9

6. Monthly mite population growth by a factor of 1.2 with well-developed mite resistance mechanisms.

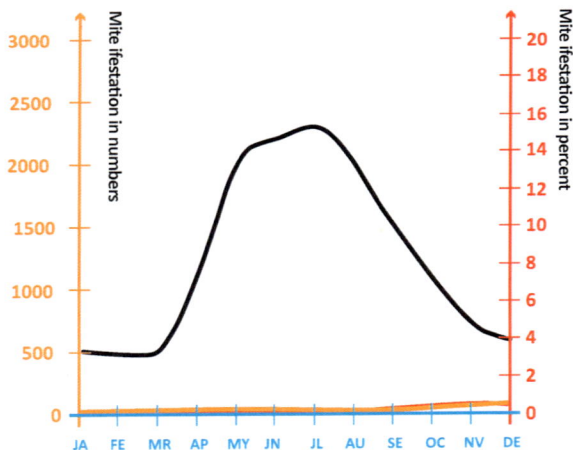

	Colony size (numbers of workers in thousands)	Number of mites	Mite infestation in percentage
JA	10	20	0,2
FE	9,5	18	0,18
MR	10,5	21	0,2
AP	25	25	0,1
MY	45	30	0,06
JN	50	36	0,07
JL	52	43	0,08
AU	45	51	0,1
SE	33	61	0,18
OC	24	73	0,3
NV	15	87	0,6
DE	12	79	0,65

━━━ Colony size (number of workers)

━━━ Varroa mite infestation in percentage

━━━ Varroa mite infestation in numbers

Social apoptosis

Apoptosis is an evolutionarily developed natural biological process involving the programmed and controlled death of cells of the organism. It can prevent abnormal development of the cells and tissues in the presence of unfavourable factors, e.g. mechanical damage, heat shock, radiation, etc.[357]. However, apoptosis occurs not only when a pathogenic or pathological external factor appears, it controls the natural development of the organism (eliminates unnecessary cells and tissues to prevent their overgrowth). Lack of apoptosis may result in congenital defects (e.g. of the vascular system or limbs). Apoptosis is therefore essential for the health of complex organisms.

Scientists have discovered that in the worker brood of Eastern honey bee (*Apis cerana*) there is a mechanism that leads to the stopping of bee development and subsequent death of the pupae if they are infested by varroa[358].

They noticed that this phenomenon is very similar to apoptosis in cells of multicellular organisms, which is why the mechanism was called "social apoptosis" (in this process, however, not the cells are dying but individual insects which constitute the larger superorganism). Although some individuals die, thus reducing the potential size of the colony, the phenomenon slows down the development of the mite population growth and significantly hinder the spread of viruses. The death of individual pupae (as well as single cells of multicellular organisms) increases the probability of survival of the bee colony.

Recently, a similar phenomenon has been discovered in some populations of western honey bees (*Apis mellifera*) characterized by increased resistance to varroa mites[359]. The research was carried out on two lines selected by American researchers, in particular Russian bees and a line known as Pol-line[360] (the control group consisted of Italian bees, not resistant to the mite). The highest degree of social apoptosis

---

357   Where, for example, damaged DNA could lead to the formation of cancerous cells.

358   P. Page, Social apoptosis, Cited.

359   K.E. Ihle, Social apoptosis in Varroa Mite Resistant Western Honey Bees (*Apis mellifera*), Journal of Insect Science, 2022, 1.

360   The Pol-line comes from a research project on the VSH trait by the USDA Honey Bee Breading, Genetics and Physiology Laboratory in Louisiana. Bees obtained during research programs were then selected for high foraging and pollination ability. These activities were carried out to provide commercial apiaries with bees with increased resistance to varroa (the name of the line comes from the word "pollination").

was observed in the Primorsky (Russian bee) colonies with had relatively low levels of mite infestation[361]. Scientists suggest when high levels of varroa infestation occur (regardless of the bee race or origin), social apoptosis may well become diluted among the many processes, both immunological and pathological, related to the high varroa and virus levels.

Researchers also believe that the evolutionary processes that shaped the Russian bees are consistent with those that developed the resistance to varroa in the Eastern honey bee species[362].

Resistance to pathogens, individual and social immunity

Contrary to the popular opinions of beekeepers, it should be stated that western honey bees have different varroa resistant mechanisms, which can be divided into individual and social[363]. Social immunity results primarily from the behaviour of individuals that make up the superorganism. This type of immunity includes the previously described various hygienic behaviours and social apoptosis, but there are also phenomena related to, for example, sick individuals leaving the hive before their death. Also, as one of those mechanisms we can consider the so-called "social fever", where the nest temperature is increased due to the presence of various pathogens[364].

The mechanisms of individual immunity of insects include physical and chemical barriers (e.g. chitin shell and epithelial cells of the digestive system)[365]. The second line of individual defence of bees is their immune

---

361 In the Pol-line bees, this feature is at a higher level than in the control group but still has a lower value than found in the Russian bees. This was maybe due to the artificial selection of the Pol-line being based on other mechanisms, primarily selection on VSH. Primorsky Russian bees, on the other hand, come from a population subjected to a quite long process of natural selection, which led to a different balance of resistance characteristics.

362 In the case of Primorsky bees, the mechanism of social apoptosis was confirmed only in connection with mite infestation. In the Eastern honey bee, this mechanism is also observed in the case of artificial injured (puncture) larvae which was induced during so called pin-test. Scientists therefore suggested that the mechanism of social apoptosis may be strong in Primorsky bees due to evolutionarily developing specific immunity to pathogens carried by mites. The initiation of apoptosis processes in the brood may also be a clear signal for the workers to clean the infected brood (VSH behaviour).

363 Some of these issues are discussed in detail (mainly in the chapter on bee health).

364 This can be also induced by the beekeeper. Spores causing chalkbrood react very badly to higher temperatures, therefore an effective treatment for this disease is to insulate the hive and tighten the nest, which makes it easier for the bees to thermoregulate.

365 M. Chmielewski, Odporność okrywy ciała, bariery anatomiczne Ii funkcjonalne chroniące owada przed zakażeniem, [Resistance of the body covering, anatomical and functional barriers protecting the insect against infection]. Pszczelarstwo 2022, 3.

reaction to the pathogens that have managed to penetrate their external defences.

SOCIAL APOPTOSIS

The trait is not present

The trait is present

The death of the mite-infested pupae may contribute to the slowing down of the growth rate of the varroa population, figure by M. Uchman.

Various scientific studies have documented the relationship between proper insect diet and higher resistance to pathogens, however, it is believed that even well-nourished honey bees may have reduced immunity when highly infested with varroa[366]. Individual immunity to pathogens would be extremely difficult to achieve and cannot be selected without the use of specialized research methods (e.g. genetic). However, research shows that in the bee population there is variability in the degree of immune response, theoretically, therefore, selecting bees in this direction would be possible. For example, bee colonies in County Gwynedd, Wales, have a very high level of DWV type B[367] infection. Some researchers suggest that it is a more benign strain, but others claim that it is more virulent and lethal to colonies than type A[368]. Meanwhile, it turns out that these bees apparently do not know that they are carriers of a deadly strain and contrary to scientists' fears they are coping perfectly well, since their survival rate remains high. Margarita Lopez Uribe's team also confirmed that wild-living bee colonies have a better immune response to pathogens and can survive even with severe infections of the DWV virus. In the case of bees infected with varroa mite, resistance to viruses is a necessary feature for survival.

A very interesting characteristics, from an evolutionary point of view, is the feature known as "suppressed *in ovo* virus infection"[369]. In eggs laid by the queen (having this feature), there are no viruses at all (vertical transmission is stopped), since the pathogen does not pass from the mother to the offspring, or there are very few of them (below the detection threshold). Queens, which are carriers of the virus, are equipped with some mechanism (not entirely understood yet) that allows the eggs to be cleansed of viruses. It turns out that this condition is transmitted genetically and therefore hereditary. Thus, honey bee colonies with this property are significantly less likely to have high levels of virus infection relative to others.

One of the mechanisms is a very important feature of bees' behaviour (social immunity) that serves their health is propolis. A property that beekeepers have unfortunately been eliminating in the breeding process.

---

366 G. DeGrandi-Hoffman, Y. Chen, Nutrition, immunity and viral infections in honey bees. Current Opinion in Insect Science, 2015, 8.

367 J.L. Kevill. DWV-A Lethal (...). Cited.

368 Z. Lipiński, Wirusy pszczoły miodnej: *Varroa destructor* i jego wirusowa gwardia przyboczna [Honey bee viruses: *Varroa destructor* and the viral army at its side], "Pszczelarstwo" 2021, 2.

369 D.C. de Graaf, Heritability estimates of the novel trait suppressed in ovo virus infection in honey bees (*Apis mellifera*), Scientific Reports, 2020, 8.

NATURAL FOOD
availability, richness, diversity

WHEATHER AND CLIMATE
conditions for colony development,
lenght of the growing season,
quality of wintering

HABITAT
insulation from external
conditions, propolis,
climate of the inside

**PRESENCE OF
SYMBIOTIC ORGANISMS**
book scorpions

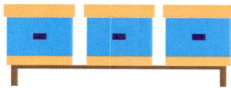

**THE INFLOW OF THE
MITES FROM OUTSIDE**
robbing, drifting,
mites on flowers

**MITE POPULATION
GROWTH DYNAMICS**
mite population growth
vs natural death rate

**ACTIVE MITE
POPULATION
REDUCING**
grooming

**REDUCING OF THE MITE
POPULATION REPRODUCING**
recapping, hygienic behaviours, shortening
of the pupa development, other influence
on mite fertility (e.g. SMR)

MITES LEAVING THE BEE HABITAT
robbing, drifting, mites on flowers

CONDITIONS OF KEEPING THE BEES
density of bee clonies, frequency of inspections,
interfering in life cycle of the bees, swarming prevention,
treatments, disinfection of bees' environment

**MICROBIOLOGICAL
ENVIRONMENT OF THE BEES**
healthy bacterial and fungal biofilms, presence of
pathogenic flora, virulence of pathogenic microbes,
presence of pathogenic flora vectors (e.g. Varroa),
immune response of the bees

**EXTERNAL AND INTERNAL
ENVIRONMENTAL
CONTAMINATION**

Many factors influence the bees' immunity and, as a result, the survival of the bee colony. These contributing factors are both the external and internal, figure by M. Uchman.

Therefore, it is worth directing the selection in such a way as to promote colonies that use a lot of propolis, and to use rough internal surfaces that stimulate the deposition of propolis[370]. The results of research by Italian scientists published in December 2021 put the issue of propolis use in a new light[371]. They showed that propolis contained in bee cells in which the larvae were raised may increase the mortality of female mites (by 20%) and reduced their reproductive success. It turns out that propolis not only helps bees in dealing with pathogenic micro-organisms but can also be a natural acaricide! It should also come as no surprise that propolis in cells reduces DWV infection in bees growing in them, despite the presence of varroa. Additionally, scientists suggest that the deliberate application of propolis by bees to an environment where there is a factor that threatens their health may be one of the factors of their behavioural immunity mechanisms, also leading to mite resistance. The presence of varroa mites and the associated viruses may stimulate an increase in the gathering of propolis by the colony and its deposition in bee cells or other parts of the nest, which is one of the mechanisms of behavioural immunity.

Wild-living bees also have their own mechanisms for preventing epidemics: they live remotely. They are characterized by a dominant vertical transmission of parasites and pathogens, which favours the co-evolution of entire ecological systems.

Local bees

The "locality" of bees is difficult to consider as a separate feature or tool of resistance, but it is impossible not to mention it. Breeding and selecting of locally adapted bees appear to have a significant impact on the population health of honey bees (and indirectly also of other bees). Many beekeepers believe that keeping locally adapted bees is impossible when foreign subspecies and ecotypes have been introduced to their country for at least a hundred years and so-called inter-racial crosses have been created. Therefore, they feel justified to continue in the full swing with the process of importing bees from various locations. Some import them from western European beekeeping institutes, others from the south (ecotypes whose

---

370   In this context, removing propolis by cleaning the hive bodies seems to be an inappropriate procedure, while beekeepers are constantly encouraged to clean or disinfect them. In my opinion, the effect will be the opposite of what was intended.

371   M. Pusceddu, Honey bees use propolis as a natural pesticide against their major ectoparasite. Proceedings of the Royal Society B, 2021, 12.

queens lay more eggs and so create bigger colonies), believing that since the racial purity of the Central European bee cannot be preserved, the idea of a local bee is a fiction. They add that climate change requires bees to face new challenges. I consider such thinking to be inappropriate and fundamentally wrong. The Polish population is dominated by the genes of two subspecies of bees; the Central European[372] and Carniolan.

LOCAL ADAPTATIONS/LOCAL BEE

Local adaptations are necessary for organisms to remain healthy. A bee adapted to one location does not necessarily do well in another, figure by M. Uchman.

Research indicates that the Central European bee genotype is more common in the north of Poland, and the further south to go the greater the share of Carniolan genes. However, because of importing bees from abroad, Polish populations included genes of many other subspecies from all over Europe and even other continents (genes of Italian and Caucasian bees are common). At the end of the 20th century and the beginning of the 21st century, inter-racial crosses based on the Italian bee with an admixture of various subspecies and ecotypes of bees from around the world became very popular. I mean Buckfast bees, whose pedigree is attributed to one of the most famous beekeepers of the 20th century, Brother Adam[373], a Benedictine monk from Buckfast Abbey in the south of England. The assumptions were that the Buckfast bee was to respond, on the one hand, to the needs of efficient apiary management, and, on the other hand, to the health problems

---

372 Research has confirmed that this genotype is common in Poland. The question is, how much has it crossbred with other subspecies?

373 Real name: Karl Kehrle. Born in 1898 in Mittelbiberach (southern Germany). When he was 12, his mother sent him to the Benedictine monastery in the English county of Devon, where he took the name Adam. Brother Adam is known primarily as an excellent practicing beekeeper and the creator of the Buckfast bee line. He was also an excellent mead maker.

of the local bee population in Great Britain, caused by the Acarine (Tracheal) mite.

Brother Adam then noticed that all the bees of the Central European subspecies died in the abbey's apiary, but crossbreeds of the Italian bee survived. This prompted him to look for crosses of bees that could be resistant to the Acarine mite and at the same time be productive. Thanks to his numerous travels, he gathered a very rich genetic background of bees, mainly from Europe and Asia Minor. However, he worked not only on efficiency, but also on resilience. In my opinion, what remains of the original assumptions of Buckfast bee breeding is the pursuit of a constant increase in bee productivity, unfortunately at the expense of their health. Today, diseases are treated using chemicals. It is worth adding that some beekeepers believe that the Buckfast bee is just a very inter-racial cross, so in a way they equate the selection work of artificial breed breeders with the genetic mix that is created locally, if only because each beekeeper imports into their own apiary, the bees he wants, regardless of the need for the bees to develop adaptive features to the local environment[374]. I am a strong supporter of breeding locally adapted bees that can cope with diseases, especially varroa, as the most serious threat[375]. The problem, however, is that when catching a swarm, we cannot be sure that it comes from the locally adapted population. The actions of beekeepers have therefore led to a paradox, which consists in minimizing the risk by obtaining non-local bees by bringing them from far outside the immediate surroundings (e.g. from conservation areas which may be hundreds of kilometres away).

It is worth returning for a moment to Buckfast breeding, because there are many misunderstandings about it. Many bees currently associate such breeding work (Buckfast method) with the need to constantly import foreign genetic material and create further crosses (after all, that's what Brother Adam did). Meanwhile, Erik Österlund sees the problem completely differently. Brother Adam was looking for bees that would be primarily resistant to the Acarine mite and he also wanted to create a line that would meet the needs of beekeeping, i.e., be gentle, fast developing, productive, and non-swarming. However, he did not intend to give up the features

---

374   From the point of view of population health and environmental adaptations, both processes are equally harmful to local bees. The difference is that breeding work is more under control and is probably more conducive to bee performance. In the book I discuss completely different processes.

375   All breeders claim that their bees are resistant to diseases (at least they are selected for this feature), but probably none of them claim that their customers may stop treating bees! Sometimes I have the impression that when breeders talk about resistance, they mean the ability to deal with diseases such as chalkbrood, as if it were a real problem in beekeeping.

of independence and resilience, i.e. those that should characterize a bee adapted to local conditions. Therefore, he decided that it would be easier to introduce into the local population features already developed in other foreign subspecies and ecotypes of bees than to undertake selection work towards a specific trait in his own apiary[376]. Such selection is often long-lasting and labour-intensive, especially when in a subspecies the trait does not have much variability[377].

However, he drew attention to the need to ensure high genetic diversity of bees and their environmental adaptations. Maybe I'm wrong, but I have the impression that many modern breeders have somehow lost these basic ideas[378]. Quite a few of them treat importing foreign genetic material as the norm (every season or two, their offer includes new crosses of bees) and relying on the so-called heterosis effect, which often promotes increased efficiency of new crosses. This type of practice, often called Buckfast breeding, in my opinion has little in common with the principles of Brother Adam's breeding.

A locally adapted bee does not have to be historically local (we can only dream of a racially pure primeval bee; it will never be possible to recreate it). A local bee is one that has been living under specific conditions for at least several generations and is adapted to them, regardless of its genetic sources.

In my opinion, the condition for full environmental adaptation is currently resistance to varroa, because in most regions of the world the mite is an integral part of the ecosystem in which the bee lives. However, even if we leave varroa aside, in some cases we can talk about bees adapted locally.

376　I am writing here about breeding work consisting in crossing various subspecies to develop an appropriate balance of features in the local population. From the perspective of natural beekeeping, this was the wrong choice (it's not just my opinion). Jonathan Powell told me during the conversation: "England apologizes for Buckfast."

377　If the local subspecies A is characterized by very little variability in the swarming trait (i.e. most bees of this subspecies are very swarming because it was evolutionarily advantageous under the given environmental conditions), then selection for lower swarming may be difficult, take a long time with no guarantee of success. Then the breeder looks for bees from subspecies B, which are characterized by low swarming behaviour (in other conditions, this was conducive to the evolutionary success of these bees). Then the breeder tries to manage the selection and crossbreeding of bees in such a way as to consolidate the low swarming feature in the mix of the A x B subspecies, while maintaining other valuable features of the A subspecies. This process, although it does not have to be easy and the result may be difficult to predict, can be faster and more effective than selection based only on the local subspecies A. The problem for local adaptation may be that subspecies B may dilute other features of subspecies' A (e.g. it may not be resistant to local pathogens or may not tolerate wintering well in a new environment).

378　In such a situation, the term "breeders" does not seem appropriate, because they are rather engaged in importing queens from other breeders and raising queens for the needs of the beekeeping market (honey production or pollination).

After all, numerous pathogens live in the environment of bees, requiring constant adaptation from the bees' immune systems. Bees also function in specific types of nest sites and specific climatic conditions to which they must adapt. They should be adapted to wintering and be able to limit brood during winter or times when nutrition is low. As numerous examples from around the world[379] show, thanks to such abilities they can, in some way, cope better even with mites.

Brother Adam in the apiary, photo by E. Österlund.

The key to local adaptation of bees is to stop bringing them from distant corners of the globe. We can work on it, starting today, and no longer remember that our predecessors turned the apiaries into a genetic mix (This harmful process is very advanced in Poland, however the problem as far as its basis, is common all over the world – especially in regions where industrial agriculture principles dominate). Meanwhile, many beekeepers,

379   Also the COLOSS research conducted in several European countries, which proved that the survival rate of bees untreated for mites was higher in the group of local colonies than in the group of bees from foreign subspecies brought to a new region. This result was statistically significant, regardless of the location. It also turns out that bees from populations with higher mite-resistance transferred to other locations somehow lost their ability to survive (M.D. Meixner et al., Effects of genotype, environment, and their interactions on honey bee health in Europe, Current Opinion in Insect Science 2015, 8). This means that bees' ability to survive is strongly correlated with local adaptations.

instead of allowing the bees to adapt to the local environment and even the conditions of a given apiary or local meadows and landscapes, repeat the historical mistake again and again. It turns out that locally adapted bees respond to more subtle environmental stimuli. Thanks to this, they can more easily adapt to current environmental conditions (e.g. weather). They are more malleable.

Any beekeeper who keeps bees without treating for varroa relies on their own local bees, that every year successfully overcome cycles of collapse and recovery[380].

The first step to the environmental adaptation of bees is to give up purchasing queen bees. An exception - subject to certain conditions - are those sources (professional and amateur beekeepers) that work on selecting for varroa resistance. However, this applies only to acquire bees for basic initial selection, multiplication and local insemination. These bees can provide our local population with resistance traits to the mite. The greater the probability of success should be, the closer the location of the breeding apiary is to ours: same climate and a similar ecological situation, e.g. the microbiological environment. Kirk Webster advises (quite rightly, in my opinion) that when such bees are imported, we should breed them immediately and allow for their daughters to free-flight insemination locally. Introducing a queen bee (even such that are recognized as "genetically resistant") to our colony will not solve the varroa problem locally.

The transition period to beekeeping without treatments

Transition period - what is it?

The experiences of people who share their own selection methods, as well as my day-to-day observations, lead to the conclusion that in continental Europe it is impossible to immediately abandon the treatment of bees without jeopardizing their high productivity and low mortality.

I consider the application of the principles of natural selection to be the best and fastest method of achieving the goal, but this path is not attractive

---

380   See: https://kirkwebster.com/collapse-and-recovery-the-gateway-to-treatment-free-beekeeping, accessed: November 20, 2022.

to beekeepers. It may also not work well – as my example shows – in the regions that are not considered to be friendly to bees, unless there is some local cooperation that will help to propagate the selected genetics. In fact, most beekeepers reject this method as senseless, un-economic and at least ethically questionable. At the same time - as one might expect, they treat their colonies as a sad necessity. Even if we ignore the issue of contamination of the honey bees' living environment, treatments involve additional costs and effort. However, this is not the kind of effort that involves looking after and observing bees, which is a source of constant satisfaction for beekeepers. It is also not an effort that brings in an income, unlike e.g. harvesting honey. Treatment also means exposure to dangerous chemicals. Beekeepers often declare that would probably agree to stop using toxic substances in apiaries if they could save the productivity of apiaries and reduce bee mortality to a minimum. Therefore, I define the transition period as the time of initial selection, the period of investment in breeding the local population of bees, which - at least hypothetically - would allow the abandonment of varroa treatments without jeopardizing their projects and goals.

A healthy apiary requires many years of selection work, and meanwhile beekeepers are looking for simple, quick, and ready-made solutions. They expect someone to simply supply them with resistant bees. At the same time, however, they almost completely ignore the offers of bees from breeders, where attention is paid to breeding mite resistance. In turn, the lack of demand discourages breeders from making efforts because the risk increases that breeding work will not bring the expected profits. It is true that bees from this type of breeding are still far from perfect; it often turns out that when moved to new places they have problems exactly alike those observed in the existing susceptible populations. Most often, however, they are equipped with basic resistance traits, thanks to which at the beginning it would be possible to at least limit mite treatments, and over time, saturate the population with resistance traits in the otherwise locally adapted bees. In my opinion, the responsibility for the health of bees and the characteristics of the population is our own, of all beekeepers as a community as well, as each individual.

I would like to emphasize it once again: **the methods of selecting bees for varroa resistance are known and the knowledge regarding the specific procedures is widely available.** A discussion of the traits of resistant bees can be found in numerous publications and studies, including scientific ones[381]. All you need to do is take the effort to read them and then translate them into your own practice. In such situations, I usually return to Erik Österlund's texts, in which he shares with the reader his own experiences, advice and the history of his selection. They can be found in the article Breeding for Resistance, where he talks step by step about the principles of his work[382]. Against the background of reports from breeders who can boast of good selection results for resistance, Österlund's publications are an exception[383]. It is difficult to find such a detailed description of methods and precise instructions in other sources. While I can imagine at least several reasons why beekeepers do not turn to natural selection, I am unable to find the answer to the question: why the selection proposed by the Swedish breeder (or other similar methods) is not used commonly?

The solution to the varroa problem, and thus to most of the health problems of honey bees currently, would be systemic selection carried out by many beekeepers to cover most bee colonies throughout a region. In my opinion, the situation will not improve until a significant part of the beekeeping community starts paying attention to the environmental adaptations of bees, their local adaptation and mite resistance. In Poland, the fragmentation of apiaries is very high, and each beekeeper has his own goals, methods, and lines of bees. The key to selecting mite-resistant honey bees should be co-operation, which does not exclude the beekeeper from running his own selection program and pursuing his own beekeeping goals. However, if each of us believes that little depends on our actions, and the abilities of our bees

---

381   Some of them are mentioned in the chapter discussing the characteristics and properties of resistant bees; see also: R. Büchler et al., Evaluation of Traits for the Selection of Apis mellifera for Resistance against Varroa destructor, Insects 2020, 9; M. Guichard, Three Decades of Selecting Honey Bees that Survive Infestations by the Parasitic Mite Varroa destructor: Outcomes, Limitations and Strategy, https://www.preprints.org/manuscript/202003.0044/ accessed 21 November 2022; J. Kefuss, Selection for resistance to Varroa destructor under commercial beekeeping conditions, Journal of Apicultural Research 2015, 5; R. Büchler, Varroa Tolerance in Honey Bees-Occurrence, Characters and Breeding, Bee World 1994, 75(2); Tenze, Evaluation of characters for testing on varroa resistance, 2, 46 International Apimondia Congress, 2019, 9; M. Buchegger, Relationships between resistance characteristics of honey bees (Apis mellifera) against Varroa mites (Varroa destructor), Journal of Central European Agriculture"2018, 19(4): R. Büchler, A. Uzunov, Selecting for varroa Resistance in German Honey Bees, Bee World 2016, 93(2).

382   E. Österlund, My first 12 years with varroa, https://www.elgon.es/resistancebreeding.html, accessed: February 7, 2025.

383   I learned e.g. that Paul Jungels, a well-known bee breeder from Luxembourg, does not perform treatments on a population of nearly 300 colonies. He also cooperates with local beekeepers on similar basis as Erik Österlund near Hallsberg in Sweden. Joseph Koller works in a similar way in the populations of Elgon bees and the Primorski bee (https://www.josefkoller.de); However, it's not that easy to find detailed guidelines on they both conduct your own selection and how to implement them in your own practice.

have no impact on the overall picture of the population, we will not move forward. **The changes depend only on us: the health of the bees will not improve until we start selection work with a joint goal.**

One of the outstanding European researchers of mite-resistance, Dr. Ralph Büchler, in a conversation with Prof. James Ellis from Florida State University (USA) said: "I'm absolutely sure the long-term future of beekeeping is going to be with resistant bees. So, it's extremely important, and it's absolutely in our hands to develop such bees. The biology is on our side. If we would not always rely on our chemical treatments, and with all our management tricks, then nature would very soon develop such resistant bees, as we can see them in so many places around the world right now. It's much more about the way we manage our bees and how we deal with varroa mite. The big question is how quickly the beekeeping industry can be turned around, with beekeepers reducing chemical treatment, and only treating above certain mite thresholds, which we need to have if we want to give the more resistant colonies a better chance of reproduction. What happens now in our industry here is that the beekeepers routinely treat two or three times a year with highly effective drugs, and in doing so, they do not even recognize the differences in susceptibility or resistance to mites in their colonies. We have colonies, which do very well without treatment even over several seasons and still produce a lot of honey. So, they are there. But the point is they are not well-identified, and they are not clearly preferred in selection programs. This is despite our breeders doing a very good job producing queens, with well-organized breeding programs. The trickier question is how can we convince beekeepers that they can make a step change in their routine in colony management and treatment? That is how I see it then, and we discuss it intensively with our beekeeper organizations. And we are looking for ways out of this spiral of treatment (...). **So it's a man-made problem. It's not the nature and it's not the biology of bees, which brought us to this point. It's simply the way how we keep our bees**"[384]

---

384   Quoted in: "Two Bees in a Podcast" episode 66, European Honey Bees & Sting Management, https://entnemdept.ufl.edu/media/entnemdeptifasufledu/honeybee/pdfs/two-bees-in-a-podcast-transcription/Episode-66_Mixdown-PROOFED_otter_ai.pdf; access: February 7, 2025.

HERE ARE SOME PRACTICAL TIPS BY ERIK ÖSTERLUND FOR OBSERVING THE BASIC CHARACTERISTICS OF COLONIES FOR SELECTION/BREEDING[385]

1. Check the level of mite infestation in the colony at least 2-3 times a season, or more often if you feel it is necessary. If the level of mite infestation on the worker bees in a colony exceeds 3%, start treating it as soon as possible. The first spring inspection is the most important, if it turns out that the degree of infestation exceeds 2%, then re-examine the infestation level after about 4 weeks.

2. Place mats (50 cm x 50 cm) in front of the hive entrances and check what is on them (at least once a week). Pay attention especially to deformed bees (infected with DWV) removed from the hive. There may also be pupae or their fragments; this usually means that the bees have some hygienic traits. If you find deformed bodies of bees on the mats, consider examining the level of mite infestation, if it is more than 3%, consider treatment.

3. Check if the bee colony is developing as planned? If not, check what is happening in the hive, perhaps it is a symptom of a large mite infestation. If the hive is filled with bees, it may be a signal that it is time for another super or to check for swarming. It is possible that this is a good time to split the colony.

4a. If you find single open cells on a comb with dense, capped brood, there is no need to worry, this is a sign that the mite infestation is minor - if your bees have the basic skills to fight the mite.

4b. If the brood on the comb is very dispersed and the capped brood constitutes no more than half of the brood cells, it may mean that the bees have not started cleaning themselves early enough, so the mite infestation and viral levels in the colony may be very high. It may also mean that the colonies have some other brood disease. Whatever the reason replacing the queen seems rational.

---

385   See: https://elgon.es/guidelines.html.

Each beekeeper has an influence on improving the genetics of bees in our region, regardless of whether we keep bees in isolation or in the vicinity of other apiaries, and regardless of the approach of other beekeepers to the mite problem. Selection can only be carried out in some of the colonies, e.g. in a separate apiary, and then selected queens (with greater level of resistance traits) can be introduced to other bee colonies. Queens should be open mated in your apiary. If you can get bees that already have some basic mite resistance, even better, it will probably make your selection easier.

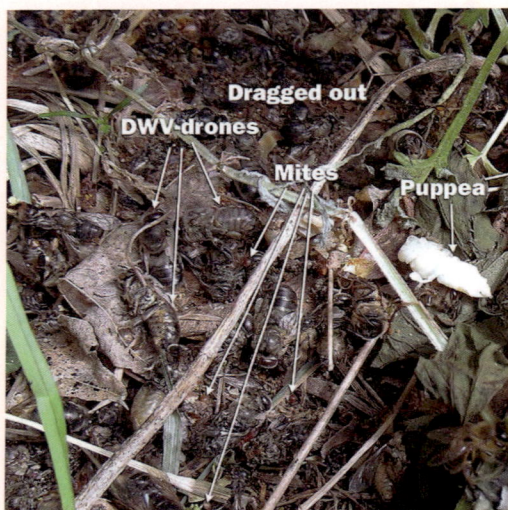

Erik Österlund analyses hive waste, which can be a rich source of information about the health of the bee colony, photo by E. Österlund.

TIPS FOR SELECTING BEES FOR IMPROVING THEIR GENETIC CHARACTERISTICS AND MITE RESISTANCE:

▸ Identify colonies that maintain the lowest mite numbers for the longest period (use your notes).

▸ Among the colonies that cope best with the mite, choose those that have yielded above average honey crop and are not defensive, raise queens from them.

▸ Remove queens from the weakest 30-50% of your colonies, destroy the queen cells after a week and give them the sealed mature queen cells.

▸ Create an additional 10-30% of colonies by making splits and giving them queen cells from your best colonies.

▸ Create queen-less nucs from colonies that can cope with the mite (better than average), are productive and not overly defensive, and let them raise their own queens. After a week, destroy the queen cells, leaving one or two next to each other (to avoid the risk of swarming).

▶ If you have only one apiary, move the colony with the queen to another place so that as many bees as possible are left in the orphaned hive. If you have several bee-yards, take the nucs to the other.

▶ Make sure to create a larger number of colonies than you want to overwinter (usually not all queens will be mated). It is better to feed young colonies (preferably with fondant) so that a lack of food does not prevent them from growing.

The transition period, during which we keep our finger on the pulse and treat the bees when they require it, is a chance for the beekeeper to better plan the first years of the development of the apiary and obtain funds for development. Probably, thanks to such selection, it is possible to limit bee losses (or at least spread them over time). Therefore, you can conduct normal apiary management. It is difficult to say whether we will not experience greater losses after complete discontinuation of treatment (I used to think that we had to take them into account, but today I am not so sure, because I have seen how good the effects of local selection are while treating). But colony losses occur even when the bees are treated. I know beekeepers who claimed that after discontinuing treatment, they did not experience any major losses for several years, and bees began to die only after many seasons. There are also such cases (as in my apiary) where bees died in the first season after discontinuing treatment. However, there are dozens of examples where beekeepers carry out mite treatments but still the colonies die off[386]. Therefore, it is difficult to formulate rules that would be objectively true for everyone. We can only talk about probability.

Treating bees during the transitional period make sense in that beekeepers are starting to trust your own methods and local bees. They can conduct normal apiary management, which is important from an economic point of view. They are therefore able to go through the selection process more gently before they decide to return the bees to the laws of nature. Perhaps some will never be ready for it. However, if they work on bees' resistance, they can have a very positive impact on the genetics of the local population.

---

386 Every year I ask the president of my local beekeeping association what the level of bee losses is in our association. My losses are always higher than average and higher than I would like and expect, yet each year at least some beekeepers that treat and still lose a higher (or the same) percentage of bee colonies than me.

How long does the transition period last? At what point can we discontinue treatment? When can colony losses reach an acceptable level, statistically the same as for treated colonies? It is difficult to give clear answers. Maybe it's worth taking advantage of Österlund's experience again. According to data from 2018, approximately a decade after the start of proper selection in the presence of the mite, the breeder treated approximately 50% of the colonies in his apiary. In 2020, it was only about 20%, because according to the adopted selection criterion (mite infection below 3%), a very high percentage of bees did not require treatment. Each year, despite treating those colonies which needed it, he lost a small number of them (usually 5-10%, which is an acceptable level of loss). 2020 was the last year the Swedish breeder did any treatments, but the low level of bee mortality remained in the following years. However, we do not know what would have happened if the breeder had not performed the treatment in earlier years in the colonies he treated at that time (that is, whether the bees would have died). Colonies can survive with a mite level higher than 3%, but they can also die with infestation rate lower than 3%. Those that have some resistant traits and higher level of tolerance, however, can also reduce the number of mites using their own efforts. If we assume that an infestation exceeding the level recommended by Österlund results in the death of the colony during the overwintering period, we can believe that after 12 years of selection (and the transition period), bee losses might be comparable to or even lower than those recorded in apiaries where bees are treated. However, everything depends on many factors, so it is difficult to say clearly when ceasing of treatments after the initial selection will allow us to enjoy low losses. The laws of nature cannot be expressed in a mathematical formula.

It should also be remembered that Österlund has been co-operating with beekeepers in the surrounding area for years, who together are selecting for varroa resistance in hundreds of bee colonies. This is a completely different situation from the one we are dealing with in Poland. The effort of a hobby beekeeper who works on the selection of bees in his own apiary may be wasted if the neighbours are guided only by the performance of the bees, without paying attention to their resistance and local adaptation. Perhaps the period in which this was achieved in Sweden would turn out to be too short under the conditions in Poland. The question of how long the apiary would have to be left in the transition period for bee mortality to stabilise at a relatively low level remains open. However, it seems that even a period of three or four years should bring the first noticeable effects. Let us list the most important tasks of the transition period:

1. Choosing a method in which selection for resistance is consistent with your existing beekeeping goals.

2. Gradual withdrawal from chemical treatments.

3. Selection of the traits needed for the bees to survive in the presence of the varroa mite

4. Shaping the bees' living environment and the ecological situation.

5. Stimulating the immune response of bees to pathogens.

Varroaresistenz-2033, i.e. Dr Ralph Büchler's method based on biotechnical procedures

In 2023 in Germany a group of beekeepers led by Dr Ralph Büchler, former head of the Bee Institute in Kirchhein, founded a group called Varroaresistenz-2033[387]. Its name relates to the goal of making the bees resistant to varroa in Europe by 2033.

The propagated approach is based on so called biotechnical methods, i.e. that is controlling the development of the colony (egg laying) through queen isolation or brood-break and brood removal.

Assumptions of the biotechnical method based on varroa and honey bee biology and host-parasite-pathogen relationship

Swarming and thus summer brood break is very beneficial if not crucial for preserving the health of a honey bee colony. It may help not only with curing brood diseases, but also with reducing varroa mite population growth (I've written about that in other parts of the book, so I won't expand on that here).

Most of the mites are in the brood – it is said that even up to 85% of the mites are in the cells, feeding on bees' pupae and reproducing, while around 15% live on adult bees. There, we can treat the bees by taking out the brood.

The goal is slowing down horizontal transmission of pests and pathogens in the late summer. What is most important in this previously discussed subject, is that in the late summer the mite infestation is very high and colonies are still strong and looking for forage, while often currently it

---

387 See: https://varroaresistenzprojekt.eu/.

is the late summer dearth. Bees therefore seek for the honey in other hives (arranged closely in the apiaries, sometimes in big numbers). This is the time that most robbing occurs and thus the most horizontal transmission of viruses and mites leading to so called re-infestation. This causes the risk of total die-off of whole apiaries, and combined with the treatments and other management methods leads to increasing virulence of the pathogens. To stop the transmission bees should be treated earlier (no/less mites = less viral infection = no/less transmission), however it is impossible with most accepted methods (chemicals), since the bees still forage and gather crops and provide a surplus for the beekeeper. That's why some other form of treatment needed to be developed, which was not based on using chemicals.

Healthy winter bees are the basis for good survival. Standard treatments (with chemicals) are usually carried out in late summer, autumn and are repeated throughout the winter. In many cases they are just too late. They are done during the development phase of so-called winter bees, i.e. those that are supposed to be wintering in the hives and raise new spring generation in the following season. In our climatic zone they are usually raised between late July and September. As the mites are killed, they don't vector DWV to the winter bees. Therefore, if we want to have winter bees raised with low mite and viral infection, the treatments should be performed early enough – but not in this period and not directly before it. Dr Büchler claims that the best results were noticed when the treatment was performed in the late June or not later than early July (in Central European conditions). The later it was done, the worse results there were, as the winter bees were more infected with viruses. Even later reinfestation was not a problem and Ralph Büchler claims that even around 500 mites in the hive should not be the problem for surviving till spring, if winter bees were raised when the infestation was low in the hive.

Treating bees once a year (one biotechnical procedure – or series of them in early summer) allows a good assessment of resistance or tolerance to the varroa mite. The colonies treated in peak season should be able to survive even if they allow regrowth of some infestation before the winter. However, if the colonies do not have some varroa resistance traits the infestation before next treatment (spring) should be showing and would be easy to be assessed. That's why in peak season it would be easy to choose breeder-queens for further propagation. Next to this, drones of colonies with a large mite population would be infested much, and so propagation of this genetics through drones would be hindered or stopped.

PRACTICAL TIPS ON THE BIOTECHNICAL METHOD

Preferably bees should be treated once a year (with one or a series of procedures) – this allows for keeping some mite pressure and makes it easy to take breeding decisions;

The treatment should be performed early enough so the winter bees are raised with low mite and viral pressure (in early summer);

The procedure should be done with no chemicals or only one chemical treatment:

a) taking out the brood and making artificial swarm on empty frames, drawn combs or foundation (with the brood we take up to 85% of all the mites);

b) isolating the queen in a cage, and after all the brood hatches making one treatment with chemical (which should be very effective in a broodless period);

c) using queen isolators (so called: trap frames) – queen only lays eggs on one or few frames – that way we can trap all the mites without taking out much of the brood from the colony;

Brood taken out of the colonies can be disposed of or used for making new colonies/nucs. In that case it should be treated with one dose of chemicals when all the brood hatches and we avoid putting chemicals to the colonies that gather honey;

Trapping mites with queen trapping frames can be repeated so that more mites are taken out of the colony (this way we can take out up to 100% of varroa mites);

Biotechnical treatment can be planned in accordance with local flowering periods – if we plan to isolate the queen before the nectar flow, with less brood in the hive, bees can gather more crop for the beekeeper (there would be less larvae to eat it);

Breeding decisions should be based on low mite infestation.

Choosing a method in which selection for resistance is consistent with your existing beekeeping goals

The right selection method is one that not only leads to the goal, but is consistent with the beekeeper's preferences, views and will not be in opposition to the previous method and goals of running the apiary. It is difficult to do something against yourself and your own beliefs. As I have already mentioned, subjecting bees to natural selection is consistent with my views. I believe much more in the abilities of nature than in my own, and production of honey is not the most important part of my beekeeping at this stage. However, if natural selection is undertaken by a person who is not convinced of such a method, it is difficult to expect that they will calmly watch the death of their bee colonies and following drop in honey production. They will rather give up quickly, having achieved nothing. I know at least a few such people. In the first years, we were united by similar experiences, and we also perceived the long-term goal in a similar way. I managed to persevere despite many drawbacks thanks to my conviction that the chosen path was the right one, but there were also those who gave up on selecting bees for their mite resistance (not only by natural selection but simply stopped paying attention to it). In their case choosing a method that was inconsistent with their beliefs resulted in retreating from the path leading to the original goal.

It is worth adding that people who keep detailed apiary records and statistics (such as the number of varroa on the hives' bottom board, the number of stings from a given colony, production analysis etc.) have completely different perspectives. This allows them to compare, for example, the degree of mite infestation in colonies, to learn about the dynamics of the mite population growth, and on this basis to make breeding decisions. It will also be easier for them to choose a criterion that will allow them to select colonies requiring treatment and determine the duration and intensity of treatment. People with a research temperament who are not afraid of performing tests or observing specific bee behaviours can conduct hygiene tests or observe the behaviours of VSH or recapping.

However, if keeping detailed records and analysing results is not your passion, it is hard to force yourselves. I say this with full conviction, because I am one of those people, no one will force me to count the number of mites on the bottom board or to examine the degree of mite infestation in my colonies! It is not a matter of laziness. I simply try to direct my efforts in a different direction, believing that natural selection will take care of the best adaptations of the population.

It is also worth remembering that selecting individual, single traits of a bee colony is unlikely to bring the expected results. Many beekeepers and researchers believe that to cope with the mites, bees need a complex set of traits. Therefore, only if we simultaneously select several traits and behaviours related to varroa resistance, the breeding effect should be satisfactory, and this could allow us to reduce the number of treatments over several seasons. If we do not believe in our own breeding abilities, we can rely on the co-operation and help of, for example, a neighbouring beekeeper.

We can also give up importing foreign genetic material, breed our own local queens and limit mite treatments. Then the mortality rate in the apiary may be higher than when conducting several treatments, but we'll probably be able to maintain most of our own population while moving forward.

If we limit the treatments to a minimum, the health status of the bees will show us which colonies are worth breeding from. Then the individual colonies will begin to differ in condition, which will allow us to easily assess which of them are independent or self-sufficient in varroa control and which cannot cope with the mite.

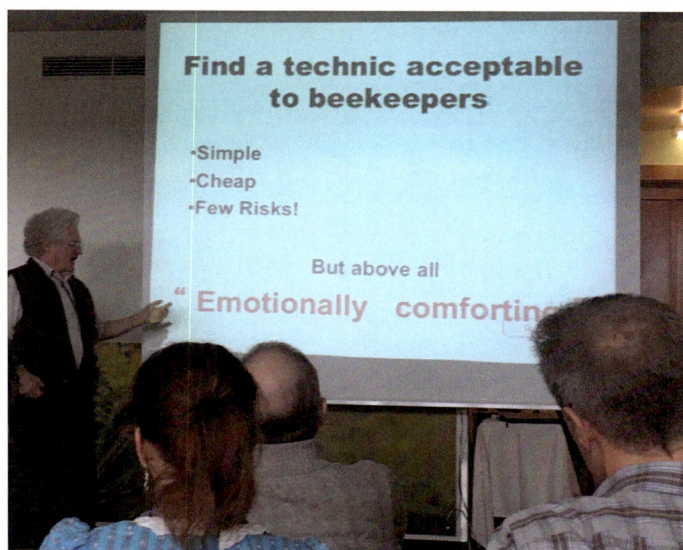

Finding a selection technique that brings inner peace to the beekeeper is one of the basic goals of the transition period. John Kefuss during a lecture in Neusiedl in 2018, photo by S. Kempf.

Gradual withdrawal from chemical treatments

During the transition period, everything should be done to limit the use of chemicals to combat the mite. Unfortunately, many beekeepers still use them for treating prophylactically in the winter, then in spring, and often in summer[388]. This is one of the biggest systemic mistakes of beekeeping in terms of bee health. If we decide to treat, we should only do it when we are

---

388 Treatments are most often not adapted to the needs of the bees, but to the seasons of the year. They are usually carried out when there is no flow (blossoming), regardless of whether there are 50 or 1000 mites in the hive. Reaching for highly toxic and dangerous chemicals, beekeepers are unlikely to know if there is a real need for that or not.

dealing with a disease[389]. If prophylactic treatments are used, we will not make even a half-step in solving the varroa problem. If we are going to stop using toxic substances, we need to start by changing our thinking. (I wonder how many beekeepers reach prophylactically for various pharmaceuticals, e.g. antibiotics, without consulting a doctor?).

Overusing pesticides because of prophylactic practices is not only the dark side of beekeeping, but also of the entire agriculture of the 21st century. At certain times of the season, we should monitor varroa and leave treatments (if we decide to perform them) only for the period when varroa infestation is high. **I think that implementing only this idea on a wide scale (universally), combined with using the results of monitoring in further breeding, would allow solving the varroa problem in under 20 years.**

There are many methods that can effectively replace chemical treatments. Using biotechnical methods[390] instead of treatment would free

---

389   The situation in bees is very specific, they are carriers of the varroa mite, which does not disrupt the normal course of the vital functions of the superorganism and does not indicate the occurrence of a real disease. The mere presence of the mite does not indicate a disease state (varroosis), if it were so, we could basically assume that in most regions of the world there is not a single healthy bee colony. According to this definition even the Eastern honey bee colonies could not be considered healthy either. In the event of a disease, its course is usually quite rapid, the colony depopulates within a few weeks and is practically impossible to save. This does not change the fact that very often colonies that do not need treatment are still treated. For the decision to carry out the procedure to become rational, monitoring of the mite infestation would be necessary. Only then would we be able to be aware of which colonies have a mite count approaching a level that threatens the health of the bee colony, and which ones possibly have resistant mechanisms.

390   There are many biotechnical methods, some are very effective in reducing mite populations, but they also have disadvantages and some also happen to be ethically questionable. Biotechnical procedures used to reduce the mite population include: a) the use of a brood-break (e.g. swarm management or de-queening he colony by making an artificial swarm/package with an old queen); b) isolation of the queen either in order to limit the number of eggs laid or to use so-called frames-traps for the mite; c) removal of brood from the colony (either the removal of a drone-brood frame or removal of all brood, this procedure can be combined with queen isolation on one or two frames and then theoretically we can remove less brood containing a greater number of mites); d) various methods of heating the brood (the mite offspring are less resistant to higher temperatures than bee brood, by heating the brood we can kill mites and return the brood to the hive). Some of these methods can be combined with limited pesticide treatments, we can use trap-frames for the queens in the apiary (when the brood hatches outside the trap frames, we can assume that a large percent of the mites will be in the brood containing in the trap frames). We can remove such frames from the hive, taking away most mites from the bee colony. We can create bee colonies from frames with capped brood, if we remove frames with capped brood just before hatching (e.g. in July, when temperatures are high and stable also at night), we can provide only a minimum number of workers for such newly created colonies. After all the brood hatches, we can perform one acaricide treatment natural or synthetic (even taking away frames to avoid contamination of the combs). We can repeat the procedure two or three times to remove as many mites as possible from each colony. In this way, we can eliminate over 90% of mites from the entire apiary, performing just a few so-called chemical treatments on no more than 20% of colonies. Of course, to eliminate chemicals, the removed brood can be disposed of or heated.

the bee environment from toxic substances and reduce chemical pressure on the bees, pathogens and the hive ecosystem.

Erik Österlund during the examination of brood, photo by E. Österlund.

At the same time, however, we must observe the specific features (properties, traits) of the bees to constantly improve their adaptation to life in the presence of varroa through breeding.

Another solution is to limit the treatment, even to one, which is carried out at a time when it is most justified and effective (e.g. in the brood-less period in autumn or after isolating the queen). However, it is important to reduce the number of treatments from year to year and use less effective methods or milder chemical substances.

Ralph Büchler from the institute in Kirchhain (Germany) postulates to abandon chemical treatments and limit the fight against the mite to biotechnical methods. The researcher recommends, for example, removing all brood, or enclosing the queen in a trap-frame and removing the brood from the trap-frame several times. According to him, these methods can be used to remove the majority (90-95%) of mites in the colony. Thus, the effectiveness is comparable to the most effective (toxic) chemical treatments.

Of course, there is a risk of mite reinvasion, but this can happen in every colony, including those in which chemical treatments were performed. Let us also remember that we are in the process of conducting selection. If we manage to remove mites in most colonies, and in a few colonies, reinvasion occurs, which will end tragically for them, we should accept it as part of the process. Bees that allow reinvasion should be eliminated by means of selection.

One of the symptoms of bee health is their ability to survive at a relatively large strength during the winter. This may also apply to treated colonies (but not where treatments are performed all the time throughout the winter), but above all to those that must deal with varroa by themselves. Comparing the strength of colonies in autumn and following spring is therefore one way to assess their condition and ability to cope with varroa. Ralph Büchler's research shows that using so-called queen isolators, i.e. isolating the queens in the summer months (and then, as I understand it, removing the brood), as the only form of mite treatment gives very good results. After such treatment was carried out in August, it turned out that the colonies in spring had an average 51% of the strength (number of bees) it had in summer. The same treatment performed in July gave an even better result, with an average of 64% strength of the colonies from before the winter, while in the control groups subjected to formic acid treatment there was only about 40% of the summer-strength in the spring. The biotechnical procedure instead of chemicals allowed not only the removal of most of the mites, but also the colonies to come out of the winter in excellent condition, better than in the case of using acid. The Büchler method was tested by several beekeepers from Westerham in Great Britain. The results were so encouraging that most of the local beekeepers were quickly convinced of it, and their results motivated them to continue selection for genetic resistance.

It seems that limiting the treatments to only one session in late autumn may also contribute to a relatively rapid pace of selection. In such a case, however, it must be considered that the most infested colonies may not live to see treatment, and some others may die due to high viral loads even if varroa mites are killed. If we started with bees that were completely non-resistant and not adapted locally, we should even consider the risk of all colonies in the apiary dying (a result of so-called mite bombs and huge virus loads). Colonies infested to a moderate degree would probably live to see treatment, but they would die in winter or early spring, or due to their very poor condition, the chances of running a successful honey production would be negligible. It is worth noting, however, that colonies infested with mites to a lesser extent would be freed from the mite in autumn, and potentially the good condition of the bees in the next season would quickly allow colonies to be identified, propagated and the appropriate genetic traits to be established in the surviving population.

In my own apiary, I stopped using treatments and procedures overnight, so I had no reason to think about a year-to-year strategy for slowly giving up treatment. However, I can use my thoughts resulting from observing the process of building resistance, as well as the experiences of other beekeepers who have developed such strategies.

In my own apiary, I do not accept isolating queens on a frame or removing brood. These procedures are at odds with my beliefs about the way bee colonies should live. However, I know that they can be a necessary evil (maybe a lesser evil). Perhaps it is a lesser evil than the regular use of chemical treatments? I also realise that most beekeepers would rather remove (kill) all brood from the hive than watch the colony collapse, leaving an empty hive behind.

On the next page, in the table, I present an alternative method for moving away from chemicals over the course of several years. This is one of the ideas that can be modified at your own discretion.

This method can be modified according to one's own beliefs (e.g. by checking the infestation using other methods), and the treatments can be carried out with other chemical compounds (than those suggested above). The dates of the practices can also be changed depending on the local calendar of honey flows. For example, the summer treatments may be adjusted to your local honey flows. If the treatment is with chemicals, it should be done after the flows. However, if you decide to use some biotechnical method (e.g. disposal of the brood) it may – or even should - be done earlier which would have a positive effect on lowering the virus infestation and healthier winter bees. In the first year after the main honey flow, instead of isolating the queen on a frame and carrying out treatment, one can put the bees onto empty frames to remove not only the mites but also all the contaminated wax from the hive. Such a one-time procedure, during which we clean the nest of the remains of any metabolites from the synthetic acaricides, will be good for the bees. The two-year period of practice is only a suggestion.

A possible method of moving away from miticides (minimizing the number of treatments) over several seasons

Years	Procedure	Deadline	Breeding activities
Year One	Pin-test (hygienic behaviour test)	Before the breeding season	Creating splits with queens raised from larvae from the colonies that perform best in the pin-test. Stimulate the rearing of as many drones as possible from these colonies.
	Isolating the queen on a frame; treat with thymol or formic acid and after treatment release the queen.	June-July	Measure of the natural mite fall; Count the mite fall after treatment.
	Treat with oxalic acid or Apivarol (one tablet)	Late autumn during the brood-less period	Count the mite fall after treatment.
Year Two	Repeat the activities from the first season, however, breeding decisions must consider the results of natural mite drop during the previous season, the number of fallen mites after treatments and the condition of the colonies in spring.		
Year Three	No treatment during the summer season, but if the colonies have high levels of natural mite drop in the early summer, isolate the queen on a frame and treat the colony after the brood hatches (as in season 1-2) or remove all the brood (so removing the mites without using chemical treatment)	Breeding season	During the season we periodically count the natural mite fall on the bottom board and propagate those colonies in which the natural mite fall (in the previous season after treatments) was the lowest and those in which during the season it is also low.
	Treat with oxalic acid or Apivarol (one tablet)	Late autumn, brood-less period	Count the mite fall after treatment.
Year Four	Repeat the activities from the third season. Again, breeding decisions consider the colonies with the lowest mite levels in the natural fall and after treatment in the previous year. Also consider colony condition in spring.		
Year Five and beyond	Stop the use of all systematic chemical treatments, only observe natural mite drops on the bottom board (or periodically examine it using the flotation method). We now begin the period of intervention treatment, i.e. we treat only those colonies (preferably after isolating the queen) that have a high infestation rate (e.g. big number of fallen mites on the bottom board), however, it must be considered that in hygienic colonies mite-drop may be periodically high due to the bees' removing infested brood - so-called mite black holes (before treatment possibly examine the fallen mites if they are injured/damaged). It may therefore turn out that after a possible treatment, the mite-drop in such a colony will be low, which means that the bees have some traits associated to the resistance to varroa.		

If after the first few years the results turn out to be unsatisfactory, the period needed to change the procedure and move on to the next phase can be extended[391].

As in the previous method, the period needed to change the procedure and move to the next stage can be extended. If we perform two treatments in the first few years, we should expect that the colony will not enter wintering in great strength. However, it is very likely that it will be freed from most mites. Experience with wintering weaker colonies tells me that the bees should winter without any problems if we allow them to raise healthy winter bees. If we believe that the second treatment is not necessary, we can skip it. We can also remove all the brood at the end of the summer or early autumn, when the winter bees have already hatched. This way we would free the colony of the remaining mites (however probably it will not be of importance for the survival of bees for the upcoming winter). We can also react as we go if we notice something disturbing, e.g. by creating artificial swarm and removing all the brood or, in extreme cases, by carrying out some treatment. If we decide to take away the brood and/or the trap frame, we must carefully choose the date of the treatments to provide the bees with the time needed to raise a generation of wintering bees.

---

391 In the table I have presented the shortest possible periods in which we can see positive changes and reduce chemical treatments in favour of interventional ones. After three or four seasons we should notice positive changes in the resilience and vigour of the bees, but it is difficult to assume that after such a period giving up the treatment would be painless (in conditions like Central-European). Many breeders who do not use treatments noticed a real change in the condition of the population after about a decade. For example, in Juhani Lunden's apiary during the first decade of selection the bee population declined (due to large losses) and only then began to recover. Also, in Österlund's apiaries after about a decade we could see a rapid increase in the percentage of colonies that do not require treatments. Danny Weaver, a breeder from Texas, USA, also talks about a period of about ten years needed for population selection. While during the first seasons we can observe colonies that have acquired resistance traits in the population, a period of ten years seems to be necessary to spread these traits throughout in the entire population. This can be seen in Randy Oliver's apiary, who, having been breeding resistant bees for several years, claims that bee colonies appeared that could be considered completely resistant, and yet there are major problems with the heredity of resistance traits, i.e. there are daughters who coped with mites, but for most of them varroa is still a big problem, "Beekeeping Today Podcast" season 4 episode 34, Scientific Beekeeping with Randy Oliver, https://www.beekeepingtodaypodcast.com/scientific-beekeeping-with-randy-oliver-54-e34, accessed: November 20, 2022. The same problems were described by the Scandinavian breeders.
It is therefore necessary to systematically disseminate resistance traits of the population or build healthy bacterial ecosystems of bees, and this takes time. The situation is similar in other sectors of other human economic activities, where the use of shorter production (and selection) periods is the norm. Changes in nature are often much longer than those that would result from people's expectations and practices. Many naturalists believe that the periods (intervals) established in forest management should be extended at least several times to be able to observe how nature (forest ecosystems) copes with its problems. Meanwhile the trees are cut when they are about 100 years old, and it is just a moment when the natural ecosystem of a mature forest only starts to develop. See podcast: "Drzazgi świata" ["Splinters of the World"] episode 25, Czwarty wymiar lasu [The Fourth Dimension of the Forest], http://www. drzazgiswiata.pl/025-czwarty-wymiar-lasu, accessed: November 20, 2022.

Below I present a method of immediately abandoning chemical treatments and using only biotechnical methods (during the initial selection period).

A possible method for conducting bee selection and keeping using only biotechnical methods

Years	Procedure	Deadline	Breeding activities
Year One	Confinement of the queen in a frame-isolator (removal of brood from the trap frame).	Before the main early summer honey flow (limiting the amount of brood may also result in a larger harvest) - June	Uncapping some of the removed sealed brood from the frame and counting the mites in the brood. Additionally, possibly during the season, observing the natural mite-fall on the bottom board.
	Making artificial swarm – putting all the mature bees onto empty frames with new wax or foundation OR taking away the capped brood.	Early August	Uncapping some of the removed brood and counting the mites.
Year Two	Queen rearing	Breeding season	Selection of queens from those colonies which had the least mites in the removed frames with brood (trap frames) and in the natural mite-drop in the previous season.
	Confinement of the queen in an isolator (removal of brood from the trap frame).	Before the main early summer honey flow.	Uncapping some of the removed sealed brood from the frame and counting the mites in the brood. Additionally, during the season, observing the natural mite-fall on the bottom board.
	Making artificial swarm – putting all the mature bees onto empty frames with new wax or foundation OR taking away the capped brood.	Early August	Uncapping some of the removed brood and counting the mites.
Years three to five	Queen rearing	Breeding season	Selection of queens from those colonies which had the least mites in the brood from the trap frame/s and consider the natural mite-drop in the previous season.
	Use a queen isolator ("trap frame") or taking all the capped brood.	Early Summer (June) – depending on your local situation	After removing the trap-frame, uncap the brood and count the mites.
Year six and beyond	Throughout the season, periodically record the natural mite-fall on the bottom board, in case of a large fall, use the trap-frame and remove the mites. Breeding decisions always based on the lowest mite-fall and infestation.		

Selection of the traits needed for the bees to survive in the presence of the varroa mite

Already more than 30 years ago, Professor Jerzy Woyke noticed that to successfully breed for varroa resistance, two fundamental questions must be answered. The first is, are there differences in resistance to varroa between different bee colonies? The second is, what are the mechanisms of resistance to varroa?[392]

It seems that we already know the answer to the first question, it can be given at least based on the examples I have given. Yes, we agree that bee colonies are characterised by different levels of varroa resistance. As for the answer to the second question, today we can provide much more details than were available more than three decades ago.

I have already described some of the best-known mechanisms, however there are still many issues that require research. It is worth observing colonies in this respect during the transition period and taking the observations into account when making breeding decisions. Selection should be conducted comprehensively since focusing on one feature may lead to losing others in the breeding process. It is also necessary to remember that populations react differently to selection, it is therefore necessary to consider the predispositions of our bees[393].

Many practicing beekeepers do not select for specific features, but rather for the level of mite infestation (infestation dynamics or parasite population growth). I have the impression that this is the more appropriate direction. Observing the degree of infestation will usually lead to the selection of those colonies that cope with the mite, although it will not necessarily allow for the identification of the specific traits that are responsible for this. Varroa infestations can be tested in several ways. Every beekeeper should choose the one they think will work best for them. Let us present the most important

---

392   J. Woyke, Hodowla..., op. cit.

393   In my own apiary I decided to be guided by natural selection, among other things, because of the difficulty in assessing which traits occur in my population that may have sufficiently high variability to be subject to selection, as well as the lack of possibility to assess which of them may be important for their survival under local conditions. I think that if I were to adopt a different breeding strategy, I would focus mainly on the hygiene (pin test), recapping (assessment of the capping from the underside of the cell), assessment of general mite infestation and infestation dynamics (I do not know how I would test this, each method has many advantages and disadvantages; the design of my hives does not allow for testing of debris on the bottom of the hive, which would probably be the least invasive and best method for me. I also have a problem with performing flotation using alcohol, because during the test we kill the bees).

of them. The first one consists in assessing the infestation based on the natural mite-fall on the bottom board (considering the natural mortality of the mite). It is assumed that during the summer season, the daily mite mortality can oscillate around 1-2% of the population. Each dead mite on the hive bottom (that died on a given day) is a signal that there are 50-100 mites living in the colony. According to the principles of modern apiary management, a summer drop of five mites per day may mean an infestation of up to 500 mites in the colony and is a recommendation to carry out an immediate treatment.

In late autumn and winter the ratio of mite-fall to living mites in the hive is different from that in the summer. In winter, mites, like bees, live longer. It is therefore assumed that in winter a fall of one mite per day means an infestation of around 500 female mites in the colony[394]. However, this method of estimating the mite infestation is difficult to use if we do not have mesh bottoms boards with inserts. Lifting the hive (especially if it is occupied by a large colony with many honey boxes) to check the mite fall is not practical. Estimating the mite infestation using this method may also be burdened with an error, because the natural mortality of the mite depends on the season or the biological condition of both the bee colony and the mite. There is also a risk that the fallen mites will be removed from the hive, e.g. by ants or other insects.

Another method of estimating the degree of mite infestation is called flotation or alcohol-wash. A sample of 300 bees (approximately 100 ml) is poured into a container with alcohol solution. The procedure, of course, kills the mites and bees. The mites fall off the bees and after straining the alcohol through a sieve can be easily counted. Supporters of this method believe that a sample of 300 bees ensures a relative accurate way to measure the mite infestation level of the adult bees. Opponents, on the other hand, claim that such a sample is not sufficient to be accurate for a large colony, especially if we sample bees from different parts of the hive (e.g. the bees in the honeycomb and the brood nest may differ in the degree of infestation). For many beekeepers, killing bees in such way is unethical and raises emotions and there is also a risk of killing the queen.

---

394 However, it must be considered that mites may fall off due to grooming. In such a situation, a drop of five mites per day on the hive bottom may mean a much smaller infestation in the colony than a few hundred mites, but it may mean that the bees have developed at least some varroa resistance characteristics. Before a potential decision on treatment, it is worth checking whether the dropped mites are damaged (mutilated) or not.

Another version of this method is to roll or shake the sample of bees in powdered sugar in containers like those used in flotation. Thanks to this, the bees do not die and can be returned to the hive after the test. However, there are opinions that such a test may be less reliable (than rinsing in alcohol), because the mites sometimes don't fall off the bees.

Some people also use carbon dioxide instead of sugar or alcohol to remove the mites since it temporarily puts both the bees and mites to sleep. As the gas doesn't kill the bees, they can return to the hive after the study.

The choice of method depends on the beekeeper. Those who know all of them claim that each of them gives a sufficiently reliable picture of the level of mite infestation in the hive (provided that a minimum diligence is exercised). However, it is important to consistently use the same method throughout the entire period, because only then are the results comparable. Regardless of the exact values, we simply assume that if the level of infestation is low, we can refrain from treatment, if it is medium, the colony must be kept under observation and if the level of infestation is high, the colony should be treated, or we should accept that it could collapse. Monitoring mite infestation allows colonies to be spared numerous treatments with toxic substances and to conduct selection effectively.

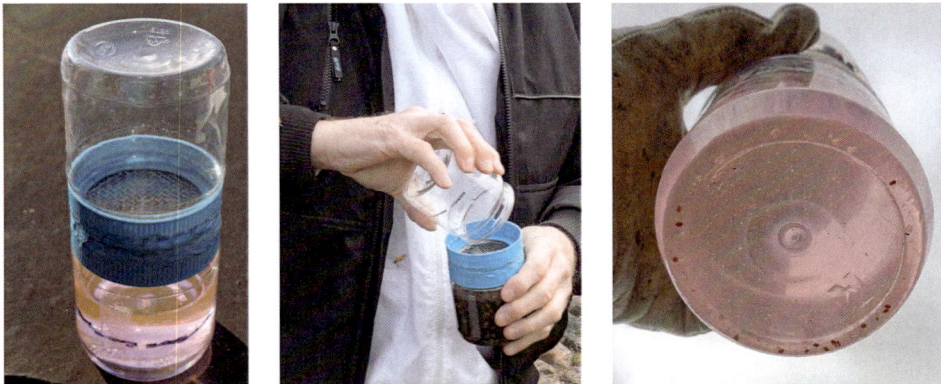

For flotation, a two-part container with a sieve is used, which retains the worker bees, but allows the much smaller mite females to pass through it, photo by E. Österlund.

It is also worth adding that phone applications have emerged that allow you to estimate the level of mite infestation in a colony (based on photos of frames occupied by bees). The fact is that most of the mites are invisible to the camera since they are located underside of the bees' body or in sealed brood cells, however, thanks to algorithms the results are apparently quite reliable and does not differ significantly from the estimates obtained using the methods described previously. I have not used these applications so far, only heard about them, but they could undoubtedly be helpful tool in bee breeding. It is also important that they are non-invasive (they practically do not interfere with the life of the bee colony as does studying the daily mite-drop), and they could be used quickly and conveniently during hive inspections[395].

During the selection process the bees' traits can be systematically strengthened. It does not have to be the case that some colonies in our apiary will be completely non-resistant, while others will be 100% resistant. Therefore, we need to breed from those bees from our population that have the most abilities to defeat the mites, while maintaining the greatest possible genetic diversity. Over time, the resistant traits should become widespread throughout the apiary. During the breeding process, it is also important to maintain a selective pressure, because without it, it is difficult to talk about selection. We need to extend the periods between treatments enough to be able to observe the differences between colonies in terms of their health. In bees that are completely non-resistant, the infestation will grow quickly, but bees that are resistant should maintain the number of mites at a low level and have lower growth dynamics. Selection also cannot be carried out if there is no differentiation of traits in the population. That is if none of the bees prove to be resistant, even to a basic degree, it is difficult to talk about the selection of resistance. Fortunately, bees are characterised by great natural variability and even in a small apiary (10-15 colonies) if we do not have any bees valuable in this respect, the necessary features may be revealed in further crosses, i.e. in subsequent generations. From experiments conducted at the Puławy Veterinary Institute they found that in one population the range of mite infestation can fluctuate within the range of several hundred to almost ten thousand mites (per bee colony),

---

395 Modern beekeeping technologies are on the rise, and various hive monitoring systems are available. Some only measure the temperature or humidity in the hive (however, this alone can provide answers to many questions about the biological condition and development of the colony). Work is also underway to study the sounds made by the bee colony to describe its biological condition (e.g. preparation for swarming). Some believe that analysing the sounds of the bee colony will help, among other things, determining the degree of mite infestation (the application could suggest when to treat). Also interesting is research on novel methods of mite control. For example, laser method of mite control has been developed recently. In the frames of the bee nest, a special "gate" is installed with sensor and the laser – whenever the sensor detects the bee with the mite on it, the laser beam ""shoots down" the mites from the bees. The mite is overheated and dies, whereas the bee does not suffer by it. See: L. Chazette, et. al. Basic algorithms for bee hive monitoring and laser-based mite control, IEEE Photonics Technology Letters, 2016, 12. The future belongs to such modern technologies. Data from some of these devices and applications could also be used in the process of selecting for mite-resistance.

the difference may be almost twenty-fold. Which ones to select for further propagation? The choice is simple.

Shaping the bees' living environment and the ecological situation

During the transition period, we must ensure that the bees' living environment becomes increasingly friendly to them. I mean both the external environment (apiary and its surroundings) and the internal hive environment. Therefore, to the extent possible, we should shape apiaries so that they correspond to the adaptations of the bees, e.g. in accordance with the principles of so-called Darwinian beekeeping, i.e. aim to eliminate horizontal transmission of pathogens (protection against drifting, robbing of weakened colonies, giving up transferring frames between colonies and combining colonies, etc.). Bee hives should correspond to the bees' needs, in that they should be constructed in such a way as to stimulate propolising of internal surfaces by bees. It is also worth detoxifying bee hives by removing old combs that may have absorbed synthetic pesticides or their metabolites. It would be extremely useful to run a foundation-free system or at least base the production of foundation on pure wax from a reliable source. In an area, attention should be paid to the richness and diversity of flowering plants and trees that are desirable to bees. As a passionate supporter of long-term solutions, I focus on planting trees and shrubs rather than sowing annual plants. But of course, there are no bad solutions here. The more flowers for the bees, the better. It's worth emphasizing again: the richness and diversity of natural diet is the key condition for building resilience and health of the bees.

Stimulating the immune response of bees to pathogens

The micro-life of the hive is something that we are not able to observe with the naked eye. In beekeeping activities. we focus on pathogens that we want to combat, doing it instead of the bees and in the process, we lose sight of the positive effects of many bacteria and fungi.

Complex systems of micro-organisms (including pathogenic flora) constantly train the immune response of bees. Their nest sites and surroundings should be constructed in such a way that there is room for organisms that shape a healthy hive ecosystem, stimulating the development of bacterial and fungal biofilms. It is therefore important, during the transition period, not only to minimize the use of toxic and biocidal substances, but also to refrain from disinfecting the bees' living environment. In the long term, this activity seems extremely harmful, yet beekeepers are still being encouraged to do it. Yes, in the short term, disinfection can bring positive effects. It allows the removal of bees' enemies, i.e. pathogens, from the hive, that the bees' immune system must deal with, which in turn can affect, for example, the speed of colony development[396]. In the long term, however, if the immune response of bees weakens, pathogens become more virulent, and the bee populations become increasingly susceptible to diseases. If our actions are directed at ensuring that the bees' environment is free of parasites and pathogens, the bees' immune system will not be stimulated. There will also be no selective pressure that leads to the evolutionary advantage in the population of bees that have a strong and appropriate immune response. For the good of the bees (a long-term process), the pressure of parasites and pathogens in the hive is therefore important.

---

396 This principle applies to agriculture in general, for example, giving animals antibiotics stimulates their growth, the problem is maintaining their health over many years. Animals suddenly deprived of pharmacological protection often develop ailments that they had not encountered before. This is also seen in beekeeping. After treating with organic acids (effectively disinfecting the bees' living environment), colonies develop more dynamically and have more vigour, but the problem remains living in a sterile environment. Short-term solutions, which have so far won over long-term ones, allow for larger honey harvests (and thus bigger profits), but cause an accumulation of problems for the future (including economical for beekeepers, as increased costs).

SUMMARY

,,
You are beekeepers, I know the last thing you want to do is to co-operate. You really want to argue, but now it's not the time. Argue when your method is better than somebody else's, not before. Find the methods that work, tell everybody (...). You've got to fight this (...). We'll lose colonies, there is no denying that. So we'll have to learn to breed bees better than we have been doing up till now.

BOB HOGGE[397]

Bee resistance to varroa is a complex issue, and selection for it takes a long time. If it were enough to find a resistant colony in the population and transfer its features via queens raised from it, we would probably be able to enjoy beekeeping without chemicals for a long time. "Breeding bees resistant to varroa is not a one-time act, conducted only during one season. Selection must be continued for many subsequent generations," Professor Jerzy Woyke pointed out already in 1988. Scientists studying bees living with varroa have no doubt that local populations of bees can become resistant with a survive rate like those that are treated. At the same time most attempts to transfer genetically resistant bees to other locations has ended in failure. The bees seem to lose the characteristics necessary for survival. To maintain their resistant traits, they may also need to survive the local adaptation process in the conditions of their nest cavity and surroundings.

To solve the problem, it is therefore necessary to implement not only local, but also regional (systemic) solutions. The principles of keeping bees in apiaries should consider the co-evolution of bees with other organisms (including pathogenic ones) that are in their environment. Honey bees, like all organisms, are not immortal, sooner or later they will die. A bees' health problems can manifest themselves in various situations. A prolonged period of hunger, insecticide spraying carried out in an orchard next to the apiary, can all contribute to weakening the bees' protective barriers. The bee colony

---

397   Bob Hogge, a beekeeper from the British island of Jersey. The quote refers to the invasion of Asian hornets on the island, not to varroa mite, http://open.spotify.com/ episode/2I8fV7P4WmH3deucxuA5av, accessed: February 8, 2025.

can then become more susceptible to diseases, including varroosis. In many cases, administering organic acids or Apivarol in such a situation will not change anything, it can only postpone the inevitable end.

I realize that deliberately not treating the organism that is under our care is controversial and for many unethical. At the same time, I have no doubt that constant care in the form that we provide to bees only weakens their ability to survive on their own. The threat is constantly present and requires constant intervention. It is impossible to eradicate varroa from our colonies once and for all and cure bees of varroosis, you can only temporarily treat its symptoms by killing the mites. Beekeeping helps individual bee colonies survive in the short-term, but in the long-term it causes an accumulation of health problems for the entire population. In my opinion, this problem must be solved by natural selection. In the approach to bee health, short-term and long-term reasons meet. Chemicals help in the short term. Although the method is attractive because it allows beekeepers to benefit from the work of bees that are productive and appear healthy, in fact they are not. A long-term solution will allow us to solve the problem systemically, once and for all, but we need to go to its sources, i.e. improve the adaptive abilities of the bees.

I would also like those who accuse others of unethical actions (harming the health of bees or the work of other beekeepers) to consider whether they themselves do not behave in a similar way. Do they pay attention to local adaptations? Or maybe they constantly import foreign and genetically non-resistant queens/bees, preventing the acquisition of population resistance of the bees and strengthening the horizontal transmission of pathogens? By accusing others of allowing the death of colonies, do they not consciously contribute to the annihilation of many bee colonies, for example using popular beekeeping practices such as colony merging or rotational beekeeping? By accusing others of spreading mites, do they not look with approval, sometimes even admiration, at beekeepers who migrate with bees for commercial honey production or pollination, where bees are massively infected with pathogens and mites? Such colonies become vectors of pathogens and mites that later reach all corners of the country. I do not want to criticize beekeeping as such. However, I certainly cannot remain indifferent to the complete lack of interest of the beekeeping community in terms of work on mite resistance. In my opinion, we beekeepers should put the natural adaptation of bees to the threat of varroosis in the foreground, because numerous examples from around the world prove that the efficiency of apiaries does not have to suffer, which translates into financial profit. We should also adopt a multi-year perspective of selection. By minimizing the number of chemical treatments in apiaries (or completely abandoning them) and implementing the principles of natural beekeeping, we should remember that some bee colonies will eventually start to have health problems that are visible to the naked eye. Some believe that this is proof of the ineffectiveness of the method, after all, they tried to do something that, according to the assumptions, was supposed to help the bees, and the effect in many colonies is the opposite. However, this means that the population begins to differentiate in terms of coping with their problems. Some colonies remain healthy, most often because they have mechanisms that allow them to endure. Others cannot withstand the pressure. Symptoms of a weakening in a certain part of population should be a selection guide for us, and not a reason to reject long-term solutions and motivate us to return to regular chemical treatments.

The essence of this problem is perfectly illustrated by a story from my childhood, my mother once told me.

Her friend, while at the zoo, was holding her son in her arms and kept repeating: "Look! Look! It's an elephant!", pointing at the animal standing right next to the fence. "Where's the elephant?" the child asked, growing impatient, seeing only a large patch of grey-beige colour in front of him. This is how important it is to maintain the right perspective when looking at the problem. If we want to see the problem of bee health, we must maintain the right distance, only then will we be able to encompass the whole issue. It seems that the fight against varroa, fought with great passion, has blinded us to the problem of the progressive loss of adaptive abilities of the western honey bee population.

BIBLIOGRAPHY

('Pszczelarstwo' is the Polish Beekeeping Magazine).

Book publications

Bush M., Practical Beekeeper. Beekeeping Naturally,
X-Star Publishing Company, 2011.

Seeley T.D., Honey bee Democracy, Princeton University Press, 2010.

Seeley T. D., The Lives of Bees. The Untold Story of Honey
Bee in the Wild, Princeton University Press, 2019.

Heaf D., Treatment Free Beekeeping, Northern Bee Books, 2021.

Published scientific research on bees living without treatment

Bila Dubaić J., Unprecedented density and persistence of
feral honey bees in urban environments of a large
SE-European city (Belgrade, Serbia), Insects 2021, 12.

Fries I., Survival of mite infested (*Varroa destructor*) honey bee
(*Apis mellifera*) colonies in a Nordic climate, Apidologie 2006, 6.

Le Conte Y., Geofigureal distribution and selection of Honey
bees resistant to *Varroa destructor*, Insects 2020, 12.

Locke B., Natural Varroa-mite surviving *Apis mellifera*
honey bee populations, Apidologie 2015, 12.

Luis A.R., Recapping and mite removal behaviour in Cuba:
home to the world's largest population of Varroa-resistant
European honey bees, Scientific Reports 2022, 9.

Rinderer T.E., Resistance to the parasitic mite *Varroa destructor*
in honey bees from far-eastern Russia, Apidologie 2001, 4.

Seeley T.D., Life-history traits of wild honey bee colonies living in forests around Ithaca, NY, USA, Apidologie 2017, 6.

Seeley T.D., Honey bees of the Arnot Forest: a population of feral colonies persisting with *Varroa destructor* in the northeastern United States, Apidologie 2007, 1-2.

Articles on immunity, natural selection and living conditions of bees

Borba R.S., Seasonal benefits of a natural propolis envelope to honey bee immunity and colony health, Journal of Experimental Biology 2015, 11.

Buchegger M., Relationships between resistance characteristics of honey bees (*Apis mellifera*) against Varroa mites (*Varroa destructor*), Journal of Central European Agriculture 2018, 19(4).

Büchler R. et al., Evaluation of characters for testing on varroa resistance, proceedings of the 46th Apimondia International Beekeeping Congress, Montreal, Canada, 2019.

Büchler R., Varroa Tolerancae in Honey Bees - Occurrence, Characters and Breeding, Bee World 1994, 75(2).

Büchler R., Uzunov A, Selecting for varroa Resistance in German Honey Bees, Bee World 2016, 93(2).

Büchler R., Evaluation of Traits for the Selection of *Apis mellifera* for Resistance against *Varroa Destructor*, Insects 2020, 9.

Cook D., Thermal Impacts of Apicultural Practice and Products on the Honey Bee Colony, Journal of Economic Entomology 2021, 4.

Chmielewski M., Odporność okrywy ciała, bariery anatomiczne I funkcjonalne chroniące owada przed zakażeniem [Resistance of the body cover, anatomical and functional barriers protecting the insect against infection], Pszczelarstwo 2022, 3.

De Graaf D.C., Heritability estimates of the novel trait 'suppressed in ovo virus infection' in honey bees (*Apis mellifera*), Scientific Reports 2020, 8.

DeGrandi-Hoffman G., Chen Y, Nutrition, immunity and viral infections in honey bees, Current Opinion in Insect Science 2015, 8.

Desai S., Curie R.W, Genetic diversity within honey bee colonies affects pathogen load and relative virus levels in honey bees, *Apis mellifera* L. Behavioural Ecology and Sociobiology 2015, 7.

Dziechciarz P., Olszewski K, Naddominowanie behawioralne [Behavioural over-dominance], "Pszczelarstwo" 2022, 12.

Fries I., Vertical transmission of American foulbrood (*Paenibacillus larvae*) in honey bees (*Apis mellifera*), Veterinary Microbiology 2006, 5.

Grindrod I., Martin S, Parallel evolution of Varroa resistance in honey bees: a common mechanism across continents? Proceedings of the Royal Society B 2021, 8.

Grindrod I., Martin S, Spatial distribution of recapping behaviour indicates clustering around Varroa infested cells, Journal of Apicultural Research 2021, 5.

Guichard M., Three Decades of Selecting Honey Bees that Survive Infestations by the Parasitic Mite *Varroa destructor*: Outcomes, Limitations and Strategy, https://www.preprints.org/manuscript/ 202003.0044/V1.

Harwood G., Social immunity in honey bees: royal jelly as a vehicle in transferring bacterial pathogen fragments between nestmates, Journal of Experimental Biology 2021, 4.

Hawkins G.P., Martin S, Elevated recapping behaviour and reduced *Varroa destructor* reproduction in natural Varroa resistant *Apis mellifera* honey bees from the UK, Apidologie 2020, 3.

Hinshaw Ch., The Role of Pathogen Dynamics and Immune Gene Expression in the Survival of Feral Honey Bees, Frontiers in Ecology and Evolution 2021, 1.

Howis M., Aspekty biologiczne rodziny pszczelej – relacje pomiędzy Apis mellifera a Varroa destructor przy stosowaniu zabiegów ograniczających populację pasożyta (rozprawa doktorska) [Biological aspects of the bee colony - relationships between *Apis mellifera* and *Varroa destructor* when applying treatments limiting the parasite population (doctoral thesis)] University of Life Sciences, Wrocław 2012.

Ihle K.E., Social apoptosis in Varroa Mite Resistant Western Honey Bees (*Apis mellifera*), Journal of Insect Science 2022, 1.

Kefuss J., Selection for resistance to *Varroa destructor* under commercial beekeeping conditions, Journal of Apicultural Research 2015, 5.

Lang S., Context-Dependent Viral Transgenerational Immune Priming in Honey Bees (Hymenoptera: Apidae), Journal of Insect Science 2022, 1.

Lopez-Uribe M., Higher immunocompetence is associated with higher genetic diversity in feral honey bee colonies (*Apis mellifera*), Conservation Genetics 2017, 2.

Łapka Ł, Mała komórka., a pszczelarskie kłopoty [A small cell and beekeeping problems], "Pszczelarstwo"; 2019, 12; 2020, 1-2.

Łapka E., O tym jak chciano powiększyć pszczołę [How they wanted to enlarge the bee], "Pszczelarstwo" 2016, 3.

Mallinger R.E, Do managed bees have negative effects on wild bees?: A systematic review of the literature, PLoS ONE 2017, 12.

Martin S., *Varroa destructor* reproduction and cell re-capping in mite-resistant *Apis mellifera* populations, Apidologie 2019, 12.

Martin S., Kryger P., Reproduction of *Varroa destructor* in South African honey bees: does cell space influence Varroa male survivorship? Apidologie 2002, 1-2.

Meixner M.D., Effects of genotype, environment, and their interactions on honey bee health in Europe, Current Opinion in Insect Science 2015, 8.

Mikheyev A.S., Museum samples reveal rapid evolution by wild honey bees exposed to a novel parasite, Nature Communications 2015, 8.

Mitchell D., Ratios of colony mass to thermal conductance of tree and man-made nest enclosures of *Apis mellifera*: implications for survival, clustering, humidity regulation and *Varroa destructor*, International Journal of Biometeorology 2016, 9.

Moeller F.E., Effect of moving honey bee colonies on their subsequent production and consumption of honey, Journal of Apicultural Research 1975, 14.

Mondet F., Evaluation of Suppressed Mite Reproduction (SMR) Reveals Potential for Varroa Resistance in European Honey Bees (*Apis mellifera* L.), Insects 2020, 9.

Neuman P., The Darwin cure for apiculture? Natural selection and managed honey bee health, Evolutionary Applications 2016, 11.

Oddie M.A.Y., Reproductive success of the parasitic mite (*Varroa destructor*) is lower in honey bee colonies that target infested cells with recapping, Scientific Reports 2021, 11.

Olszewski K., Wpływ plastrów o małych komórkach na cechy pszczół, rodzin pszczelich oraz odporność na *Varroa destructor* [The influence of small-cell combs on the characteristics of bees, bee colonies and resistance to *Varroa destructor*], "Pszczelarstwo" 2022, 1.

Page P., Social apoptosis in honey bee superorganisms, Scientific Reports 2016, 6.

Palmer K.A., Oldroyd B.P, Evidence for intra-colonial genetic variance in resistance to American foulbrood of honey bees (*Apis mellifera*): further support for the parasite/pathogen hypothesis for the evolution of polyandry, Naturwissenschaften 2003, 5.

Peck D.T., Seeley T.D, Mite bombs or robber lures? The roles of drifting and robbing in *Varroa destructor* transmission from collapsing honey bee colonies to their neighbours, PLOS ONE 2019, 6.

Pohorecka K., Oczekiwania versus rzeczywistość, czyli skuteczność zwalczania inwazji Varroa destructor a liczebność populacji roztoczy w następnym sezonie [Expectations versus reality, i.e. the effectiveness of combating *Varroa destructor* invasion and the size of the mite population in the next season], Pszczelarstwo 2020, 10.

Pohorecka K., Bober A, Wszystko o nosemozie [Everything about nosema], Pszczelarstwo 2021, 12; 2022, 1; 2022, 3; 2022, 4.

Pusceddu M., Honey bees use propolis as a natural pesticide against their major ectoparasite, Proceedings of the Royal Society B 2021, 12.

Seeley T.D., Darwinian Beekeeping: An Evolutionary Approach to Apiculture, American Bee Journal 2017, 3.

Seeley T., Progress report on three years of treatment free beekeeping, including a test of three types of queen: wild colony, Webster Russian, and VSH Italian, American Bee Journal 2020, 8.

Simone-Finstrom M., Propolis Counteracts Some Threats to Honey Bee Health, Insects 2017, 4.

Taber S., The effect of disturbance on the social behavior of the honey bee colony, American Bee Journal 1975, 103(8).

Tautz J., Behavioural performance in adult honey bees is influenced by the temperature experienced during their pupal development, Proceedings of National Academy of Sciences 2003, 5.

Wilczyńska G., Aby zrozumieć świat pszczół miodnych, musimy badać je w naturze, rozmowa z prof. J. Tautzem [To understand the world of honey bees, we must study them in nature, interview with prof. J. Tautz], "Pszczelarstwo" 2022, 11.

Woyke J., Hodowla pszczół odpornych na warrozę [Breeding bees resistant to varroa], "Pszczelarstwo" 1988, 11.

Publications on the diet of bees

Ament S.A., Mechanisms of stable lipid loss in a social insect, The Journal of Experimental Biology 2011, 11.

Bajda M., et al, Rewersja procesu starzenia u pszczół miodnych (*Apis mellifera*)? [Reversion of the aging process in honey bees (*Apis mellifera*)?], Medycyna Weterynaryjna, 2013, 1.

Mirjanic G., Impact of different feed on intestine health of Honey Bees, XXXXIII International Apicultural Congress, Apimondia, Kyiv, Ukraina, 2013.

Noordyke E.R., Ellis J, Reviewing the Efficacy of Pollen Substitutes as a Management Tool for Improving the Health and Productivity of Western Honey Bee (*Apis mellifera*) Colonies, Frontiers in Sustainable Food Systems 2021, 11.

Wheeler M.M., Robinson G.E, Diet-dependent gene expression in honey bees: honey vs. sucrose or high fructose corn syrup, Scientific Reports 2014, 4.

Wu W, Responses of two insect cell lines to starvation: autophagy prevents them from undergoing apoptosis and necrosis, respectively, Journal of Insect Physiology, 2011, 6.

Publications on the role of micro-organisms in the bees' and its environment

Engel P., The Bee Microbiome: Impact on Bee Health and Model for Evolution and Ecology of Host-Microbe Interactions, ASM Journals" 2016, 4.

Forsgren E., Novel Lactic acid bacteria inhibiting *Paenibacillus larvae* in honey bee larvae, Apidologie 2010, 1.

Fredrick J.L., Saccharide breakdown and fermentation by the honey bee gut microbiome, Environmental Microbiology 2014, 6.

Fries I., Camazine S, Implications of horizontal and vertical pathogen transmission for honey bee epidemiology, Apidologie 2000, 5-6.

Gilliam M., Identification and roles of non-pathogenic microflora associated with honey bees, FEMS Microbiology Letters 1997, 10.

Lipiński Z., Wojtacka J, Przełom w poznaniu biologiznej roli symbiotycznych bakterii przewodu pokarmowego pszczół miodnych [A breakthrough in understanding the biological role of symbiotic bacteria in the digestive tract of honey bees], Pszczelarstwo 2016, 2.

Powell J.E., Field-Realistic Tylosin Exposure Impacts Honey Bee Microbiota and Pathogen Susceptibility, Which Is Ameliorated by Native Gut Probiotics, ASM Journals 2021, 6.

Vasques A., Symbionts as Major Modulators of Insect Health: Lactic Acid Bacteria and Honey bees, PLOS ONE 2012, 3.

Publications on varroa (co-evolution of bees with the mites/ increased virulence/other)

Anderson D.L., *Varroa jacobsoni* (Acari: Varroidae) is more than one species, Experimental and Applied Acarology 2000, 3.

Beaurepaire A.L., Population genetics of ectoparasitic mites suggest arms race with honey bee hosts, Scientific Reports 2019, 8.

DeGrandi-Hoffman G., Are Dispersal Mechanisms Changing the Host-Parasite Relationship and Increasing the Virulence of *Varroa destructor* (Mesostigmata: Varroidae) in Managed Honey bee (Hymenoptera: *Apidae*) Colonies? Environmental Entomology 2017, 8.

Dynes T.L., Assessing virulence of *Varroa destructor* mites from different honey bee management regimes, Apidologie 2020, 4.

Frey E., Invasion of *Varroa destructor* mites into mite-free honey bee colonies under the controlled conditions of a military training area, Journal of Apicultural Research 2015, 4.

Frey E., Rosenkranz P., Autumn Invasion Rates of *Varroa destructor* (Mesostigmata: Varroidae) Into Honey Bee (Hymenoptera: Apidae) Colonies and the Resulting Increase in Mite Populations, Journal of Economic Entomology 2014, 4.

Millán-Leiva A., Mutations associated with pyrethroid resistance in Varroa mites, a parasite of honey bees, are widespread across the USA, Pest Management Science 2021, 3.

Morfin N., Surveillance of synthetic acaricide efficacy against *Varroa destructor* in Ontario, Canada, The Canadian Entomologist 2022, 4.

Morro A., Host-Parasite Co-Evolution in Real-Time: Changes in Honey Bee Resistance Mechanisms and Mite Reproductive Strategies, Insects 2021, 1.

Rosenkranz P., Biology and control of *Varroa destructor*, Journal of Invertebrate Pathology 2010, 1.

Ramsey S.D., *Varroa destructor* feeds primarily on honey bee fat body tissue and not haemolymph, Proceedings of National Academy of Sciences 2019, 1.

Ramsey S.D., Foreign Pests as Potential Threats to North American Apiculture *Tropilaelaps mercedesae*, *Euvarroa* spp. *Vespa mandarinia*, and *Vespa velutina*, Veterinary Clinics of North America: Food Animal Practice 2021, 3.

Ramsey S.D., Kleptocytosis: A Novel Parasitic Strategy for Accelerated Reproduction via Host Protein Stealing in *Varroa destructor* [preprint patrz:https://www.biorxiv.org/content/10.1101/2022.09.30.509900v1.article-info].

Sammataro D., The resistance of varroa mites
(Acari: *Varroidae*) to acaricides and the presence of
esterase, International Journal of Acarology 2005, 31.

Wagoner K, Stock-specific chemical brood signals are induced
by Varroa and Deformed Wing Virus, and elicit hygienic
response in the honey bee, Scientific Reports 2019, 6

Publications on Deformed Wing Virus (DWV) and other viruses

Brettel L.E., Martin S.J, Oldest Varroa tolerant honey bee
population provides insight into the origins of the global
decline of honey bees, Scientific Reports 2017, 4.

Dalmon A., Evidence for positive selection and recombination
hotspots in Deformed wing virus (DWV), Scientific Reports 2017, 1.

Geffre A.C., Honey bee virus causes context-dependent changes in host
social behavior, Proceedings of the National Academy of Sciences 2020, 4.

Hybl M., Evaluating the efficacy of 30 different essential oils against *Varroa
destructor* and Honey Bee Workers (*Apis mellifera*), Insects 2021, 11.

Kevill J.L., DWV-A Lethal to Honey Bees (*Apis mellifera*):
A Colony Level Survey of DWV Variants (A, B, and C) in England,
Wales, and 32 States across the US, Viruses 2019, 5.

Lanzi G., Molecular and biological characterization of deformed wing
virus of honey bees (*Apis mellifera* L.), ASM Journals 2006, 5.

Lipiński Z., Wirusy pszczoły miodnej: Varroa destructor i jego
wirusowa gwardia przyboczna [Honey bee viruses: *Varroa
destructor* and the viral army at its side], Pszczelarstwo 2021, 2.

Manley R., Knock-on community impacts of a novel vector:
spillover of emerging DWV-B from Varroa-infested honey
bees to wild bumblebees, Ecology Letters 2019, 8.

McMahon D.P., Elevated virulence of an emerging viral genotype as a
driver of honey bee loss, Proceedings of the Royal Society B 2016, 6.

Mordecai G.J., Superinfection exclusion and the long-term survival of
honey bees in Varroa-infested colonies, The ISME Journal 2015, 10.

Natsopoulou M.E., The virulent, emerging genotype B of Deformed wing virus is closely linked to overwinter honey bee worker loss, Scientific Reports 2017, 7.

Posada-Flores F., Pupal cannibalism by worker honey bees contributes to the spread of deformed wing virus, Scientific Reports 2021, 4.

Roberts J.M.K., Multiple host shifts by the emerging honey bee parasite, *Varroa jacobsoni*, Molecular Ecology 2015, 4.

Roberts J.M.K., Tolerance of Honey Bees to Varroa Mite in the Absence of Deformed Wing Virus, Viruses 2020, 5.

Publications on pesticides, cleanliness of the bee's environment and the impact of chemical substances on humans

Böhme F., From field to food, will pesticide contaminated pollen diet lead to a contamination of royal jelly? Apidologie 2017, 8.

Bykowy M., Chorbiński P. Kwas szczawiowy w zwalczeniu *Varroa destructor* [Oxalic acid in the control of *Varroa destructor*], Pszczelarstwo 2021, 6.

Chauzat M.P., Faucon J-P, Pesticide residues in beeswax samples collected from honey bee colonies (*Apis mellifera* L.) in France, Pest Management Science 2007, 11.

Dai P., Chronic toxicity of amitraz, coumaphos and fluvalinate to *Apis mellifera* L. larvae reared in vitro, Scientific Reports 2018, 4.

Desneux N., The Sublethal Effects of Pesticides on Beneficial Arthropods, The Annual Review of Entomology 2007, 52.

Dhooria S., Agarwal R, Amitraz, an underrecognized poison: A systematic review, Indian Journal of Medical Research 2016, 9.

Gregorc A., Cell death in honey bee (*Apis mellifera*) larvae treated with oxalic or formic acid, Apidologie 2004, 9-10.

Hawthorne D.J., Killing Them with Kindness? In-Hive Medications May Inhibit Xenobiotic Efflux Transporters and Endanger Honey Bees, PLOS ONE 2011, 11.

Hernandez M.R., Short term negative effect of oxalic acid in *Apis mellifera iberiensis*, Spanish Journal of Agricultural Research 2007, 4.

Higes G., Negative long-term effects on bee colonies treated with oxalic acid against *Varroa jacobsoni* Oud., Apidolgie 1999, 4.

Howis M., et al., Uszkodzenia mechaniczne i pozycja Varroa destructor na dennicy ula po zastosowaniu rónych środków warroabójczych [Mechanical damage and the position of *Varroa destructor* on the bottom of the hive after the use of various varroacidal agents], Medycyna Weterynaryjna, 2012, 10.

Johnston P., et al., The plight of bees. Analysis of pesticide residues in bee bread and pollen collected from honey bees (*Apis mellifera*) in 12 European countries, Greenpeace Technical Report, Greenpeace Poland Foundation, 2014, 3.

Johnson R.M., Acaricide, Fungicide and Drug Interactions in Honey Bees (*Apis mellifera*), PLOS ONE 2013, 1.

Johnson R.M., Pesticides and honey bee toxicity in the USA, Apidologie 2010, 5-6.

Karwan D., *Varroa destructor* temat rzeka. Jak nie skazić środowiska ula? [*Varroa destructor* – never ending story. How not to contaminate the hive environment?] Pszczelarstwo 2017, 1-2.

Lipiński Z., Wojtacka J, O amitrazie i nie tylko [About amitraz and more], Pszczelarstwo 2015, 11.

Mullin Ch. A., High Levels of Miticides and Agrochemicals in North American Apiaries: Implications for Honey Bee Health, PLOS ONE 2010, 3.

Rademacher E., Effects of Oxalic Acid on *Apis mellifera* (Hymenoptera: Apidae), Insects 2017, 8.

Wu J.Y., Sub-Lethal Effects of Pesticide Residues in Brood Comb on Worker Honey Bee (*Apis mellifera*) Development and Longevity, PLOS ONE 2011, 2.

Additional publications (beekeeping and non-beekeeping issues)

Darwin Ch., The Origin of Species, P. F. Collier & Son, 1909, New York,

Dogantzis K., Thrice out of Asia and the adaptive radiation
of the western honey bee, Science Advances 2021, 12.

Jaroński J., Hodowca pszczół świadczy usługi dla pszczelarzy.
Musi odpowiadać na ich potrzeby oraz nowe wyzwania, rozmowa
z Przemysławem Szeligą [A beekeeper provides services to
beekeepers. He must respond to their needs and new challenges,
interview with Przemysław Szeliga], Pszczelarstwo 2022, 4-5.

Mazur P., et al., Dziedziczenie grzechów
[Inheritance of sins], Wiedza i Życie, 2012, 8.

Orzyłowska-Śliwińska O., Pilnuj genów
[Watch your genes], Wiedza i Życie 2016, 1.

Strachecka A., Ciało tłuszczowe, tkanka odpowiedzialna za metabolizm
[Fat body, tissue responsible for metabolism], Pszczelarstwo 2022, 11.

Szaciło A. et al., Kyzys zapyleń a pszczoła miodna? Lek na całe zło czy
niekoniecznie? [Pollination crisis and the honey bee? A cure for all evil
or not necessarily?] Kosmos. Problemy Nauk Biologicznych, 2020, 2.

Wohlleben P., The Hidden Life of Trees, Greystone books, 2016

Wulf A., The Invention of Nature. The Adventures of Alexander von
Humboldt, the lost hero of Science, John Murray Publishers, 2015.

Useful Internet Addresses Pages in Polish:

Author's blog: http://pantruten.blogspot.com/

Łukasz Łapka's blog: http://llapka.blogspot.com/

"Bee Brotherhood" website: https://bractwopszczele.pl/eng/eng.html

Website of the "Bartnicy Sudetów" project (Sudety Treebeekeepers): https://bartnicy-sudetow.pl/pl/strona-glowna/

"Fort Knox" project website: https://bees-fortknox.pl/

Jakub Jaroński's podcast (Radio Varroa) and website: https://warroza.pl/

Website addresses of beekeeping organizations

Varroaresistenz-2033 Project: https://varroaresistenzprojekt.eu/

AGT (Germany): https://www.toleranzzucht.de/en/home.

Arista Bee Research Foundation: https:// aristabeeresearch.org.

COLOSS (Switzerland): https://coloss.org.

Free the Bees (Switzerland): https://freethebees.ch/en.459

Natural Beekeeping Trust (UK): https://www.naturalbeekeepingtrust.org.

Resistant bees (Canary Islands): http://english.resistantbees.es/?p=61.

Website addresses of practicing beekeepers

Michael Bush (USA): http://www.bushfarms.com/bees.htm.

Sam Comfort (USA): https://anarchyapiaries.org.

Phil Chandler (UK): https://www.biobees.com.

Norbert Dorn (Austria): https://www.treatmentfree-beekeeping.com.

David Heaf (UK/Walia): http://www.bee-friendly.co.uk.

Juhani Lunden (Finland): https://naturebees.wordpress.com

Randy Oliver (USA): http://scientificbeekeeping.com.

Erik Österlund (Sweden): https://www.elgon.es.

Treatment Free Beekeeping YT channel:
https://www.youtube.com/treatmentfreebeekeeping.

Alois Wallner (Austria): http://www.voralpenhonig.at

Kirk Webster (USA): https://kirkwebster.com

Selected lectures by beekeepers/scientists (YouTube)

Margarita Lopez-Uribe, Are Feral Honey Bee Colonies Healthier
https://www.youtube.com/watch?v=7zqy5toLV8k.

Thomas D. Seeley, Darwinian Beekeeping, An Evolutionary Aproach
to Apiculture, https://www.youtube.com/watch?v=fRj9n7Vma48

Ralph Büchler, Environmental Adaptation of Honey Bees
https://www.youtube.com/watch?v=4DVm_L7Fkqc&t=33s.

FMichael Bush, our Simple Steps to Healthier Bees,
https://www. youtube.com/watch?v=03qniBf7_Uo&t=299s.

Roger Patterson, Honey Bees in the Wild - PART
1 - What can we learn from them?
https://www.youtube.com/ watch?v=txGzuVJhJ_Y&t=371s.

Thomas Seeley, Honey Bees in the Wild - PART 2 -
What do we know about how they live?
https://www.youtube.com/ watch?v=T7CB8E7jKBc&t=4s.

Ralph Büchler, Varroa Resistance Characters and Selection Protocols,
https://www.youtube.com/watch?v=KwuR3uMkMFo&t=1179s.

9 781919 200408